HIGH PERFORMANCE DEVICES

SELECTED TOPICS IN ELECTRONICS AND SYSTEMS

Editor-in-Chief: **M. S. Shur**

Published

Vol. 18: Sensitive Skin
eds. *V. Lumelsky, M. S. Shur and S. Wagner*

Vol. 19: Advances in Surface Acoustic Wave Technology, Systems and Applications (Two volumes), volume 1
eds. *C. C. W. Ruppel and T. A. Fjeldly*

Vol. 20: Advances in Surface Acoustic Wave Technology, Systems and Applications (Two volumes), volume 2
eds. *C. C. W. Ruppel and T. A. Fjeldly*

Vol. 21: High Speed Integrated Circuit Technology, Towards 100 GHz Logic
ed. *M. Rodwell*

Vol. 22: Topics in High Field Transport in Semiconductors
eds. *K. F. Brennan and P. P. Ruden*

Vol. 23: Oxide Reliability: A Summary of Silicon Oxide Wearout, Breakdown, and Reliability
ed. *D. J. Dumin*

Vol. 24: CMOS RF Modeling, Characterization and Applications
eds. *M. J. Deen and T. A. Fjeldly*

Vol. 25: Quantum Dots
eds. *E. Borovitskaya and M. S. Shur*

Vol. 26: Frontiers in Electronics: Future Chips
eds. *Y. S. Park, M. S. Shur and W. Tang*

Vol. 27: Intersubband Infrared Photodetectors
ed. *V. Ryzhii*

Vol. 28: Advanced Semiconductor Heterostructures: Novel Devices, Potential Device Applications and Basic Properties
eds. *M. Dutta and M. A. Stroscio*

Vol. 29: Compound Semiconductor Integrated Circuits
ed. *Tho T. Vu*

Vol. 30: Terahertz Sensing Technology — Vol. 1
Electronic Devices and Advanced Systems Technology
eds. *D. L. Woolard, W. R. Loerop and M. S. Shur*

Vol. 31: Advanced Device Modeling and Simulation
ed. *T. Grasser*

Vol. 32: Terahertz Sensing Technology — Vol. 2
Emerging Scientific Applications and Novel Device Concepts
eds. *D. L. Woolard, W. R. Loerop and M. S. Shur*

Vol. 33: GaN-Based Materials and Devices
eds. *M. S. Shur and R. F. Davis*

Vol. 34: Radiation Effects and Soft Errors in Integrated Circuits and Electronic Devices
eds. *R. D. Schrimpf and D. M. Fleetwood*

Selected Topics in Electronics and Systems – Vol. 35

HIGH PERFORMANCE DEVICES

Proceedings of the 2004 IEEE Lester Eastman Conference on High Performance Devices

Rensselaer Polytechnic Institute 4 – 6 August 2004

Editor

Robert E Leoni III

Raytheon RF Components, USA

World Scientific

NEW JERSEY • LONDON • SINGAPORE • BEIJING • SHANGHAI • HONG KONG • TAIPEI • CHENNAI

Published by

World Scientific Publishing Co. Pte. Ltd.

5 Toh Tuck Link, Singapore 596224

USA office: 27 Warren Street, Suite 401-402, Hackensack, NJ 07601

UK office: 57 Shelton Street, Covent Garden, London WC2H 9HE

British Library Cataloguing-in-Publication Data
A catalogue record for this book is available from the British Library.

HIGH PERFORMANCE DEVICES
Proceedings of the 2004 IEEE Lester Eastman Conference

ISBN 981-256-196-X

Editor: Tjan Kwang Wei

This book is printed on acid-free paper.

Printed in Singapore by Mainland Press

PREFACE

This special issue of the *International Journal of High Speed Electronics and Systems* (IJHSES) presents state-of-the-art works from top academic and research institutions in the areas of high performance semiconductor materials, devices, and circuits. The special issue will also be reprinted in book form, providing a volume for the *2004 Proceedings of the Lester Eastman Conference on High Performance Devices*. This conference, which was originated in 1967 by Professor Lester F. Eastman as the Biennial IEEE-Cornell Conference on Microwave Semiconductors, shares the IJHSES charter: to promote engineering education by advancing interdisciplinary science between electronics and systems and to explore high performance technology in electronics and photonics.

This issue of IJHSES features a broad coverage of topics relating to high performance devices and circuits. There are 46 contributed papers covering a wide range of materials, device types, and applications. These papers describe the results of ongoing research in three general areas: high speed technologies for advanced mixed signal and terahertz applications, advanced technologies for high performance optical links and light sources, and high power density and high efficiency technologies for next generation microwave front ends and power electronics.

Papers in the area of high speed / terahertz technologies describe developments in material, device, and circuit level research in the antimonide, indium phosphide and silicon material systems. The antimonide system provides a wide range of desirable electronic characteristics that enable extremely fast, low power consumption devices in the terahertz range. The antimonide contributions in this issue of IJHSES describe core material and device level accomplishments and demonstrations. The indium phosphide contributions describe advancements in scaling of heterostructure bipolar transistors and integration of tunnel diodes to realize high performance mixed signal circuitry. These technologies will ultimately enable low power consumption, high dynamic range circuitry for direct digital synthesis and analog-to-digital conversion at microwave frequencies. Finally, the silicon papers present innovative solutions for high speed and terahertz needs through the use of an extremely mature technology.

Technological advancements in the areas of UV light sources and high speed optical links are represented in the optoelectronics manuscripts. Current research in UV LEDs are fueled by the need to significantly reduce the cost and footprint of future lighting, medical diagnostic systems, and biological / chemical agent detection systems. UV LEDs are also being pursued for covert, low power consumption communication systems. The contributed UV LED related papers describe new understandings and observations of core material and device level physics that will lead to enhanced LED and LED based system performance. The high speed optical link contributions describe methods to achieve enhanced link performance through the use of novel passive structures and optoelectronic integration techniques.

High power density and high efficiency technologies for next generation microwave front ends and power electronics are currently rich areas of research which are fueled by the gallium nitride (GaN) and silicon carbide (SiC) material systems. Manuscripts in this special issue describe material innovations (diamond, gallium-indium nitride, GaN growth on GaN substrates) and new understanding of material issues (trap centers, impact of material parameters on performance, etc.). The impacts of device geometry (field plates, recessing, etc.) and device configuration (FETs, BJTs, SITs, P-I-N diodes) are also explored. Also, the significant circuit-level gains that are made possible by these new materials and devices are demonstrated in a number of contributed manuscripts.

Acknowledgements

The editor would like to thank Professor Michael Shur of Rensselear Polytechnic Institute for initiating this special issue and book volume. As editor, he is also indebted to the many contributors to this volume. He is indebted to the Lester Eastman Conference Committee members for their assistance in collecting and reviewing contributed manuscripts. The committee members include Dr April Brown, Dr. Greg DeSalvo, Prof. Alan Doolittle, Prof. Lester Eastman, Dr. Thomas Jenkins, Prof. Erhard Kohn, Dr. Christie Marrian, Prof. Umesh Mishra, George Narode, Dr. Laura Rea, Dr. Joan Redwing, Dr. Robert Sadler, Prof. Fredrick Schubert, Prof. Alan Seabaugh, Prof. Michael Shur, Dr. Marko Sokolich, Prof. Micheal Spencer, Prof. Robert Trew, Prof. Gary Wicks, Prof. Jerry Woodall, and Dr. John Zolper.

2004 IEEE LESTER EASTMAN CONFERENCE ON HIGH PERFORMANCE DEVICES

Committee List

General Chair:
John C. Zolper
DARPA/MTO
3701 N. Fairfax Drive
Arlington, VA 22203
Email: jzolper@darpa.mil

Program Chair:
James Kolodzey
Department of Electrical Engineering and Computer Engineering
216 Evans Hall
University of Delaware
Email: kolodzey@ee.udel.edu

Local Arrangements:
Michael Shur
Rensselaer Polytechnic Institute
ECSE and Physics
CII9017, RPI, 110 8th Street,
Troy, NY 12180
Email: shurm@rpi.edu

Treasurer:
George Narode
Rensselaer Polytechnic Institute
110 8th Street, Troy, NY 12180
Email: narode@ecse.rpi.edu

European Rep. For Germany, England, Sweden, France:
Erhard Kohn
University of Ulm
Dept. of Electron Devices and Circuits
Albert-Einstein-Allee 45
D-89081 Ulm, Germany
Email: dept-ebs@sunrise.e-technik.uni-ulm.de

Proceedings and Publications:
Robert Leoni
MEC/L2 N7
Raytheon RF Components
362 Lowell Street
Andover, MA 01810
Email: rleoni@raytheon.com

Technical Committee Members

April Brown
Duke University
Durham, NC

Greg DeSalvo
Northrop Grumman
Baltimore, MD

Alan Doolittle
Georgia Institute of Technology
Atlanta, GA

Tom Jenkins
AFRL/SNDM
Wright-Patterson AFB, OH

Erhard Kohn
University of Ulm
Ulm, Germany

Christie Marrian
DARPA/MTO
Arlington, VA

Umesh Mishra
University of California
Santa Barbara, CA

Laura Rea
Wrigth Patterson Air Force Base
Dayton, OH

Joan Redwing
Penn State University
University Park, PA

Robert Sadler
RF Micro Devices
Charlotte, NC

Fred Schubert
Rensselaer Polytechnic Institute
Troy, NY

Alan Seabaugh
University of Notre Dame
Notre Dame, IN

Marko Sokolich
HRL Laboratories
Malibu, CA

Michael Spencer
Cornell University
Ithaca, NY

Robert Trew
Virginia Tech
Blacksburg, VA

Gary Wicks
University of Rochester
Rochester, NY

Jerry Woodall
Yale University
New Haven, CT

2004 IEEE LESTER EASTMAN CONFERENCE ON HIGH PERFORMANCE DEVICES

August 4 – 6, 2004
Rensselaer Polytechnic Institute, CII, Troy, NY 12180

Wednesday, August 4, 2004

Session I: Plenary
Session Chairs: Prof. Jim Kolodzey and Dr. John Zolper

8:30 – 9:00	WELCOME *J. Zolper, DARPA*
9:00 – 9:30	SiGe, SOI, AND BEYOND *G. Shahidi, IBM*
9:30 – 10:00	POLYMER ELECTRONICS *J. Sturm, Princeton University*
10:00 – 10:30	TERAHERTZ *P. Siegel, JPL*
10:30 – 11:00	BREAK
11:00 – 11:30	BIOELECTRONICS *M. Stutzmann, Walter Schottky Institute*
11:30 – 12:00	InN *W. Schaff, Cornell University*
12:00 – 12:30	InP *B. Brar, Rockwell Scientific*
12:30 – 2:00	LUNCH

Session II: High Speed Devices and Circuits
Session Chairs: Prof. Alan Seabaugh and Prof. Jerry Woodall

2:00 – 2:20	VERTICAL SCALING OF TYPE I InP HBT WITH F_T > 500 GHz *J. Lai, University of Illinois*
2:20 – 2:40	NUMERICAL INVESTIGATION OF THE EFFECT OF DOPING PROFILES ON THE HIGH FREQUENCY PERFORMANCE OF InP/InGaAs SUPER SCALED HBTS *D. Veksler, Rensselaer Polytechnic Institute*
2:40 – 3:00	HIGH PERFORMANCE InP /InGaAs/InP DHBTs WITH PATTERNED SUBCOLLECTOR FABRICATED BY ELEVATED TEMPERATURE N+ IMPLANT *M. Chen, HRL*

3:00 – 3:20 HIGH SPEED 4 BIT ACCUMULATORS
S. Turner, University of Maine
3:20 – 3:50 BREAK
3:50 – 4:10 TUNNEL DIODE/TRANSISTOR DIFFERENTIAL COMPARATOR
Q. Liu, University of Notre Dame
4:10 – 4:30 NOISE AND THz RECTIFICATION CHARACTERISTICS OF ZERO-BIAS QUANTUM TUNNELING Sb-HETEROSTRUCTURE DIODES
A. Luukanen, NIST
4:30 – 4:50 TEMPERATURE DEPENDENCE OF TERAHERTZ EMISSION FROM SILICON DEVICES DOPED WITH BORON
R. T. Troeger, University of Delaware

Poster Session and Reception
Session Chairs: Prof. Morris Washington, Prof. Michael Shur, and Mr. George Narode
6:00 – 8:20 POSTER SESSION

Thursday, August 5, 2004

Session III: AlGaN / GaN HEMTs
Session Chairs: Dr. Greg DeSalvo, Prof. Bob Trew, and Prof. Umesh Mishra
8:30 – 8:50 MBE GROWN AlGaN/GaN HEMTs ON SiC
S. Rajan, University of California at Santa Barbara
8:50 – 9:10 STABLE HIGH POWER GaN-ON-GaN HEMT
P. Chao, BAE Systems
9:10 – 9:30 UNSTRAINED InAlN/GaN HEMT STRUCTURE
M. Neuberger, University of Ulm - Germany
9:30 – 9:50 A NEW FIELD PLATE GaN HEMT STRUCTURE WITH IMPROVED POWER & NOISE PERFORMANCE
H. Xu, University of California at Santa Barbara
9:50 – 10:20 BREAK
10:20 – 10:40 GaN MOS-HEMT USING ATOMIC LAYER DEPOSITION Al_2O_3 AS GATE DIELECTRC AND SURFACE PASSIVATION
P. Ye, Agere Systems
10:40 – 11:00 TEMPERATURE DEPENDENT I-V CHARACTERISTICS OF AlGaN / GaN HBTs AND GaN BJTs
H. Xing, University of California at Santa Barbara
11:00 – 11:20 ANALYSIS OF HIGH DC CURRENT GAIN STRUCTURES FOR GaN/InGaN/GaN HBTs
J. Li, University of California at San Diego
11:30 – 1:00 LUNCH

Session IV: AlGaN/GaN HEMTs - part 2
Session Chairs: Dr. Greg DeSalvo, Prof. Bob Trew, and Prof. Umesh Mishra

1:00 – 1:20 SELECTIVE DRY ETCHING OF GaN OVER AlGaN IN BCl_3/SF_6 MIXTURES
D. Buttari, University of California at Santa Barbara

1:20 – 1:40 NOISE CHARACTERISTICS OF FIELD PLATED GaN HEMTs
Y. Wu, Cree Santa Barbara Technology Center

1:40 – 2:00 HIGH LINEARITY GaN HEMT POWER AMPLIFIER WITH PRE-LINEARIZATION GATE DIODE
S. Xie, University of California at Santa Barbara

2:00 – 2:30 BREAK

Session V: Nanotechnology and Innovative Concepts
Session Chairs: Dr. Tom Jenkins, Dr. Christie Marrian, Prof. April Brown, and Prof. Alan Doolittle

2:30 – 3:00 CARBON NANOTUBE DEVICES
J. Appenzeller, IBM

3:00 – 3:30 CARBON NANOTUBE NETWORKS
E. Snow, NRL

3:30 – 3:50 HIGH-VOLTAGE DIAMOND SCHOTTKY RECTIFIERS
W. Huang, Rensselaer Polytechnic Institute

3:50 – 4:10 HIGH SPEED 4 BIT ACCUMULATORS
S. Turner, University of Maine

4:10 – 4:30 LEAKY SURFACE ACOUSTIC WAVES IN SINGLE-CRYSTAL AlN SUBSTRATE
G. Bu, Rensselaer Polytechnic Institute

6:00 – 8:00 CONFERENCE DINNER

Friday, August 6, 2004

Session VI: SiC
Session Chairs: Prof. Paul Chow and Dr. Robert Sadler

8:30 – 8:50 2-kV 4H-SiC DMOSFETs FOR LOW-LOSS, HIGH-FREQUENCY SWITCHING APPLICATIONS
S. Ryu, Cree

8:50 – 9:10 2.5-kV, 17-mW*cm^2 NORMALLY-ON 4H-SiC VERTICAL JFET DEVICES
S. Krishnaswami, Cree

9:10 – 9:30 SIMULATION STUDY ON BREAKDOWN BEHAVIOR OF FIELD-PLATE SiC MESFETs
H. Cha, Cornell University

9:30 – 9:50	ION-IMPLANTED SiC STATIC INDUCTION TRANSISTOR WITH 107 W OUTPUT POWER AND 59% POWER-ADDED EFFICIENCY UNDER CW OPERATION AT 750 MHz *P. Esker, Northrop Grumman Corp.*
9:50 – 10:10	HIGH POWER, DRIFT-FREE 4H-SiC PiN DIODES *M. Das, Cree*
10:10 – 10:40	BREAK

Session VII: Optoelectronics

Session Chairs: Prof. Fred Schubert and Dr. Shuji Nakamura

10:40 – 11:10	GaN LEDS *A. Fischer, SNL*
11:10 – 11:30	GaInN LIGHT-EMITTING DIODES WITH $RuO_2/SiO_2/Ag$ OMNI-DIRECTIONAL REFLECTOR *F. Schubert, Rensselaer Polytechnic Institute*
11:30 – 12:00	UV LED CHARACTERIZATION *P. Wrabeck, ARL*
12:00 – 12:20	LIFETIME OF NONEQUILIBRIUM CARRIERS IN AlGaN EPILAYERS WITH HIGH Al MOLAR FRACTION *J. Mickevicius, Rensselaer Polytechnic Institute*
12:20 – 12:35	CLOSING REMARKS

CONTENTS

International Journal of High Speed Electronics and Systems
Vol. 14, No. 3 (2004) 625–631

VERTICAL SCALING OF TYPE I InP HBT WITH F_T > 500 GHZ

J.W. LAI, W. HAFEZ, AND M. FENG
University of Illinois, 208 N. Wright Street, Urbana, IL 61801, USA
e-mail: jlai@uiuc.edu Tel#: (217) 333-4054 Fax#: (217) 244-3662

We have fabricated the high-speed InP/InGaAs-based single heterojunction bipolar transistors (SHBTs) with current gain cutoff frequency, fT from 166GHz to over 500GHz by the approach of vertical scaling. Collector thickness is reduced from 3000Å to 750Å and the peak current density is increased up to 1300kA/cm2. In this paper, device rf performance has been compared with respect to materials with different vertical dimensions. The scaling limitation is also studied by analytical approach. The extracted physical parameters suggest that the parasitic emitter resistance is the major limit on further enhancing ultra-scaled HBT intrinsic speed due to the associated RECBC delay. The cut-off frequency of a 500Å collector SHBT has been measured and the results indicate a dramatic drop on fT, supporting the conclusion projected by model analysis. It is also commented that for deeply downscaled HBTs, impact ionization could be another degrading mechanism limits device bandwidth.

1 Introduction

InP/InGaAs heterojunction bipolar transistors (HBTs) have been considered as one of the most promising technology to realize ultra wideband applications due to its wide bandwidth, high current driving capability, good linearity, and good uniformity of threshold voltage distribution for circuit integration. These reasons give the motivation to improve the performance and investigate the limits of HBTs.

Several technologies have already demonstrated the breakthroughs on bipolar transistor speed. Over 300GHz fT was achieved by InP/InGaAs heterojunction bipolar transistor [1][2]. 350GHz silicon bipolar transistors fabricated by SiGe technology [3] were demonstrated. Type-II GaAsSb DHBTs with fT over 300GHz were also reported [4]. HBT bandwidth can be increased by reducing electron transit time and RC charging time. The electron transit time is primarily improved by downscaling the layer thickness, and the increased parasitic capacitance is reduced through the dimensional shrinking of devices. Therefore, when the epitaxial layer structure is downscaled to enhance the device intrinsic bandwidth, promising HBT performance has to be accomplished by proper lateral scaling also with the support of well-developed process and well-designed device layout. By doing so, one unavoidably increases certain parasitic resistances and trade-offs have to be made through device scaling. In this paper, we compare and analyze device characteristics of InP/InGaAs single heterojunction bipolar transistors under the vertical scaling on material layer structure design. By model extraction and transit time component comparison, our study points out that the RECBC poses the scaling limit at ultra-scaled devices.

2 Device Fabrication

Materials with In0.53Ga0.47As collector layer thickness varied from 750Å to 3000Å are grown by MBE. Collector doping concentration is fixed at 1×1016 cm-3 for all materials. 300Å compositionally graded base is grown through all materials under comparison, with In0.5Ga0.5As at the emitter end to In0.53Ga0.47As at the collector end. An internally built-in electrical field is formed to introduce maximum electron velocity in the base layer for assistance to electron transport. Emitter thickness is highly reduced to eliminate the adverse effect of parasitic emitter resistance. As the discussions are going to be made, the emitter resistance is going to be the main limit on achieving high speed/high performance InP HBTs. With the considerations of our manufacturing capability, 800Å total thickness is chosen. Device manufacturing techniques involve e-beam lithography to define <0.5μm wide emitter. Self-aligned sub-micron HBT process has been developed, which applies BCB etch-back technique to avoid contact via opening and allows interconnection metal with size larger than emitter mesa. The device also features air-bridge connection between the base metal, which surrounds emitter mesa, and the base post that is located outside the intrinsic device. By this way, the base interconnection can be made by larger interconnection components without increasing base-collector capacitance and therefore maximizes device cut-off frequency. The cross-section of the fabricated device is shown in Fig.1.

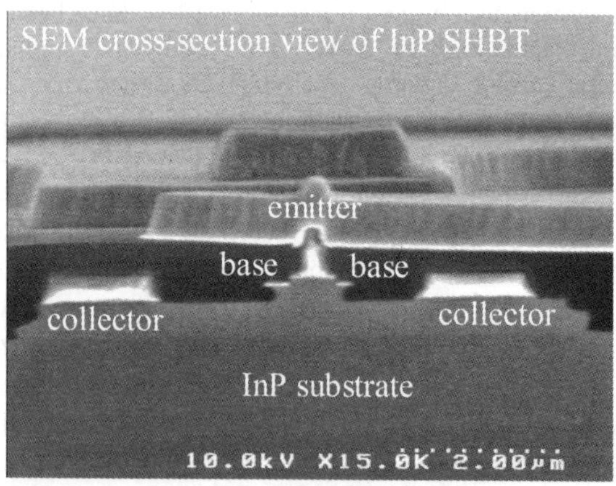

Fig.1. The SEM image of the fabricated sub-micron SHBT.

3 Measurement Results

Fig. 2 shows the measured frequency dependent current gain (H21) and unilateral power gain (U) of various materials at peak fT. Measurement data are calibrated by SOLT standards through 0.5GHz to 50GHz by 8510 network analyzer. fT enhancement has been observed from 166GHz to 509GHz with the collector thickness reducing from 3000Å to 750Å. The corresponding decrease of fMAX is expected due to the increases of base-collector capacitance in these devices. The trend can be clearly observed by cut-off frequency versus current density plot shown in Fig.3, which also shows that the maximum current density posed by Kirk effect proportionally increases with the shrinking of collector layer thickness.

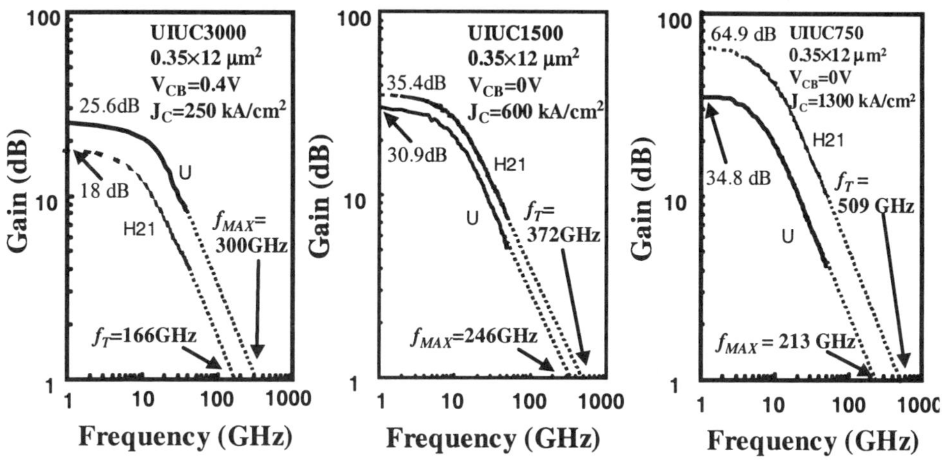

Fig.2. The measured rf response of 0.35×12μm² SHBTs with collector thickness t_C = 3000Å (left, UIUC3000), 1500Å (center, UIUC1500) and 750 Å (right, UIUC750) at peak f_T.

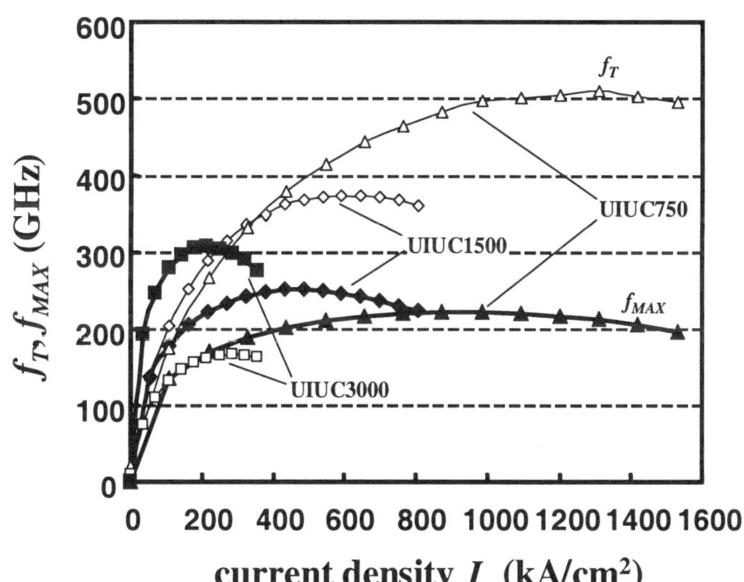

Fig.3. f_T / f_{MAX} versus J_C plots of 0.35×12μm² SHBTs. The measured results of UIUC750 (collector layer thickness =750Å), UIUC1500 (1500Å) and UIUC3000 (3000Å) are shown.

TABLE.I. The extracted physical parameters of UIUC InP SHBTs.

Coefficients (units)	descriptions	values
ε (F/m)	Dielectric constant of collector	1.26×10^{-10}
C_{je} (fF/μm^2)	B-E junction capacitance	12.9
ρ_C (Ω-μm^2)	Specific base contact resistance	180
R_S(Ω/□)	Base sheet resistance (t_B=300Å)	970
r_E (Ω-μm^2)	Emitter resistance	15.8
R_C(Ω)	Collector resistance (l_E=12μm)	0.927
L_T (μm)	Transfer length of base contact	0.24
Jc,max (kA/cm^2)	Maximum current density (t_C=3000Å)	250
D_n (cm^2/s)	Effective diffusion coefficient	49.9
V_{eff} (V_B) (cm/s)	Effective electron velocity	4.2×10^7
ΔE (ev)	Base graded energy	0.026
T (k)	Junction temperature	500
Θ (ps/Å)	Normalized coefficient	1.5×10^{-4}

4 Device Analysis

The measured data are fit in small signal T model [5], and the corresponding components are converted into the conventional cut-off frequency relation:

$$\frac{1}{2\pi f_T}=\tau_F+\frac{kT}{I_C}C_{je}+(R_C+R_E+\frac{kT}{I_C})C_{BC},$$

(1)

which can be further expanded to be [6]

$$\frac{1}{2\pi f_T}=\frac{kT}{D_n\Delta E}[1-\frac{kT}{\Delta E}(1-e^{-\frac{\Delta E}{kT}})]\,t_B^2+\frac{kT}{v_B\Delta E}(1-e^{-\frac{\Delta E}{kT}})\,t_B+\Theta\, t_B+\frac{t_C}{2v_{eff}}+\frac{kT}{J_C}c_{je,dep}+\frac{\varepsilon\, r_E w_C}{w_E t_C}+R_C\frac{\varepsilon\, w_C l_C}{t_C}+\frac{\varepsilon\, kT w_C}{J_C w_E t_C}.$$

(2)

By the s-parameters measured from 1500Å and 3000Å devices, these physical parameters are extracted and listed in Table I.

Comparing the small signal parameters, it is interesting to note that the delay portion contributed by emitter resistance significantly increases due to the vertical scaling. Fig.4 shows a summary of the extracted electron transit time τF and RECBC. For 3000Å collector design, electron transit time across base and collector layers (τF) occupies 75% of the total delay. For 750Å collector design, electron transition term τF drops and

RECBC increases to 24% of τF. Furthermore, theoretically projected results indicate that if collector layer reduced to 500Å, the portion of RECBC increases up to more than 40%. The overall trend of fT versus collector thickness is shown in Fig.5. The solid and dash line are derived from the data shown in Table I, which is extracted only from 3000Å and 1500Å devices. Good fitting of these lines to 1000Å and 750Å devices is achieved indicating the accuracy of the data measurement and model extraction. Analysis shown in Fig.5 also indicates that further scaling will lead to the overall drop of fT if collector layer thickness lower than 650Å. As a result, the shortening transit time brought by vertical scaling cannot improve HBT speed if the emitter resistance is not scaled accordingly. To verify this observation, a 500Å collector layer material is grown without changing other material properties. As marked on Fig.5, this material only shows fT at 320GHz, indicating the limit of vertical scaling on making higher speed devices. Comparing to the theoretical value, the measured fT is lower and can be explained as the result of the enhanced carrier multiplication within thin collector, which poses excess delay on consuming regenerated carriers. Enhanced carrier multiplication is observed at the zero IB on IC-VCE curves when collector is reduced to 500Å (Fig.6(a)). Also an excess drop on BVCEO is measured (Fig.(b)), suggesting that the effect of impact ionization keeps increasing in the deeply downscaled HBTs. Since the mentioned avalanche-type regenerative mechanism is not conventionally considered in the small signal fT estimation, this effect is not included in (2).

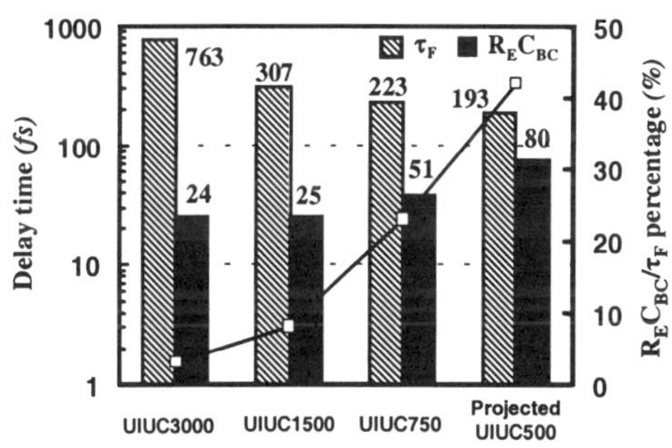

Fig.4. Comparison of the delay components in 0.35x12μm² SHBTs. Data shown in this figure are measured and extracted from small signal model except UIUC500 (500Å collector layer thickness), which is projected theoretically.

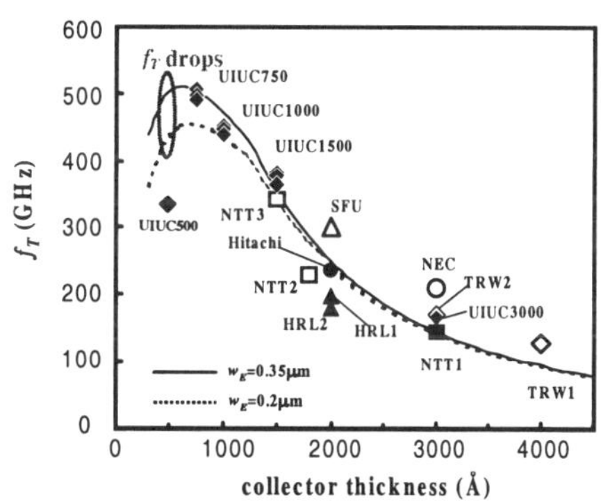

Fig.5. Scaling trends on f_T with various collector thicknesses. SHBTs are marked by solid markers, and DHBTs are marked by empty markers.

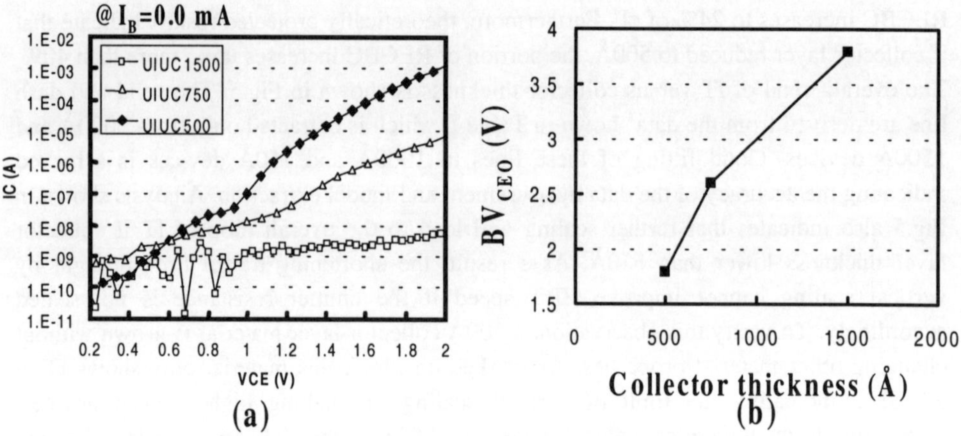

Fig.6. DC characterization of deeply downscaled HBTs versus collector thickness: (a) the collector dark current measured at I_B=0.0mA and (b) The measured BV_{CEO} of each material.

5 Conclusion

In conclusion, material can be tailored to produce over 500GHz InP SHBT by vertical scaling. Vertical scaling effectively reducing electron transit time is shown through small signal model analysis, and finite emitter resistance shows the possibility to prohibit effective speed enhancement or even degrade HBT speed if devices are scaled down deeply. To overcome such limit, an even thinner emitter cap layer may cause device yield issues. Either increasing the doping level, or replacing conventional InGaAs emitter cap by narrow bandgap low resistance material such as graded InAs, will be effective to reduce sheet resistance and contact resistance, indicating a promising way to increase device speed further.

Acknowledgement

The authors would like to thank Mr.Frank Strolli from BAE System for the DARPA-TFAST program management support, as well as DARPA-TFAST program manager Dr. John Zopler and ARL contract manager Dr. Alfred Hung for program support.

References

[1] M. Ida, K. Kurishima, and N. Watanabe, "High-speed InP/InGaAs DHBTs with a thin pseudomorphic base," *Digest of IEEE GaAs IC Sym.*, pp.211-214, 2003.

[2] W. Hafez, J.W. Lai, and M. Feng, " Record FT and Fmax performance of InP/InGaAs Single Heterojunction Bipolar Transistors, " *Electronics Lett.*, vol. 39 , no. 10 , pp.811 – 813, 2003

[3] J.S. Rieh, B. Jagannathan, H. Chen, K.T. Schonenberg, D. Angell, A. Chinthakindi, J. Florkey, F.Golan, D.Greenberg, S.J.Jeng, M.Khater, F.Pagette,

C.Schnabel, P. Smith, A.Stricker, K.Vaed, R.Volant, D.Ahlgren, G.Freeman, K.Stein, and S. Subbanna, "SiGe HBTs with Cut-off Frequency of 350GHz," *2002 IEDM Technical Digest*, pp. 771-774, 2002.

[4] C. Bolognesi, M.W. Dvorak, N. Matine, O.J. Pitts, and S.P. Watkins, "Ultrahigh Performance Staggered Lineup ("Type-II") InP/GaAsSb/InP NpN Double Heterojunction Bipolar Transistors," *Jpn. J. Appl. Phys.*, Vol. 41, no. 2B, pp. 1131-1135, 2002.

[5] M. Hattendorf, D. Scott, Q.Yang, M.Feng, "Method to determine intrinsic and extrinsic base-collector capacitance of HBTs directly from bias-dependent S-parameter data", *IEEE Electron Devices Lett.*, vol.22, no.3, pp.116-118, 2001.

[6] M.J.W.Rodwell, M.Urteaga, T.Mathew, D.Scott, D.Mensa, Q.Lee, J.Guthrie, Y.Betser, S.C.Martin, R.P.Smith, S.Jaganathan, S.Krishnan, S.I.Long, R.Pullela, B.Agarwal, U.Bhattacharya, L.Samoska, and M.Dahlstrom, "Submicron scaling of HBTs," *IEEE Trans. Electron Devices*, vol. 48, no. 11, pp. 2606-2624, 2001.

International Journal of High Speed Electronics and Systems
Vol. 14, No. 3 (2004) 632–639

World Scientific
www.worldscientific.com

NUMERICAL INVESTIGATION OF THE EFFECT OF DOPING PROFILES ON THE HIGH FREQUENCY PERFORMANCE OF InP/InGaAs SUPER SCALED HBTs

DMITRY VEKSLER AND MICHAEL S. SHUR
Department of Electrical, Computer, and Systems Engineering Rensselaer Polytechnic Institute, Troy NY 12180-3590, USA

V.E. HOUTSMA, N.G. WEIMANN, AND Y.K. CHEN
Lucent Technologies, Bell Laboratories, 600 Mountain Ave., Murray Hill NJ 07974, USA

Our simulations show that the emitter and collector doping profile engineering is very important for the transistor optimization, in particular, adjusting the low doped emitter section to the depletion length resulted in the decrease of the emitter series resistance and increased f_t and f_{max}; decreasing of the collector doping concentration and shrinking the collector thickness reduced the collector transit time. Accounting for the lateral diffusion of hot electrons in the device with submicron emitter was found to be important in the transistor optimization process. This effect determines the effective thickness of the emitter finger and the value of the push-out current.

Introduction

Submicron scaling of semiconductor devices is a conventional way to increase the operating frequencies of transistor-based integrated circuits. During the last decade, record heterojunction bipolar transistor (HBT) cutoff frequencies have been achieved by thinning the base and collector layers, shrinking the emitter width, and using the transferred substrate technology [1-8]. Epitaxial and lateral scaling of InP-based double heterostructure transistors (DHBTs) resulted in f_t = 370 GHz and f_{max} = 459 GHz for devices with the 0.6x7 μm^2 emitter–base junction area [9] and f_t = 347 GHz and f_{max} = 492 GHz devices with the 0.6x5 μm^2 emitter junction area [10]. The highest f_t = 509 GHz was achieved in InGaAs/InP single heterostructure SHBT structure [11]. InGaAs/InP material system has a number of advantages compared to Si/Ge and GaAs/AlGaAs. High electron mobility and velocity in InGaAs allow achieving high f_t and f_{max} values. Also, InGaAs has a lower surface recombination velocity, which makes scaling to small dimensions easier; finally, InP substrates have a better thermal conductivity than GaAs substrates.

In this paper, we address the following issues related to further optimization of InGaAs/InP DHBTs: understanding the effects of the emitter, collector and base doping profiles on the InGaAs/InP DHBTs performance; investigating the two dimensional effect of hot electron spreading in the base and collector regions, and developing the simple analytical model of the transistor based on the device physics and on the results of the numerical simulation.

Simulations

The cross-section of the simulated device structure is shown in Figure 1. The DHBT structure consists of a 0.25 x 4 μm^2 InP emitter, 40 nm InGaAs base and a 178 nm collector. The 30 nm section of the emitter region directly on top of the base was doped at $N_d = 5 \times 10^{17}$ cm^{-3}. Base doping concentration was $N_a = 4 \times 10^{19}$ cm^{-3}. One kT grading in the energy gap of the base was assumed in the simulations. The distance between the base contact and the emitter edge was equal to 0.1 μm. The collector structure consisted of the following layers: 18 nm layer of InGaAs, 30 nm InGaAsP grading layer, and 130 nm InP layer (all three layers were doped at $N_d = 3 \times 10^{16}$ cm^{-3}) followed by the InP implanted subcollector ($N_d = 2 \times 10^{19}$ cm^{-3}). (The implanted subcollector reduced the collector-base capacitance, C_{BC}.) The contacts to the semiconductor were assumed to have the following specific contact resistances: emitter metal – 8 x 10^{-8} ohm-cm^2; base metal - 10^{-6} ohm-cm^2; collector metal – 8 x 10^{-7} ohm-cm^2.

We used the 2D device simulation software DESSIS from ISE. This software has a hydrodynamic model for the charge carriers flow, which accounts for electron heating, device self heating, and velocity overshoot, i. e. the key electron transport features for the submicron devices.

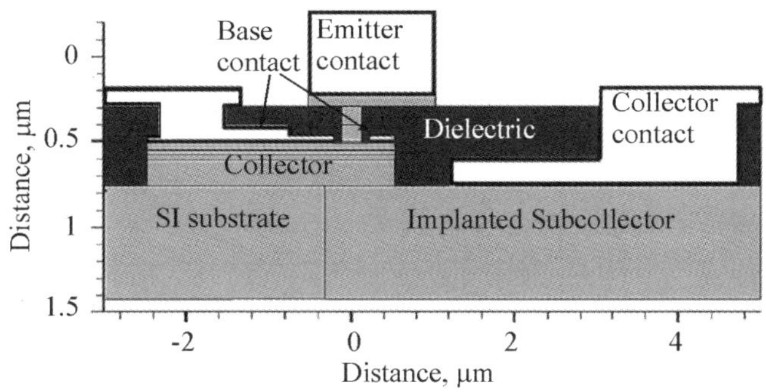

Figure 1. Cross-section of the device under investigation

Analytical f_t and f_{max} model

Figure 2 shows the simulated dependences of f_t and f_{max} as a function of the collector current density. f_t and f_{max} first rise with an increasing current density and then decrease at the point where the current is sufficiently high for base push out effect to occur.

We suggest the following expression for f_t accounting for the base push out effect:

$$\frac{1}{2\pi f_t} = \tau_B + \tau_C + \frac{kT}{qI_C}\left(1 + \left(\frac{I_C}{I_{CRIT}}\right)^m\right)(C_{EB} + C_{BC}) + (R_E + R_C)C_{BC}\,, \qquad (1)$$

where R_E and R_C are the parasitic emitter and collector resistances (including the contact resistances), C_{BC} and C_{EB} are the collector junction and emitter junction capacitances, I_C is the collector current, I_{CRIT} is a critical current for the base push out, τ_B and τ_C are the neutral base layer and the collector depletion region transit times, and m is a fitting parameter to be extracted from the 2D simulations. In the limit of low collector currents ($I_C << I_{CRIT}$), Equation (1) coincides with the expression for f_t derived using small signal equivalent circuit transistor model [12].

The maximum frequency of oscillations can be calculated using the conventional formula [12]:

$$f_{\max} = \sqrt{\frac{f_t}{8\pi R_{BB} C_{bc}}}, \tag{2}$$

where R_{BB} is a parasitic base resistance and C_{bc} is the intrinsic part of the base-collector capacitance.

Figure 2 demonstrates that Eq. (1) and Eq. (2) agree well with the simulated f_t and $f_{\max}$. The parameters used for the fit are: $(C_{EB} + C_{BC})$ = 6.55 fF/μm, I_{CRIT} = 0.014 A, $(R_E + R_C)C_{BC} + \tau_B + \tau_C$ = 0.375 ps, $R_{BB}C_{bc}$ = 0.0835 ps and m=6.45.

Figure 3 shows the fit of experimental f_t and f_{max} dependences versus collector current. Once again, we obtain a good fit. As seen from Eq. (1), f_t and f_{max} are determined by several characteristic time constants. Our simulations examined the possibilities of decreasing of each of these time constants.

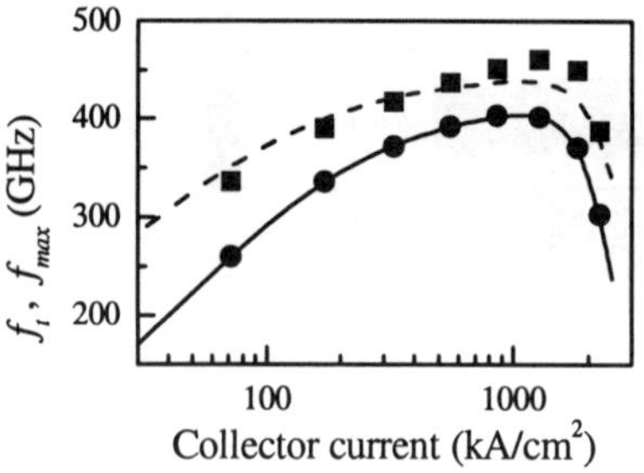

Figure 2. Simulated f_t and f_{max} dependences vs. collector current for InGaAs/InP DHBT biased at V_{CE} = 2.5 V. Circles correspond to ft and Squares stand for fmax. Solid line is a f_t theoretical fit. Dashed line is a fit of f_{max}.

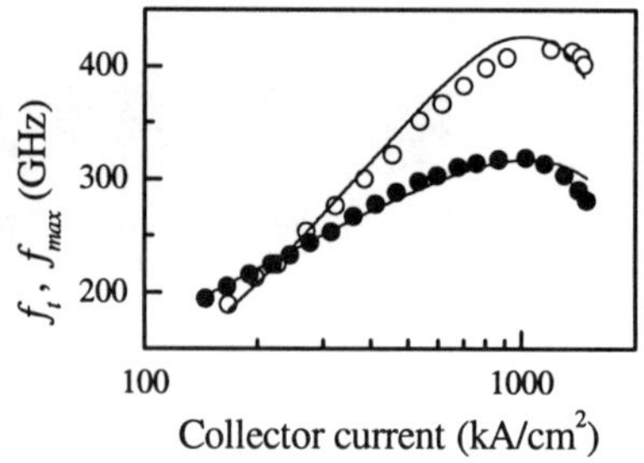

Figure 3. Experimental f_t (open circles) and f_{max} (solid circles) against collector current density for 0.35x4 μm^2 HBT from Ref. [11]. Lines correspond to theoretical fit.

Base doping profile

Figure 4 illustrates the decrease of the base transit time by engineering the doping profile in the base. Grading of the base doping level produces an additional electric field accelerating electrons in the base region. Effect of the linear and exponential doping grading was studied theoretically in the Refs. [13,14] and experimentally in Refs. [15,16]. There is also a number of papers (see refs. in [17,18]) devoted to study of the effect of emitter-base and base-collector interfaces δ-doping on an HBT performance, but, to the best of our knowledge, the influence of delta-doping on the base transit time has not been considered. Three base doping profiles employed in simulations were: flat doping profile with N_a= 6 x 10^{19} cm^{-3}, delta doping at the emitter-base junction (in addition to flat doping the top 2 nm - portion of the base from the emitter side was doped at N_a= 5 x 10^{20} cm^{-3}) and the model of a linear doping profile falling off from the emitter side (N_a=8 x $10^{19}cm^{-3}$) to the collector side (N_a= 5 x 10^{19} cm^{-3}) of the base. Both types of grading increase f_t and f_{max}. The delta doping increases f_t by approximately 9 GHz and the linear doping profile increases f_t by 5 GHz.

Delta-doping of the base-collector junction also reduces the base transit time (see Fig. 5). However, the base transit time reduction is counter balanced to a certain degree by an increased base spreading resistance: high additional *n*-type doping at the base – collector interface significantly increases the width of the depletion region in the base reducing the base thickness. Also, the δ-doping reduces the electron velocity in the collector region, increasing the transit time of collector depletion region.

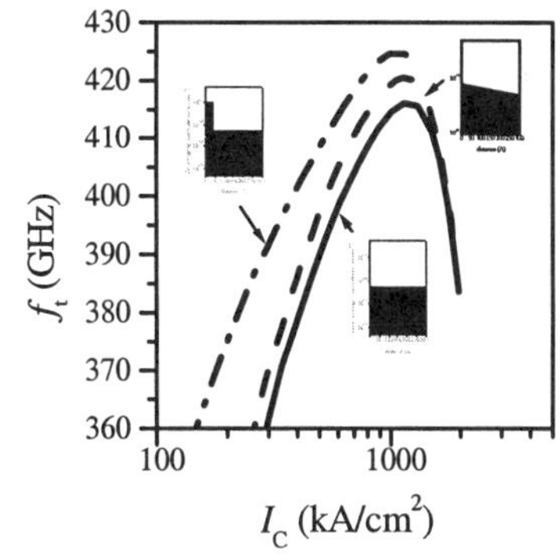

Figure 4. Calculated f_t dependences vs. collector current at different base doping profiles. Inserts in the figure show the profiles corresponding to: $f_t(I_C)$ curves. (here base thickness is 30 nm).

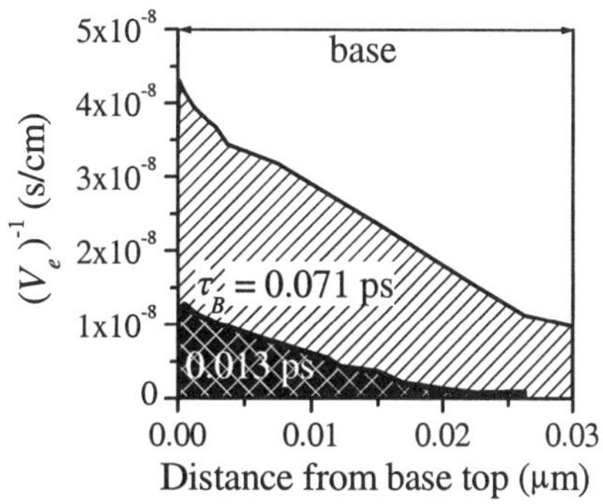

Figure 5. Inverse electron velocity profile in the base cross-section. Cross-section goes from emitter side of the base to collector one through the middle of emitter. Black area plot corresponds to δ-doping at base-collector junction. Black line is related to the flat doping profile. Numbers in figure represent the area under the related graph, i.e. base transit times. V_{CE} = 2.2 V.

Collector doping

As seen from Fig. 6, a decrease of the collector doping concentration increases f_t and f_{max} but reduces the gain. Fig. 7 illustrates the effect of the doping profile effect on transistor high frequency performance. This figure shows the simulated electron velocity profiles in the central cross-section of the collector at two different doping levels of the InP collector layer. The electrons enter the collector region with a velocity several times higher than the saturation velocity (due to the overshoot effect). Then the charged impurity scattering in the depleted collector decreases the velocity. The lower collector doping level leads to a smaller collector transit time.

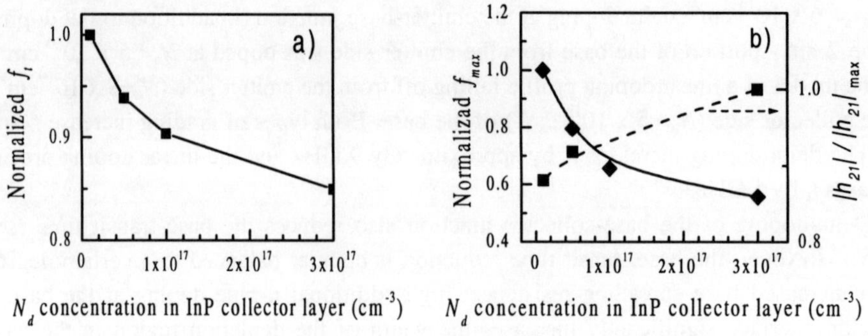

Figure 6. f_t (a), f_{max} and gain (b) dependence vs. collector doping. Squares and dashed curve correspond to the current gain. f_t is normalized by 250 GHz, and f_{max} is normalized by 548 GHz and $|h_{21}|_{max}$ = 135. These calculations were performed assuming V_{CE} = 1.5 V and neglecting the contact resistances.

On the other hand, low collector doping concentration makes the critical current for the base push out effect smaller. The crude estimates for the critical current density for N_d = $5 \cdot 10^{16}$ cm $^{-3}$ yield the value of 800 kA/cm^2:

$$I_{CRIT} = 2 \cdot q \cdot N_d \cdot v_{eff}, \tag{3}$$

where v_{eff} = 5·x 10^7 cm/s is the average electron velocity in the collector.

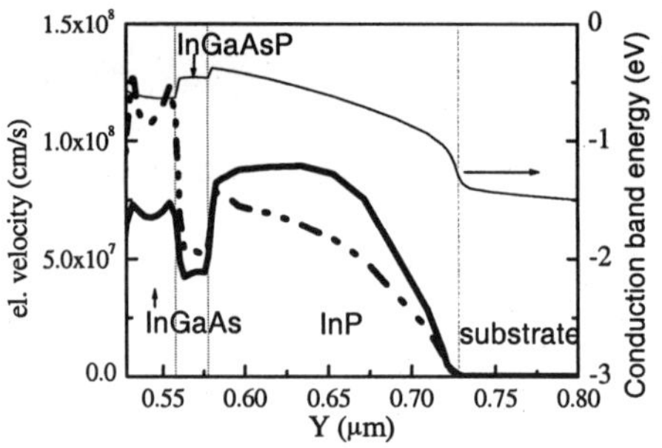

Figure 7. Electron velocity profile in the collector region at two different doping concentrations in InP layer. Solid line corresponds to $N_d = 5 \cdot 10^{16}$ cm^{-3}. Dashed line is for $N_d = 10^{17}$ cm^{-3}. Thin line represents the conduction band energy. V_{CE} = 1.5 V.

Figures 8 and 9 show the electron temperature and concentration distributions, respectively. As seen, the hot electron effects play an important role and have to be accounted in the process of the transistor optimization. Electrons injected from the emitter into the base spread out in the lateral direction increasing the effective width of the emitter finger. The extent of this lateral spread is on the order of the Debye radius of the hot electrons injected into the base and is close to 0.1 micron for the simulated device. The lateral spread of hot electrons might be even more important in the collector region, where electrons are hotter. This affects the value of the push-out current.

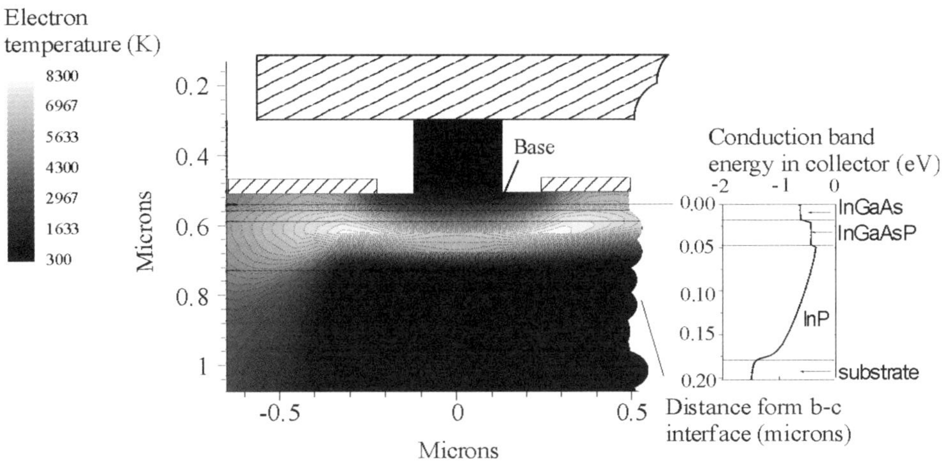

Figure 8. Temperature distribution in the HBT cross-section. $V_{CE} = 2.5$ V. $V_{BE} = 1.0$ V, corresponding to the critical collector current. The insert demonstrates the band diagram of the collector heterostructure at zero bias.

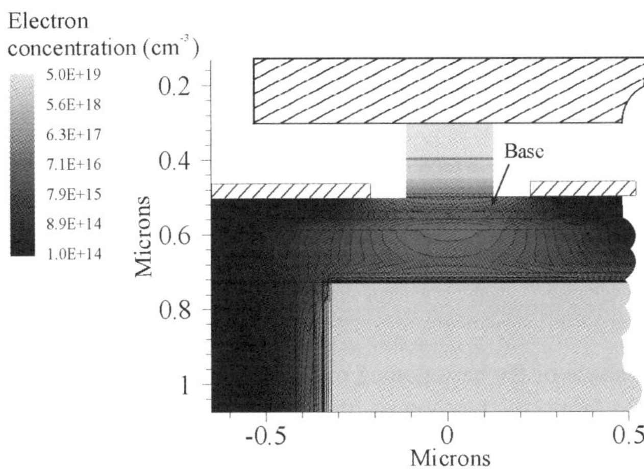

Figure 9. Electron concentration distribution in the HBT cross-section. $V_{CE} = 2.5$ V. $V_{BE} = 1.0$ V.

Emitter optimization

The emitter charging time $C_{EB}kT/qI_C$, and $R_E C_{BC}$ delay time contribute to the overall emitter to collector delay, which is proportional to the inverse of f_t (see Eq. (1)) The emitter-base junction capacitance can be reduced by increasing the depletion layer thickness. This can be achieved by reducing the doping concentration in the emitter region at base side. At the same time, lowering of the emitter doping increases the emitter series resistance. Also, the effect of emitter narrowing occurs due to the surface charges at the emitter sides. Hence, it is possible to find the optimal doping profile in the emitter. We propose to reduce the length of the low doped emitter to the value of the depletion length. In this case, an optimum tradeoff will be achieved between the series resistance and junction capacitance. Different types of emitter profiles, evaluated in the simulations and simulated f_t and f_{max} are shown in Fig. 10. (The emitter depletion width at this bias voltage was estimated to be about 20 nm. The contact resistances were not taken into account in these simulations.)

Fig. 10 demonstrates an increase in the transistor speed due to the optimization of the emitter low doped region thickness.

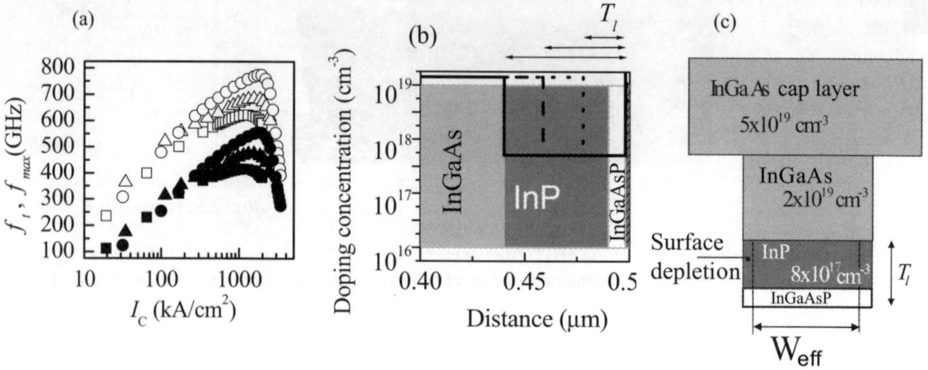

Figure 10. (a) f_t (filled symbols) and f_{max} (open symbols) of devices with different emitter doping profiles. Squares correspond to the low-doped emitter portion length Tl = 58 nm, Triangles stands for T_l = 40 nm, and Circles are related to T_l = 20 nm. All the simulations were performed neglecting the contact resistances. Base thickness was 300 Å. There was no filling of the emitter-collector space with dielectric. Base doping level was 6×10^{19} cm^{-3}. (b) doping profiles employed in simulations; (c) emitter cross-section. T_l is the thickness of the low doped emitter portion. W_{eff} is the effective width of emitter.

Conclusions

Our simulations show that the shape of the base doping profile has relatively little effect on the HBT high frequency performance. In contrast, the emitter and collector profile engineering was found to be very important for the transistor optimization. Adjusting the low doped emitter section to the depletion length resulted in the decrease of the emitter series resistance and increased f_t and f_{max}. The decrease of the collector doping concentration and shrinking the collector thickness (to about 180 nm) reduced the collector transit time. Hot electron effects play a big role in the submicron device. Lateral diffusion of hot electrons ejected into the base leads to the increase of the effective width of the emitter finger. The length of this lateral spread is close to 0.1 micron for the simulated device. The spread of hot electrons might be even more important in the collector region, where electrons are hotter. This allows the lower doping concentrations in the collector region for the same value of the base push-out current.

References

1. M. Dahlstrom, X.-M. Fang, D. Lubyshev, et. al., *IEEE El. Dev. Lett.*, **24**, 433 (2003).

2. M. Kahn, S. Blayac, M. Riet, P. Berdaguer, V. Dhalluin, et. al., *Proc. Int. Conf. on Indium Phosphide and Related Materials* (2003), p 134.
3. Yongsik Jeong, Yongjoo Song, Sunkyu Choi, Myounghoon Yoon, Kyounghoon Yang; *Proc. Int. Conf. on Indium Phosphide and Related Materials* (2003), p 22.
4. Scott, D.W.; Wei, Y.; Dong, Y.; Gossard, A.C.; Rodwell, M.J., *IEEE El. Dev. Lett.*, **25**, 360 (2004).
5. Y. Yang, J. Frackoviak, C.T. Liu, C.J. Chen, et. al., *IEEE GaAs IC Symp. Tech. Dig*, 269 (2002).
6. Daekyu Yu, Kyungho Lee, Bumman Kim, D. Ontiveros, K. Vargason, J. M. Kuo, Y. C. Kao, *IEEE El. Dev. Lett.*, **24,** 384 (2003).
7. S. Lee, M. Urteaga, Y. Wei, Y. Kim, M. Dahlstrom, S. Krishnan, M. Rodwell, *Device Research Conference (2002). 60th DRC. Conf. Digest*, p 107.
8. J.-S. Rieh, B. Jagannathan, H. Chen, et. al., *IEDM Tech. Dig.*, p 771 (2002).
9. Z. Griffith, M. Dahlström, M. Urteaga, M. J. W. Rodwell, X.-M. Fang, D. Lubyshev, Y. Wu, J. M. Fastenau, and W. K. Liu, *IEEE El. Dev. Lett.,* **25**, 250 (2004).
10. M. Ida, K. Kurishima, K. Ishii, and N. Wantanabe, *IEEE GaAs IC Symp. Tech. Dig.*, p. 211 (2003).
11. W. Hafez, J.W. Lai, and M. Feng, *Electronics Letters*, **39**, 1475 (2003).
12. M. J. Rodwell, M. Urteaga, Y. Betser, T. Mathew, et. al., *Int. J. of High Speed Electronics and Systems*, **11**, 159 (2000).
13. K. Suzuki, *IEEE Transactions on El. Dev.*, 38, 2128 (1991).
14. Y. Zebda, A. Elnagar, A. Hussein, *IEE Proc. Circuits, Devices and Systems*, **144,** 375 (1997).
15. D. C. Streit, M. E. Hafizi, D. K. Umemoto, et. al., *IEEE El. Dev. Lett.*, **12**, 194 (1991).
16. M. Dahlstrom, X.-M. Fang , D. Lubyshev, M. Urteaga, et. al., *IEEE El. Dev. Lett.* **24**, 433 (2003).
17. H. R. Chen, C. Y. Chang, C. P. Lee, C. H. Huang, J. S. Tsang, K. L. Tsai, *IEEE El. Dev. Lett.* **15,** 336 (1994).
18. W.C. Wang, S.Y. Cheng, W.L. Chang, H.J. Pan, Y.H. Shie and W.C. Liu, *Semicond. Sci. Technology*, **13,** 630 (1998).

International Journal of High Speed Electronics and Systems
Vol. 14, No. 3 (2004) 640–645

TUNNEL DIODE/TRANSISTOR DIFFERENTIAL COMPARATOR

QINGMIN LIU, SURAJIT SUTAR, AND ALAN SEABAUGH
Department of Electrical Engineering, University of Notre Dame
Notre Dame, IN 46556, USA

A new tunnel diode/transistor circuit topology is reported, which both increases speed and reduces power in differential comparators. This circuit topology is of special interest for use in direct digital synthesis applications. The circuit topology can be extended to provide performance improvements in high speed logic and signal processing applications. The circuits are designed based on InP/GaAsSb double heterojunction bipolar transistors and AlAs/InGaAs/AlAs resonant tunneling diodes. A self-aligned and scalable fabrication approach using nitride sidewalls and chemical mechanical polishing is outlined.

1. Introduction

As minimum feature sizes are reached in transistor technologies, circuit performance can be expected to saturate. Therefore, it is important to find new ways to augment the performance of integrated circuit technology. Integrated tunnel diodes enable design options which can result in a decrease in power dissipation and an increase in speed and function per area.[1]

Since 1957, the tunnel diode has been widely investigated for high-speed circuit applications, due to its intrinsic high switching speed, and its "N-shaped" current-voltage (I–V) characteristic featuring negative differential resistance (NDR).[2,3] Today, methods for integrating tunnel diodes with transistors in compound semiconductor technologies are well known.[4,5,6] In silicon technology both Si and SiGe p^+n^+ tunnel diodes have been fabricated using CMOS-compatible processes.[7,8] Monolithic integration of a Si-based tunnel diode and a heterojunction bipolar transistor (HBT) have also been demonstrated.[9]

In this paper, we report on a circuit topology, utilizing high speed tunnel diodes integrated with transistors to lower the static power dissipation in a differential comparator and simultaneously increase the circuit speed. A novel scalable fabrication approach using nitride sidewalls and chemical mechanical polishing is also outlined for fabricating this tunnel diode/transistor (TDT) differential comparator.

2. Circuits Design

The schematic of the TDT differential comparator is shown in Fig. 1(a). Two tunnel diode pairs, D_1-D_3 and D_2-D_4, are connected to the outputs of the differential pair Q_1 and Q_2. Emitter followers Q_3 and Q_4 are used to buffer the output signal.

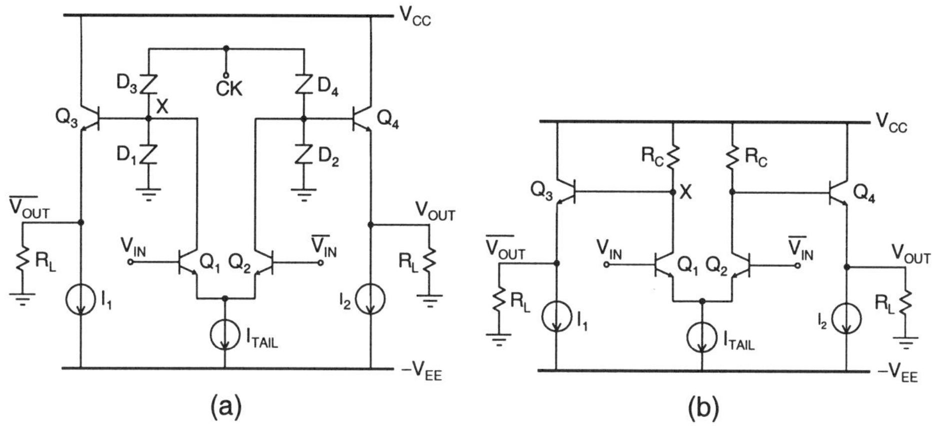

Fig. 1. Schematic diagrams of (a) tunnel diode/transistor (TDT) and (b) conventional bipolar transistor differential comparator.

When the clock is at a low level, the tunnel diode pair is in a monostable state. In this state, no matter what the applied input signal, the output voltage is low. When the clock switches to a high level, the tunnel diode pair switches to a bistable state. The output voltage latches to either a high or low voltage state, determined by the tunnel diode peak currents and the transistor collector current occurring at the rising edge of the clock. The peak currents, I_{P3} and I_{P4}, of diodes, D_3 and D_4, are designed to be greater than the peak currents, I_{P1} and I_{P2}, of diodes, D_1 and D_2, with the following relationships $I_{P3} < I_{P1}+I_{TAIL}$ and $I_{P4} < I_{P2}+I_{TAIL}$. If the input to transistor Q_1 is low, the tail current, I_{TAIL}, flows through transistor Q_2. When the clock switches from low to high, the output voltage at node X switches to the high level, because diode D_1 has a smaller peak current and switches first. If, on the other hand, the input to transistor Q_1 is high, the tail current, I_{TAIL}, flows through Q_1. When the clock switches from low to high, the output voltage at node X remains low, because $I_{P3} < I_{P1} + I_{TAIL}$ and diode D_3 switches instead of D_1. This circuit functions as an edge-triggered return-to-zero (RZ) D flip-flop with complementary input and output.

The TDT circuit, Fig. 1(a), has smaller output open-circuit time constant of the differential pair and faster switching speed than a conventional transistor-only current-mode-switching differential amplifier,[10] Fig. 1(b). In transistor-only comparator circuits, speed improvements are usually obtained by increasing the tail current and decreasing the pull-up collector resistor. The tunnel diodes enable the differential pair current-switching at a lower tail current than the transistor-only differential pair. The tunnel diodes, therefore, lower the static power dissipation and simultaneously increase the circuit speed.[10]

3. Direct Digital Synthesis

This TDT differential comparator is of special interest for use as a switching element in direct digital synthesis applications. A recently developed algorithm, based on list decoding, for DDS provides an improved signal-to-noise ratio (SNR), compared with conventional $\Sigma\Delta$ approaches.[11,12] In both list decoding and $\Sigma\Delta$ approaches, a specially designed digital bitstream is output through a single-bit digital-to-analog converter (DAC) and a reconstruction filter to form the analog signal. The Fourier spectrum of the desired signal is embedded in the pattern of the bitstream. A high-speed and high-linearity single-bit DAC is required to achieve high SNR, because the severe distortion of the output signal is usually generated from the nonlinearity of the DAC. The circuits, proposed in this paper, are designed for use in such DDS applications.

To compare the power dissipation and linearity of both TDT and transistor-only circuits, we simulated both circuits (Fig. 1) in Agilent ADS. The transistor SPICE model we used in the simulation is for a high speed InP-based heterojunction bipolar transistor (HBT) with f_T/f_{MAX} of 140/340 GHz. Broekaert's RTD model[13] is used for an InP-based RTD with a speed index of 316 mV/ps. The input bitstream has a 100 Gbps bit rate, and an amplitude of 400 mV with rising and falling time of 1 ps. The synthesized passband signal frequency is approximately 37.3 GHz. The simulated spectrum of the synthesized signal in both TDT and transistor-only circuits is shown in Fig. 2. Approximately 60 dBc spur free dynamic range (SFDR) is obtained for both circuits, showing these two circuits have approximately the same linearity. The power dissipation of both circuits are also compared and shown in Table 1. The power dissipation is reduced by approximately 4× in the TDT differential pair and 1.6× in the full TDT circuit.

Table 1. Comparison of the power dissipation in the TDT and conventional HBT differential comparators of Fig. 1.

Power dissipation	TDT (mW)	HBT (mW)
Differential pair	3	12.5
Clock	0.5	0
Emitter followers	25.5	34
Total	29	46.5

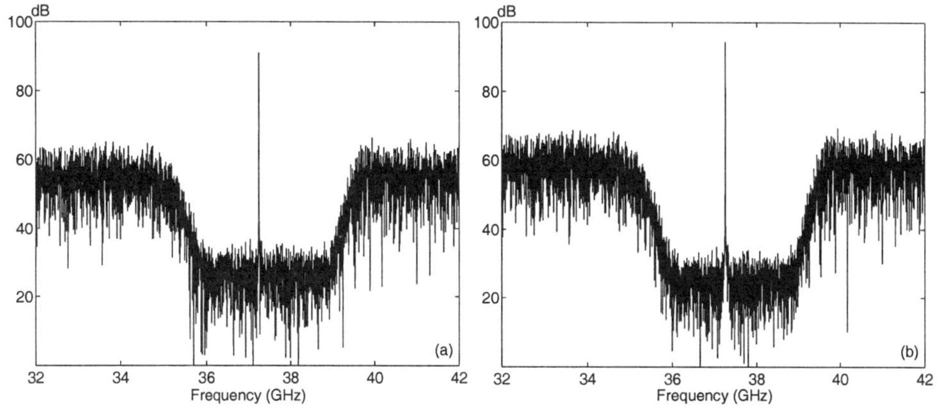

Fig. 2. Simulated output spectrum of the synthesized passband signal for (a) tunnel diode/transistor and (b) transistor-only differential comparator showing 60 dBc SFDR around 37.3 GHz. This simulation uses high speed InP-based HBT and RTD models.

4. Fabrication Approach

A fabrication approach using AlAs/InGaAs/AlAs RTDs and InP/GaAsSb HBTs is described. The material layer structure is grown by molecular beam epitaxy (MBE). The energy band diagram of the structure is shown in Fig. 3.

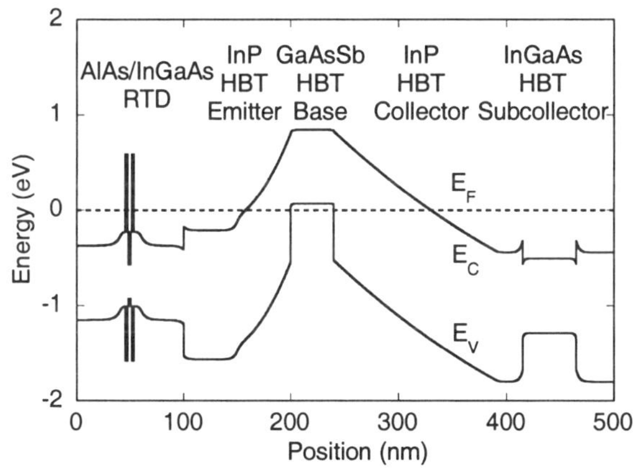

Fig. 3. Simulated energy band diagram of InP-based RTD/HBT structure, computed using the Schrödinger-Poisson solver, BandProf, of W. R. Frensley.

The process uses silicon nitride sidewalls and chemical mechanical polishing (CMP) to get scalable self-aligned contacts for the RTD and HBT. A schematic cross section of the RTD/HBT fabrication process is shown in Fig. 4. The RTD and

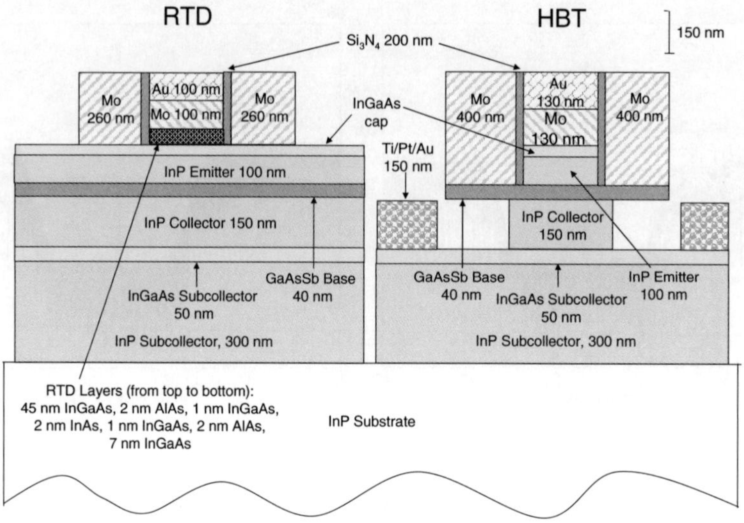

Fig. 4. Integrated resonant tunneling diode and heterojunction bipolar transistor fabrication approach, drawn to scale in the y-direction.

HBT emitter contacts are formed by lift-off. After defining the RTD and HBT mesas using reactive ion etching (RIE), silicon nitride sidewalls are formed by depositing a plasma-enhanced chemical vapor deposition (PECVD) Si_3N_4 film, followed by an RIE process using CHF_3. A blanket deposition of Mo is used to form self-aligned RTD collector and HBT base contacts. Chemical mechanical polishing is employed to polish the molybdenum and electrically isolate the HBT emitter and base contacts, as well as the RTD emitter and collector contacts. The isolation is provided by the silicon nitride sidewall. An RIE process with CF_4/O_2 is used to etch the molybdenum and define the RTD collector and HBT base contacts. The HBT collector contact is formed by lift-off. The silicon nitride sidewalls can be scaled to minimize the HBT base resistance and RTD series resistance and enhance the speed of the devices.

5. Conclusions

A new tunnel diode/transistor differential comparator is described for lowering power dissipation while simultaneously increasing speed. A fabrication approach for InP/GaAsSb HBTs and AlAs/InGaAs/AlAs RTDs using a novel scalable process, featuring nitride sidewalls and chemical mechanical polishing is also outlined. This circuit topology can be applied to all materials systems which can integrate a resonant or Esaki tunnel diode and therefore is applicable to Si and SiGe BiCMOS technologies, GaAs, InP, InAs, and GaN-based transistor technologies.

Acknowledgments

We would like to thank the Office of Naval Research under contract N00014-02-1-0924 for supporting this work. We would also like to thank Patrick Fay, Oliver Collins, and Ajay Gupta for valuable advice and discussion.

References

1. A. Seabaugh, X. Deng, T. Blake, B. Brar, T. Broekaert, R. Lake, F. Morris, and G. Frazier, "Transistors and tunnel diodes for analog/mixed signal circuits and embedded memory", in *1998 Int. Electron Dev. Mtg. Tech. Dig.*, December 1998, pp. 429–432.
2. W. F. Chow, *Principles of Tunnel Diode Circuits*, John Wiley, NY, 1964.
3. J. O. Scanlan, *Analysis and Synthesis of Tunnel Diode Circuits*, John Wiley, NY, 1966.
4. J. C. Yen, Q. Zhang, M. J. Mondry, P. M. Chavarkar, E. L. Hu, S. I. Long, and U. K. Mishra, "Monolithic integrated resonant tunneling diode and heterostructure junction field effect transistor circuits", *Solid-State Electron.* **39** (1996) 1449–1455.
5. T. Waho, K. J. Chen, and M. Yamamoto, "Resonant-tunneling diode and HEMT logic circuits with multiple thresholds and multilevel output", *IEEE J. Solid-State Circ.* **33** (1998) 268–274.
6. A. Seabaugh, B. Brar, T. Broekaert, F. Morris, P. van derWagt, and G. Frazier, "Resonant-tunneling mixed-signal circuit technology", *Solid-State Electron.* **43** (1999) 1355–1365.
7. J. Wang, D. Wheeler, Y. Yan. J. Zhao, S. Howard, and A. Seabaugh, "Silicon tunnel diodes formed by proximity rapid thermal diffusion", *IEEE Electron Dev. Lett.* **24** (2003) 93–95.
8. L. -E. Wernersson, S. Kabeer, V. Zela, E. Lind, J. Zhang, W. Seifert, T. Kosel, and A. Seabaugh, "SiGe Esaki tunnel diodes fabricated by UHV-CVD growth and proximity rapid thermal diffusion", *Electron. Lett.* **40** (2004) 83–85.
9. S. -Y. Chung, N. Jin, P. R. Berger, R. Yu, P. E. Thompson, R. Lake, S. L. Rommel, and S. K. Kurinec, "Three-terminal Si-based negative differential resistance circuit element with adjustable peak-to-valley current ratios using a monolithic vertical integration", *Appl. Phys. Lett.* **84** (2004) 2688–2690.
10. Q. Liu and A. Seabaugh, "Design approach using tunnel diodes for lowering power in differential comparators", submitted, *IEEE Trans. Circ. Syst. II* (2004).
11. A. K. Gupta and O. M. Collins, "A new interpretion and extension of $\Sigma\Delta$ modulation", in *Proc. IEEE Int. Symp. Info. Theory*, Washton, D.C., 2001, pp. 194.
12. A. K. Gupta and O. M. Collins, "Viterbi decoding and $\Sigma\Delta$ modulation", in *Proc. IEEE Int. Symp. Info. Theory*, Lausanne, Switzerland, 2002, pp. 292.
13. T. P. E. Broekaert, B. Brar, J. P. A van der Wagt, A. C. Seabaugh, F. J. Morris, T. S. Moise, E. A. Beam, and G. A. Frazier, "A monolithic 4-Bit 2-Gsps resonant tunneling analog-to-digital converter", *IEEE J. Solid-State Circ.*, **33** (1998) 1342–1349.

International Journal of High Speed Electronics and Systems
Vol. 14, No. 3 (2004) 646–651

BENCHMARK RESULTS FOR HIGH-SPEED 4-BIT ACCUMULATORS IMPLEMENTED IN INDIUM PHOSPHIDE DHBT TECHNOLOGY

STEVEN EUGENE TURNER and DAVID E. KOTECKI
Department of Electrical & Computer Engineering
University of Maine, Orono, ME 04469-5708, USA

High-speed accumulators are frequently used as a benchmark of the high-speed performance and ability to yield large scale circuits in InP double hetereojunction bipolar (DHBT) processes. In previous work, we reported test results of an InP DHBT 4-bit accumulator with 624 transistors operating at 41 GHz clock frequency with a power consumption of 4.1W. In this work, we report on modifications that allow the circuit to operate at a lower supply voltage and a corresponding lower power consumption. Simulation results for this modification indicate that a 16% power reduction can be obtained, while maintaining a high-speed operating frequency of 40 GHz.

1. Introduction

High-speed accumulator circuits are a critical component of direct digital synthesizers (DDS), which are useful for generating frequency-agile waveforms with complex modulation. To allow direct generation of these waveforms at radio frequencies up to X-band, the accumulator circuit must operate at clock rates ≧30 GHz. The accumulator must also have a wide bit width in order to provide adequate frequency resolution, thus requiring transistor counts approaching 5000 devices. The accumulator is a good benchmark of the ability to implement fast circuits with high integration levels in a given semiconductor process. For purposes of this research, we report accumulator benchmarks based on the InP DHBT process reported by Vitesse.[1] These devices have f_t and f_{max} both over 300 GHz, and use a self-aligned, highly manufacturable 0.35 μm device structure.

2. Accumulator Architecture

The accumulator architecture[2] utilizes 2-bit adder blocks, which are cascadable to any 2N-bit width as shown in Fig. 1. The cascaded architecture allows for wide bit-width accumulators with high-speed performance. As the bit-width increases, the total number of accumulators increases linearly as Eq. (1), while the total number of registers increases in a quadratic fashion as Eq. (2). As bit-widths increase, the power consumption of the registers becomes a dominant factor.

$$\# \; of \; accumulators = \frac{\# \; of \; bits}{2} \tag{1}$$

$$\# \; of \; registers = \frac{\# \; of \; bits^2}{8} - \frac{\# \; of \; bits}{4} \tag{2}$$

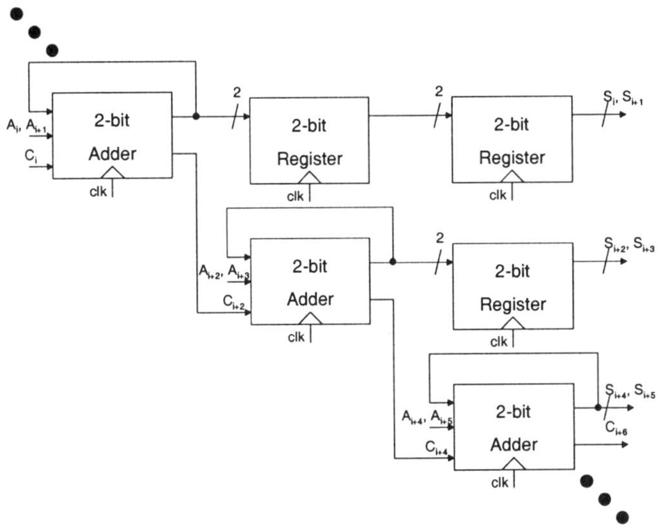

Fig. 1. Schematic of the cascadable accumulator architecture.

The individual 2-bit adder block shown in Fig. 2 contains internal pipelining and an architecture that merges the logic and latching functions.[3,4] The carry and sum blocks contain both logic functions and latches, thus the clock inputs control these internal latches. The left and right sides of the adder are driven by opposite clock phases, resulting in the computation and latching of a full 2-bit add operation in a single clock cycle.

3. Accumulator with Modified Carry Circuit

In prior work,[3,4] the cascoded carry and sum circuits both required four series-gated levels. The registers, whose architecture only requires two series-gated levels, were forced to add two series-gated levels in the form of diodes for compatibility with the sum and carry. These extra levels translated into unnecessary power consumption in the registers, which could be reduced in a design requiring a lower power supply voltage. An initial attempt at power reduction[5] gave rise to the modified carry circuit shown in Fig 3. Instead of using four-level series-gated logic, this design uses single-level parallel-gated logic as a step towards a lower supply voltage. While this circuit could theoretically enable reduction of the power supply by two diode

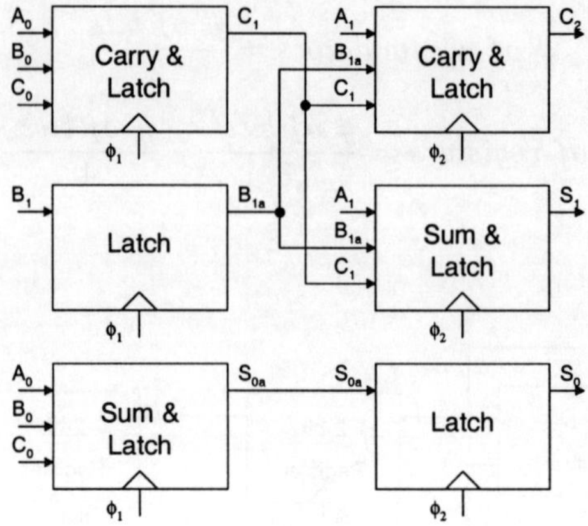

Fig. 2. Schematic of the 2-bit adder block.

drops from 5.5 V to 3.6 V, additional diodes are needed in the carry circuit for compatibility with the four-level series-gated sum circuit.

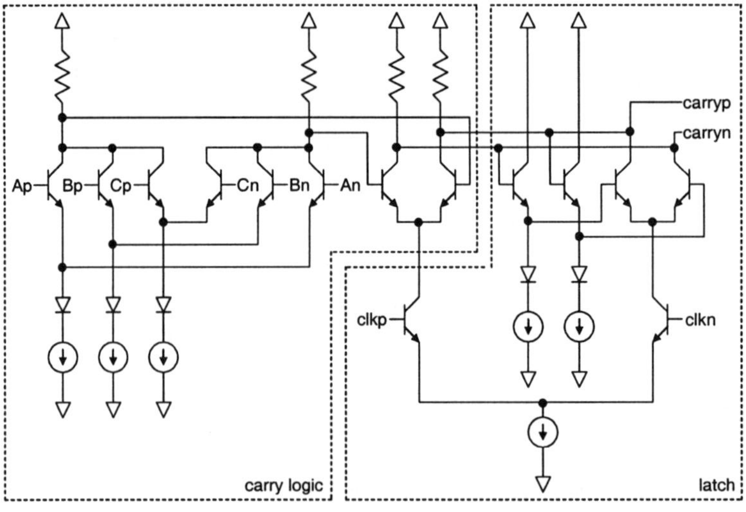

Fig. 3. Schematic of single-level parallel-gated carry circuit.

Although the four-level series gated sum circuit prevents the design from taking advantage of potential power reduction, the 4-bit accumulator constructed with the modified carry circuit showed promising results. Using InP DHBT technology with f_t and f_{max} both over 300 GHz,[1] we were able to realize a 4-bit accumulator test circuit with 624 transistors at a maximum clock frequency of 41 GHz and an overall power consumption of 4.1W. Note that this power consumption figure is for the whole test circuit, including supporting circuitry such as the DAC used for output. The 4-bit accumulator core consisting of two 2-bit adders, one 2-bit register, and the clock tree, requires 2.86 W.

4. Accumulator with Modified Carry and Sum Circuits

In order to reduce the power consumption of the design, the three-level series-gated sum circuit shown in Fig. 4 has been developed. This sum circuit separates the logic function into two three-level portions, instead of using a single stack of logic with four-levels.

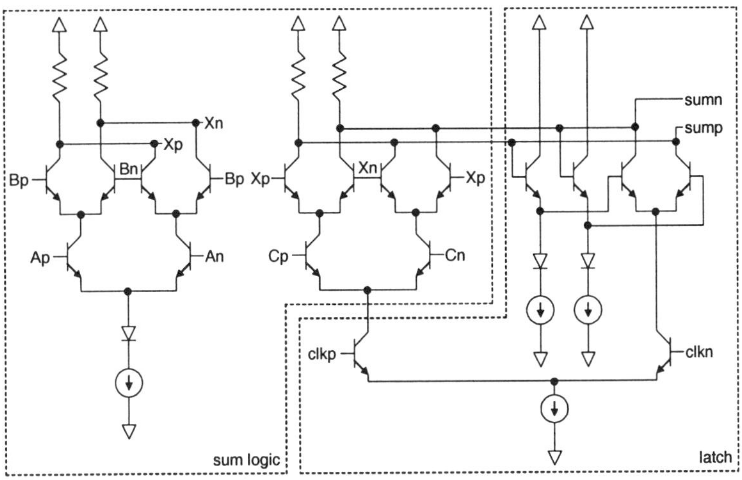

Fig. 4. Schematic of three-level series-gated sum circuit.

In the first stage sum circuit, the 'A' (accumulation increment) inputs change only at low frequency, and the 'B' inputs change only when 'clkn' is active. In this configuration, the 'X' outputs of the first stage are settled before the logic cascoded with 'C' and 'clkp' in the front-end of the second stage becomes active. As a result, the additional first stage has no impact on propagation delay. The second stage of the sum has only three cascaded levels as opposed to four in the previous design, giving it a lower overall propagation delay. While the new sum circuit is faster, the carry circuit, which is common to both designs, dominates the critical timing path. The net effect is that both designs have roughly the same maximum clock frequency.

Support for four series-gated levels in the sum circuit was the only factor constraining the previous accumulator to a 5.5 V power supply. The new sum circuit allows for a reduction of the power supply to 4.5 V by the removal of one diode drop from the remaining circuitry in the test chip, such as the carry circuit of Fig. 3. The lower power supply voltage results in an accumulator core power reduction of 16%.

A 4-bit accumulator test circuit with a single-level parallel-gated carry circuit and a three-level series-gated sum circuit has been designed and taped-out for fabrication. Simulations of a 4-bit accumulator test circuit, including extracted parasitics, showed a maximum clock frequency of 40 GHz. The test circuit includes an on-chip DAC that allows for observation of a single output using a high-speed sampling oscilloscope. Figure 5 shows a simulation of the DAC output of the test circuit operating at 40 GHz with an accumulation increment of seven. The digital values corresponding to the analog output are superimposed on the figure. Power consumption was simulated at 3.7 W for the whole test circuit, which included a DAC with higher power than the previous version. For the accumulator core consisting of two 2-bit adders, one 2-bit register, and the clock tree, the power was simulated as 2.38W. This represents a 16% core power reduction from the previous design.

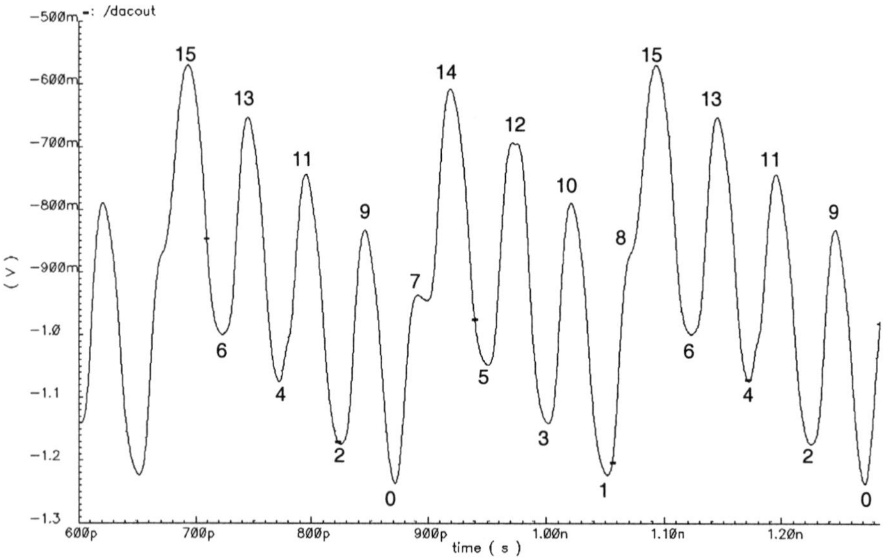

Fig. 5. Extracted simulation of the DAC output of the accumulator test circuit with an accumulation increment of seven, operating at 40 GHz. Digital values corresponding to the analog output are superimposed on the figure.

5. Conclusion

In our previous work, we were able to demonstrate the inherent speed and yield of the InP DHBT process by demonstrating an accumulator test circuit operating at a 41 GHz clock frequency with over 600 transistors. By modifying the sum circuit and reducing the power supply from our previous design, we were able to simulate a reduction in the core power consumption of over 16% while maintaining high frequency operation at 40 GHz. This new circuit is currently in fabrication and will be tested at a later date.

Acknowledgments

This work was supported by the U.S. Army Research Lab and Defense Advanced Research Projects Agency (DARPA) under contract DAAD17-02-C-0115. The authors would like to thank Dr. John Zolper at DARPA, Dr. Alfred Hung at ARL, and Mr. Frank Stroili at BAE Systems for supporting this work. We would also like to thank Mr. Richard Elder Jr. and Mr. Douglas Jansen at BAE Systems for design advice and Mr. Minh Le at Vitesse Semiconductor for fabricating our test chips.

References

1. G. He, J. Howard, M. Le, P. Partyka, B. Li, G. Kim, R. Hess, R. Bryie, R. Lee, S. Rustomji, J. Pepper, M. Kail, M. Helix, R. B. Elder, D. S. Jansen, N. E. Harff, J. F. Prairie, E. S. Daniel, and B. K., "Self-aligned InP DHBT with ft and fmax both over 300 GHz in a new manufacturable technology," *IEEE Electron Device Letters* **25** (2004) pp. 520-522.
2. C. G. Eckroot and S. I. Long, "A GaAs 4-bit adder-accumulator circuit for direct digital synthesis," *IEEE JSSC* **23** (1988) pp. 573–580
3. T. Mathew, S. Jaganathan, D. Scott, S. Krishnan, Y. Wei, M. Urtega, M. Rodwell, and S. Long, "2-bit adder carry and sum logic circuits clocking at 19 GHz clock frequency in transferred substrate HBT technology," *in Proc. of IEEE International Conference on Indium Phosphide and Related Materials*, Nara, Japan, May 2001, pp. 505–508.
4. T. Mathew, S. Jaganathan, D. Scott, S. Krishnan, Y. Wei, M. Urtega, M. J. W. Rodwell, and S. Long, "2-bit adder: carry and sum logic circuits at 19 GHz clock frequency in InAlAs/InGaAs HBT technology," *Electronics Letters* **37** (2001) pp. 1156–1157.
5. S. E. Turner, D. S. Jansen, and D. E. Kotecki, "4-bit adder-accumulator at 41 GHz clock frequency in InP DHBT technology," *IEEE Microwave and Wireless Components Letters* (2004), submitted for publication.

International Journal of High Speed Electronics and Systems
Vol. 14, No. 3 (2004) 652–657

ATOMICALLY FLAT III-ANTIMONIDE EPILAYERS GROWN USING LIQUID PHASE EPITAXY

ANIKA KUMAR, SUJATHA SRIDARAN, P. S. DUTTA†
Electrical, Computers and Systems Engineering Department
Rensselaer Polytechnic Institute, Troy, NY 12180, USA

A novel process has been developed which allows the growth of device grade ultra-smooth epitaxial layers of antimonide based III-V compounds with a controlled thickness using Liquid Phase Epitaxy (LPE). GaSb epilayers (with thickness in the range of 20-50 m) on GaSb single crystalline substrates have been grown with a root mean square surface roughness of less than 1 nm over an area of 5 §←5 m.

1 Introduction

Device grade, single crystal substrates of only a few binary compounds (such as GaAs, GaSb, InP, CdTe) with discrete band gaps and lattice constants are commercially available. Ternary or quaternary based devices are fabricated on thin epitaxial layers grown on binary substrates. To grow these compound semiconductor epilayers, various epitaxial techniques such as molecular beam epitaxy (MBE), metalorganic chemical vapor deposition (MOCVD), and liquid phase epitaxy (LPE) have been in use. However for antimonide based materials, the extrinsic doping levels possible using MOCVD and MBE are in a limited range due to high concentrations of native defects in the material [1]. LPE growth takes place under near equilibrium conditions at low temperatures and therefore has low native defects enabling a wide range of doping levels. LPE also has an additional advantage of high growth rate compared to MOCVD and MBE. However, LPE suffers from the problems of non- uniform interface and rough morphology which limits it application as an industrial epitaxy technique for a large number of devices. In this paper, we have investigated the fundamental growth mechanism of the epilayers and engineered the growth process enabling ultra-smooth layers with predictable thicknesses. GaSb epi-growth on GaSb substrate has been used as a model material in this work. The process consists of isothermal growth following an initial supercooling. Data of surface roughness and layer morphology will be presented in this paper.

†$_\psi$duttap@rpi.edu

2 Experimental details

LPE growth was carried out in a single-zone transparent vertical Bridgman furnace in hydrogen atmosphere using the dipping technique. The growth setup is shown in figure 1. A multi-bin crucible made from high purity graphite and a quartz wafer carrier was used.

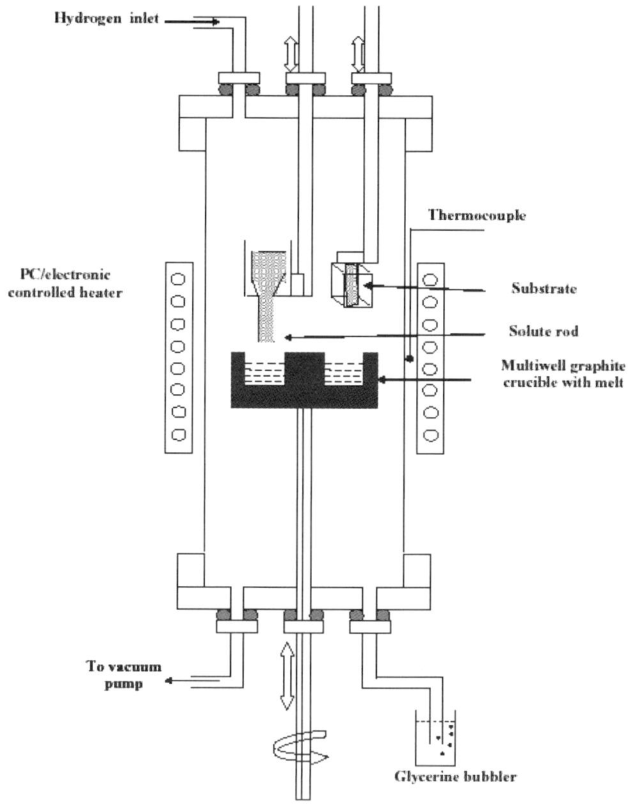

Figure 1: A schematic illustration of the epilayer growth setup.

Epitaxial layers were grown on the (100) surface of single crystalline Te-doped GaSb wafers from Wafer Tech. Corp. The wafers were degreased in warm xylene, acetone and methanol and the surface oxide was removed by a dip in HCl and a solution of Cr_2O_3:HF:H_2O (34gm:5ml:120ml). Gallium was used as the growth solution. Prior to growth, it was necessary to baked the gallium in ultra-pure hydrogen flow at 850 °C for 5-10 minutes in order to eliminate the oxide scum on the top of the melt. During this baking, the GaSb wafer and the GaSb solute rod (used for supersaturating the growth solution) were kept inside the growth chamber at about 200 °C. After the high temperature baking, the furnace was cooled to the growth temperature (in 30 minutes) and allowed to stabilize for 2-3 hours. The growth solution was prepared by dissolution of the GaSb solute rod in the Ga-liquid for 2-5 minutes at the final growth temperature. In initial growth runs, it was observed that during solute dissolution, an oxide scum was formed on the top of the solution that led to surface contamination of the wafer surface and inhibited proper nucleation during growth. Hence an oxide desorption Ga-solution was separately put in one of the graphite bins. Prior to dissolution, the surface of the GaSb rod was cleaned in the Ga-solution for 30-60 seconds. This process was also repeated for the actual growth wafer. Two separate oxide desorption solutions were used for the solute rod and the GaSb wafer to avoid cross contamination of the wafer surface with oxide scum. Without the surface oxide desorption process, the morphology of the GaSb epilayers were extremely rough and thicknesses inhomogeneous. After cleaning the GaSb wafer in a wafer oxide desorption melt, it was inserted in the growth solution. We have used two different temperature-time profiles for the epigrowth as shown in figure 2 and will be discussed in section 3. After the growth, the surfaces were cleaned using hot HCl to remove the excess gallium and the surface morphology of the layers were examined by a Nikon microscope and Scanning Electron Microscope (SEM). The root mean square roughnesses of the layers were analyzed using an Atomic Force microscope (AFM) and the layers were optically characterized by a Thermo-Nicolet Fourier Infra-red Transform. (FTIR) spectroscope. The thickness of the layers was measured by a profilometer.

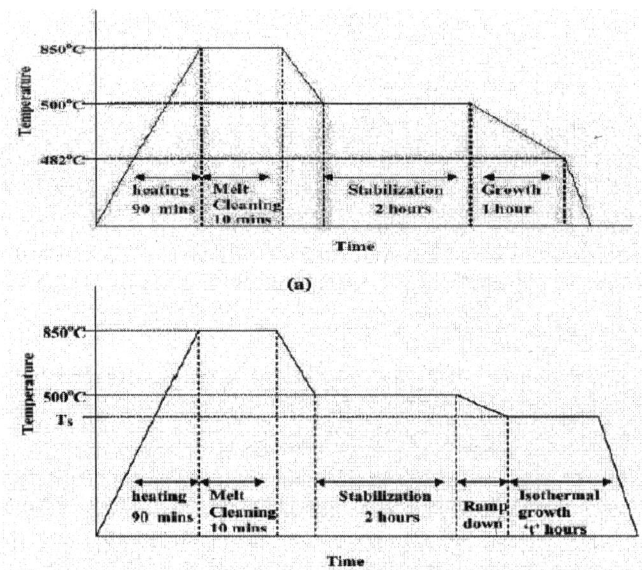

Figure 2 : The temperature-time profiles used for (a) 2-D epilayer growth using ramp- cooling and (b) isothermal growth with initial supercooling.

3 Results and Discussions

3.1 2-Dimensional Epilayer Growth using Ramp-cooling

The temperature profile shown in figure 2(a) was used in these experiments. To study the evolution of 2-D island growth, the duration of growth was varied from 30 minutes to 90 minutes, while the cooling rate of the growth solution was kept constant at 0.3 °C/min for all the experiments. It should be noted that the cooling rate used by us is much lower than that used in literature (e.g. 2-5 °C/min [3])[4,5,6]. This was intentionally done to enable the 2-D growth. The growth morphologies and SEM pictures can be seen in figure 3. Rectangular facets could be seen in all the layers.

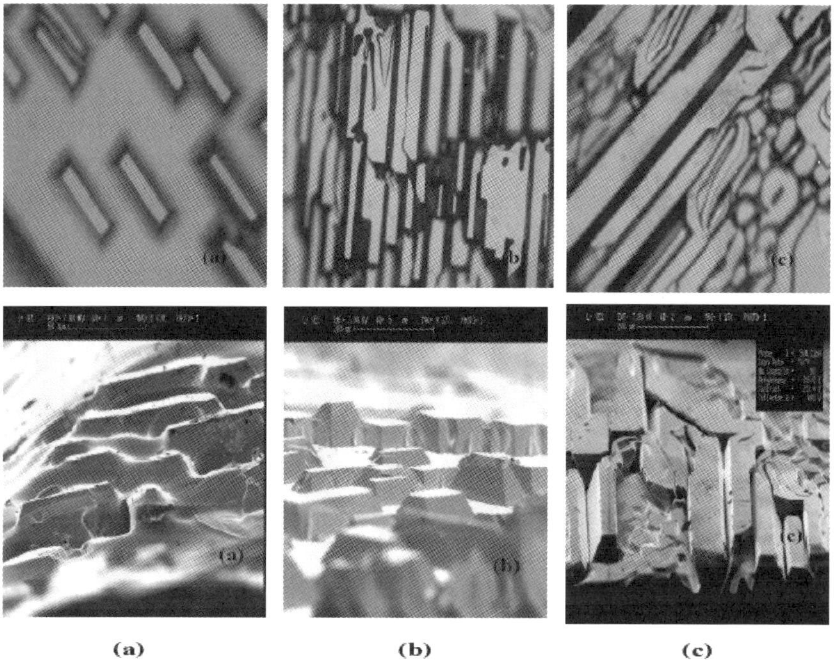

Figure 3: Optical and SEM micrographs of GaSb/GaSb layers grown with a cooling rate of 0.3°C/min for growth times of (a) 30 minutes (b) 45 minutes and (c) 60 minutes. The top row shows the optical micrographs (Magnification = 60X). The bottom row shows SEM pictures of corresponding samples.

The top faces of the facets were extremely smooth. An increase in the time of growth leads to a steady increase in the lateral dimensions of the facets implying different growth rates along different directions. The heights of the facets are approximately 50 μm for the layers grown for 45 minutes. AFM measurements show a root mean square roughness of the surface of these facets to be an average of 0.9 nm for a scan of 5 x 5 μm^2 area. After a growth time of 90 minutes, the facets merge. However, complete coverage of the substrate is not achieved at the same time. The occurrence of these facets can be explained by the fact that {100} and {111} planes have interfacial energy densities which are the highest and the lowest respectively[7]. Hence the {100} plane will tend to breakdown and produce {111} planes. This leads to the formation of facets instead of uniform terraces under non-equilibrium conditions.

3.2 *Isothermal growth with initial supercooling*

Even though the above process gave extremely flat layers, complete coverage of the substrate by the individual islands *at the same time* could not be controlled.

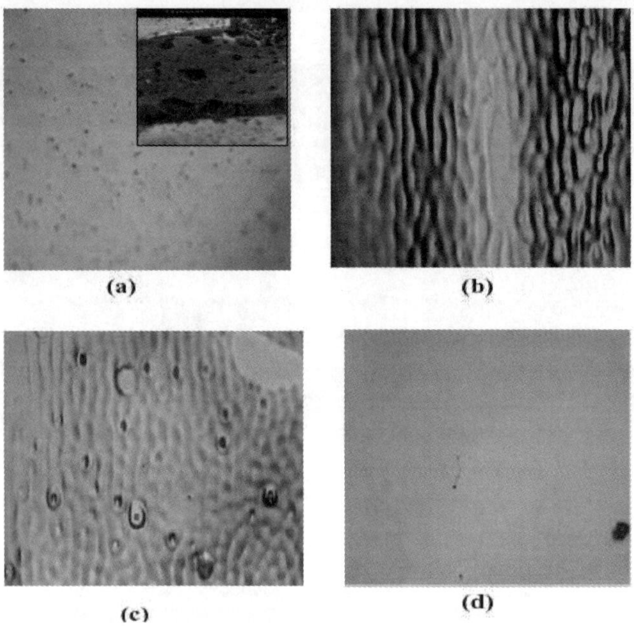

Figure 4: Optical micrographs of GaSb/GaSb layers grown for (a)$T_{s\psi}$= 2°C, ramp rate = 0.3°C/min, t= 2 hours; (inset)SEM picture of the layer (b)$T_{s\psi}$= 4°C, ramp rate = 0.3°C/min t= 2 hours (c) $T_{s\psi}$= 2°C, ramp rate = 0.3°C/min, t= 4 hours (d) $T_{s\psi}$= 2°C, ramp rate = 0.1°C/min , t= 2 hours; (Magnification = 60X)

Due to the fact that upon supersaturation, the solute takes time to diffuse through the boundary layer before it can attach to the substrate, it is intuitive that a very low supersaturation coupled with long growth duration is required to achieve a uniformly thick layer using LPE [2]. Hence, a new temperature profile as shown in figure 2(b) was used in subsequent experiments. In this process, the growth melt was cooled by an amount T_S = 2-4°C from the saturation temperature at a rate of 0.1-0.3 °C/min followed by an isothermal growth time, t= 2-4 hours. The growth wafer was inserted in the melt prior to the cooling cycle and removed at the end of the isothermal growth period. Continuous layers with complete substrate coverage could be grown using this strategy. It was observed that for a higher cooling rate, lots of pin holes were present in the layer (see Figure 4a). By decreasing the cooling rate, pinholes were eliminated (as shown in figure 4(d)). AFM measurements show that the rms roughness of the surface in figure 4(d) is 0▷88 nm. For comparison, the rms roughness of a typical GaSb starting substrate is in the range of 0.3-0.5 nm. Achievement of atomically flat surfaces (contrary to what is expected in LPE grown layers) is due to the etch-back and re-growth process as a result of the long isothermal sequence used in our temperature-time profile during growth. It should be noted that the initial surface oxide desorption process is crucial in ensuring uniform nucleation during growth. It is possible that during the oxide desorption process in gallium solution, etch pits get exposed which could form nucleation centers for the 2-D islands. The transition from 2-D faceted growth to a continuous layer takes place by a continuous etch-back and re-growth mechanism.

4 Conclusion

A new 2-D island dissolution and re-growth process has been shown to result in atomically flat epilayers by liquid phase epitaxy under isothermal growth conditions. Extremely smooth layers over large area up to 1 x 1 cm^2 have been obtained. Using this process, it is anticipated that epilayers with controlled thicknesses could be grown by changing the initial supercooling level in the solution and the growth times.

Acknowledgments

The authors would like to acknowledge financial support from National Science Foundation (NSF) through Faculty Early Career Development Award ECS-0093706 and Rensselaer Seed Research Grant. Thanks to Mr. Jeff Langer for the SEM pictures and Mr. Vinay Bhagwat for AFM scans.

References

1. P.S. Dutta, H.L. Bhat and V. Kumar, *J. Appl. Phys.*, **81** (1997) 5821-5870.
2. H.J. Scheel, *J. Crystal Growth,* **211** (2000) 1-12.
3. P.S. Dutta, K.S.R. Koteswara Rao, H.L. Bhat, K. Gopalkrishna Naik, V. Kumar, *J. Crystal Growth*, **152** (1995) 14-20 .
4. I. Crossley, M.B. Small, *J. Crystal growth*, **19** (1973) 160-168.
5. C.Woelk, K.W. Benz, *J. Crystal Growth*, **27** (1974) 177-182.
6. N.K. Udayshankar, H.L. Bhat, *Bull. Mater. Sci.*, **26(7)** (2003) 685-692.
7. B.L. Mattes, R.K. Route, *J. Crystal Growth,* **27** (1974) 133-141.

International Journal of High Speed Electronics and Systems
Vol. 14, No. 3 (2004) 658–663

NATIVE DEFECT COMPENSATION IN III-ANTIMONIDE BULK SUBSTRATES

ROBINSON PINO, YOUNGOK KO, and PARTHA S. DUTTA[†]
Electrical, Computer and Systems Engineering Department
Rensselaer Polytechnic Institute, Troy, NY 12180-3590, USA
[†] *duttap@rpi.edu*

As-grown, undoped III-antimonide bulk substrates contain high concentration of native defects resulting in high residual carrier density. In this paper, we have demonstrated that native defects can be compensated in bulk substrates of GaSb, InSb, and $Ga_{1-x}In_xSb$ via impurity doping and low temperature growth from nonstoichiometric melts and solutions. Decrease in residual carrier concentration up to one order of magnitude at 300 K and three orders of magnitude at 77 K have been achieved.

1. Introduction

III-Antimonide compound semiconductors are of particular interest amongst III-V semiconductors since the band gap can be tuned from approximately 0.17 to 1.6 eV. Furthermore, the possibility for band gap engineering in the 6.1 to 6.5 Å lattice constant range offers extraordinary opportunities for the development of novel band aligned devices for high speed electronic and photonic applications operating between the 2 to 14 μm spectral region.[1−6] Substrate properties play a significant role in the performance of devices. Antimonide based substrates are undermined by the high concentration of native defects (i.e. vacancies and antisites). For example, as grown undoped GaSb, InSb, and $Ga_{1-x}In_xSb$ exhibit high concentration of native defects and free carrier absorption.[6−8] In this work, two methodologies have been employed to compensate native defects in vertical Bridgman grown bulk substrates of GaSb, InSb, and $Ga_{1-x}In_xSb$.

GaSb substrates are readily available from commercial vendors. However, they are for the most part not optically transparent in the regions of interest and they are also highly conductive.[9] These properties severely limit the performance of optical and electronic devices grown on GaSb substrates. As grown undoped GaSb is always p-type in nature irrespective of growth technique and conditions.[7] Work over the last 3 decades has been devoted mainly for understanding the origin of the residual acceptors that are the limiting factors for both fundamental studies and device applications.[7] The residual acceptors with concentration of $\sim 10^{17}$ cm^{-3} have been found to be related to gallium vacancies (V_{Ga}) and gallium in antimony site (Ga_{Sb}) with doubly ionizable nature.[10]

Native defects in InSb have been studied by Kendall and Huggins.[11] they studied the self-diffusion in InSb using radiotracers and refined sectioning techniques. They proposed the In:Sb divacancy as the defect primarily responsible for self-diffusion of both components in InSb. In addition, they reported that the diffusion coefficients of neither In nor Sb seemed to depend on the ambient Sb pressure. Kendal and Huggins[11] showed that the native defect concentration in InSb can be greatly suppressed if the material is grown at low temperatures. However, to the best of our knowledge, no work has been reported in the low temperature bulk growth of InSb substrates.

The electrical properties of $Ga_{1-x}In_xSb$ have been investigated in detail by Joullie and co-workers.[12] As-grown undoped $Ga_{1-x}In_xSb$ is n-type in nature above $x = 0.5$ and p-type for x below 0.5.[12] Therefore, it would be expected that at a composition $x = 0.5$, $Ga_{1-x}In_xSb$ would exhibit near intrinsic properties as the native p-type like defects from GaSb would compensate for the n-type like native defects inherited from InSb.

In this work, we have successfully achieved a high level of native defect compensation in GaSb, InSb, and $Ga_{1-x}In_xSb$ bulk substrates by implementing two specific approaches during the vertical Bridgman growth process. First, compensation of the residual native acceptor concentration in GaSb and $Ga_{1-x}In_xSb$ bulk crystals was achieved via tellurium doping of the melt during growth. Second, InSb bulk crystals were grown at temperatures below the eutectic melting point, ~525 °C, from indium rich melt compositions. The electrical and optical properties of the grown crystals are presented here.

2. Experimental Details

For the charge synthesis of bulk GaSb and InSb crystals, 7N pure Ga and 6N pure Te, In and Sb were used as-received without any chemical treatment. The typical melt height was between 2 and 4 cm. The temperature gradient imposed by the furnace near the melt-solid interface was approximately 5 °C/cm. Prior to the growth of GaSb, InSb, and $Ga_{1-x}In_xSb$ bulk crystals, the following growth procedure was employed. First, the starting source materials were elevated to a temperature 50 °C above the melting point of the binary material (712 °C for GaSb and 525 °C for InSb). Then, the melt was homogenized for 1 h by rotating the crucible at 120 rpm while a static stirrer remained in the melt within 0.5 cm from the bottom (where growth initiated) of the crucible. After homogenization, the crucible was lowered at a constant vertical translational rate of 1 cm/h (0.5 cm/h for $Ga_{1-x}In_xSb$) and rotated continuously at 120 rpm, while the static stirrer remained in the melt within 0.5 cm of the solid-liquid interface at all times to keep the melt homogeneous. After the entire ingot reached outside the hot zone of the furnace, the crystal was cooled to room temperature over a period of 12 h by appropriately programming the furnace temperature-time profile.

For native defect compensation in bulk GaSb crystals, considering that undoped GaSb has between (1 to 2)$\times 10^{17}$cm^{-3} acceptors and the segregation coefficient of Te is 0.37,[13] the initial Te concentration in the melt was set at 1×10^{18} cm^{-3} to obtain the highest level of compensation at approximately midway along the growth direction. Native defects in pure bulk GaSb crystals have segregation coefficient close to unity.[7] Hence all the pure GaSb samples exhibited similar optical and electrical properties, substrate GS0. Several wafers were extracted from the region of highest compensation of the Te doped GaSb ingot, the mid region along the growth direction. The extracted wafers were labeled GS1 and GS2 respectively. Substrates GS1 and GS2 are located approximately 8 and 18 mm from bottom of the boule. Native defect reduction in InSb was achieved by the growth of the material from indium rich melts. As mentioned previously, Kendall and Huggins[11] demonstrated that the concentration of native defects in InSb can be suppressed if the material is grown at low temperatures. Therefore, InSb crystals were grown from stoichiometric and indium rich melts. Thus, low temperature grown InSb crystals were obtained from temperatures varying from below the eutectic point, 525 °C, down to very close to the melting point of elemental indium, 150 °C, according to the InSb phase diagram.[14] From the grown stoichiometric and nonstoichiometric InSb boules substrates IS1 and IS2 were obtained from approximately 4 and 18 mm from the bottom of the boules. Similarly, $Ga_{1-x}In_xSb$ bulk crystals were grown with an approximately 1×10^{18} cm^{-3} Te doping concentration and with a starting melt composition of $x = 0.75$. $Ga_{1-x}In_xSb$ substrates with varying compositions x were obtained namely, substrates IGS1, IGS2, and IGS3. All substrates were polycrystalline, without inclusions, and their properties are tabulated in Table 1.

3. Results and Discussion

Table 1 displays the electronic properties of GaSb and InSb substrates for different growth conditions at 300 and 77 K. From the table it is observed that the properties for each substrate material vary depending on the growth conditions. If we look at the GaSb properties, we can observe that the carrier concentration of the undoped pure GaSb substrate (GS0) is almost an order of magnitude higher than that observed in the Te doped GaSb substrates (GS1 and GS2). The mobility of substrate GS0 is approximately 5 times larger than in substrate GS1 and 15 times larger than in substrate GS2. However, the resistivity observed in the Te compensated substrates are approximately 15 (GS1) and 300 (GS2) times larger than the resistivity observed for the undoped GaSb substrate (GS0). The effects of impurity compensation in GaSb have been observed and explained previously in the literature. Baxter et al.[15] studied the behavior of the electron mobility in Te compensated GaSb samples at 77 K as well as in lithium diffused Te compensated GaSb samples at 300 and 77 K. They attributed the monotonic variation of mobility with carrier concentration to the dominance of impurity scattering under conditions of high compensation. Similarly, for InSb, in Table 1, it is observed that

Table 1. Electrical properties of GaSb, InSb, and $Ga_{1-x}In_xSb$ bulk substrates at 300 and 77 K for various growth conditions.

Material: dopant	Sample Name	Type	Temp. (K)	Carrier Conc. (cm^{-3})	Mobility (cm^2/Vs)	Resistivity (Ω-cm)
GaSb	GS0 [a]	p	300	1.92×10^{17}	559.15	0.061
GaSb	GS0 [a]	p	77	2.38×10^{16}	1,973.90	0.13
GaSb:Te	GS1 [b]	p	300	2.66×10^{16}	129.35	1.83
GaSb:Te	GS1 [b]	p	77	3.50×10^{13}	48.15	3,720.00
GaSb:Te	GS2 [c]	n	300	1.16×10^{16}	39.65	16.40
GaSb:Te	GS2 [c]	n	77	1.16×10^{14}	38.40	323.96
InSb	IS1 [d]	n	300	2.59×10^{17}	4.79×10^{4}	3.12×10^{-4}
InSb	IS2 [e]	n	300	3.58×10^{16}	5.64×10^{4}	3.25×10^{-3}
InSb	IS2 [e]	n	77	2.81×10^{16}	7.72×10^{4}	3.12×10^{-3}
$Ga_{0.68}In_{0.32}Sb$:Te	GIS1 [f]	n	300	2.93×10^{17}	7.03×10^{3}	3.41×10^{-3}
$Ga_{0.66}In_{0.34}Sb$:Te	GIS2 [g]	n	300	3.76×10^{17}	8.48×10^{3}	2.06×10^{-3}
$Ga_{0.58}In_{0.42}Sb$:Te	GIS3 [h]	n	300	4.92×10^{17}	1.46×10^{4}	8.70×10^{-4}

Note: Growth condition and sample thickness
[a]Unintentionally doped pure GaSb, 730 μm thickness.
[b,c]Te doped GaSb, (b) 390 and (c) 410 μm thickness.
[d]Unintentionally doped pure stoichiometric grown InSb, 620 μm thickness.
[e]Low temperature grown (~275 °C) InSb from indium rich melt, 600 μm thickness.
[f,g,h]Te doped $Ga_{1-x}In_xSb$, (f) 620, (g) 600, and (h) 740 μm thickness.

the displayed electrical properties for the low temperature grown substrates (IS2) are better than those observed in the high temperature grown InSb substrates (IS1). For example, the low temperature substrate IS2 shows lower carrier concentration, higher mobility, and higher resistivity compared to the high temperature grown substrate IS1. These data correlate with the expected native defect concentration results obtained by Kendall and Huggins[11] where low native defect concentration is expected for low temperature grown InSb. For the Te doped $Ga_{1-x}In_xSb$ substrates (GIS1, GIS2, and GIS3), we can observe from Table 1 that the electrical properties for $x < 0.42$ are very similar to the properties displayed by the low temperature grown InSb substrates (IS2). On the other hand, the electrical properties of the Te doped $Ga_{1-x}In_xSb$ substrates (GIS1, GIS2, and GIS3) do not compare well with the properties displayed by the Te doped GaSb substrates (GS1 and GS2). As we can see from Table 1, the carrier concentration exhibited by the Te doped $Ga_{1-x}In_xSb$ substrates (GIS1, GIS2, and GIS3) are about one order of magnitude higher and the resistivities are about three orders of magnitude lower relative to the Te doped GaSb substrates (GS1 and GS2). However, the electron mobility exhibited by the $Ga_{1-x}In_xSb$ substrates (GIS1, GIS2, and GIS3) are between two to three orders of magnitude higher compared to the Te doped GaSb substrates (GS1 and GS2) which would be advantageous for high speed electronic and photonic applications.

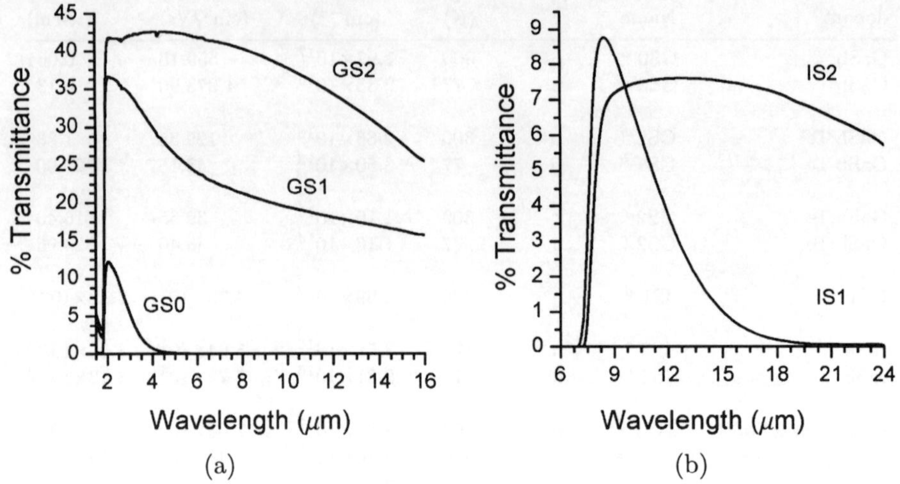

Fig. 1. Near- and mid-IR transmission measurements of pure and Te compensated GaSb and InSb substrates at 300 K, in atmospheric ambient. (a) Sample GS0 corresponds to pure GaSb sample and samples GS1 and GS2 correspond to Te compensated GaSb samples. (b) Sample IS1 corresponds to pure InSb grown at the eutectic point, ∼525 °C, and sample IS2 corresponds to low temperature grown InSb from an indium rich nonstoichiometric melt.

Figure 1 displays the near- and mid-IR optical transmission properties of all GaSb and InSb substrates. From Fig. 1a, it is observed that the transmission properties exhibited by the Te compensated GaSb substrates, GS1 and GS2, are far more transparent and broad compared to the properties observed for the undoped GaSb substrates (GS0). Also, Fig. 1b shows that the transmission properties of the low temperature grown InSb substrates (IS2) are more transparent and broad compared to the transmission properties observed for the high temperature grown InSb substrates (IS1). Also, note the difference in the observed band edge between IS1 and IS2 which can be predicted by the Burstein-Moss shift.[16] This further proves that indeed we have suppressed the free carrier density in InSb via low temperature growth. In fact, Fig. 1 helps to corroborate the electrical properties shown in Table 1. Table 1 and Fig 1 show a substantial decrease in carrier concentration, free carrier absorption, and an increase in resistivity for the Te compensated GaSb (GS1 and GS2) and low temperature grown InSb substrates (IS2) compared to the undoped GaSb (GS0) and InSb (IS1) substrates.

4. Conclusions

Suppression of native defects has been achieved in Bridgman grown bulk substrates of GaSb, InSb and $Ga_{1-x}In_xSb$ via impurity doping and low temperature growth.

Superior optoelectronic properties have been observed in the obtained substrates namely lower residual carrier concentration and free carrier absorption as well as higher resistivity compared to the undoped stoichiometric grown substrates.

Acknowledgments

This research was supported in part by the Office of Naval Research under contract No. N00014-02-1-0458.

References

1. J.B Boos, W. Kruppa, B.R. Bennett, D. Park, S. W. Kirchoefer, R. Bass, and H. B. Dietrich, "AlSb/InAs HEMT's for low–voltage, high–speed applications," *IEEE Trans. Electron Devices* **45** (1998) 1869.
2. D.H. Chow, H.L. Dunlap, W. III Williamson, S. Enquist, B. K. Gilbert, S. Subramaniam, P. -M. Lei, and G. H. Bernstein, "InAs/AlSb/GaSb resonant interband tunneling diodes and Au–on–InAs/AlSb–superlattice Schottky diodes for logic circuits," *IEEE Electron Device Lett.* **17** (1996) 69.
3. F. Fuchs, U. Weimer, W. Pletschen, J. Schmitz, E. Ahlswede, M. Walther, J. Wagner, and P. Koidl, "High performance InAs/$Ga_{1-x}In_xSb$ superlattice infrared photodiodes," *App. Phys. Lett.* **71** (1997) 3251.
4. W.W. Bewley, C.L. Felix, I. Vurgaftman, D.W. Stokes, E.H. Aifer, L.J. Olafsen, and J.R. Meyer, M. J. Yang, B. V. Shanabrook, H. Lee, R. U. Martinelli, A. R. Sugg, "High–temperature continuous–wave 3-6.1 μm 'W' lasers with diamond–pressure–bond heat sinking," *App. Phys. Lett.* **74** (1999) 1075.
5. P.R. Hammar, B. R. Bennett, M. J. Yang, and Mark Johnson, "Observation of Spin Injection at a Ferromagnet–Semiconductor Interface," *Phys. Rev. Lett.* **83** (1999) 203.
6. B.R. Bennett and B.V. Shanabrook in "Thin Films: Heteroepitaxial Systems", Chapter 9, Edited by A.W.K. Liu and M.B. Santos, *World Scientific Publishing Co.* (2000).
7. P.S. Dutta, H.L. Bhat and V. Kumar, "The physics and technology of gallium antimonide: An emerging optoelectronic material," *J. Appl. Phys.* **81** (1997) 5821.
8. "Antimonide related strained layer heterostructures" Edited by M.O. Manasreh, Optoelectronic properties of semiconductors and superlattices series, vol. 3, *Gordon and Breach Science Publishers* (1997).
9. C. Mailhiot and D.L. Smith, "Long–wavelength infrared detectors based on strained $InAsGa_{1x}In_xSb$ type–II superlattices," *J. Vac. Sci. Technol. A* **7**, 445 (1989).
10. R.D. Baxter, R.T. Bate, and F.J. Reid, "Ion–Paring Between Lithium and the residual Acceptors in GaSb," *J. Phys. Chem. Solids* **26**, 41 (1965).
11. Don L. Kendall and Robert A. Huggins, "Sefl–Diffusion in Indium Antimonide," *J. Appl. Phys.* **40** (1969) 2750.
12. A. Joullie, J. Allegre, and G. Bougnot, *Mater. Res. Bull.* **7**, 1101 (1972).
13. P.S. Dutta and A.G. Ostrogorsky, "Segregation of tellurium in GaSb single crystals and associated diffusion coefficient in the solute layer," *J. Crystal Growth* **197** (1999) 749.
14. Massalski, T. B., "Binary Alloy Phase Diagrams," Vol 2. *ASM International*, Ohio, January 1986.
15. R.D. Baxter, F.J. Reid, and A.C. Reer, "Electron Mobility in GaSb at 77 K," *Phys. Rev.* **162** (1967) 718.
16. E. Burstein, "Anomalous Optical Absorption Limit in InSb," *Phys. Rev.* **93** (1954) 632 and T.S. Moss, *Proc. Phys. Soc. London Sec.* **B76** (1954) 775.

International Journal of High Speed Electronics and Systems
Vol. 14, No. 3 (2004) 664–669

NOISE AND THZ RECTIFICATION CHARACTERISTICS OF ZERO-BIAS QUANTUM TUNNELING SB-HETEROSTRUCTURE DIODES

ARTTU LUUKANEN¶*
ERICH N. GROSSMAN§
National Institute of Standards and Technology†
325 Broadway, Boulder, CO 80305, USA
¶ *luukanen@boulder.nist.gov*
§ *grossman@boulder.nist.gov*

HARRIS P. MOYER♯
JOEL N. SCHULMAN ♠
HRL Laboratories LLC, 3011 Malibu Canyon Road, Malibu CA 90265-4797, USA
♯ *hpmoyer@hrl.com*
♠ *schulman@hrl.com*

The Sb-heterostructure quantum tunneling diode, fabricated from epitaxial layers of InAs and AlGaSb, is a recently proposed device for direct detection and mixing in the submillimeter wavelength range. These diodes exhibit especially high curvature in the current-voltage characteristic that produces the rectification or mixing without bias. Operation without bias is a highly desirable feature as the device does not suffer from large $1/f$ noise, a major shortcoming in other devices such as Schottky barrier diodes or resistive room temperature bolometers. In this paper we present the noise characteristics of the diode as a function of the bias voltage. At room temperature and zero bias, the device demonstrates a Johnson noise limited intrinsic noise equivalent power of 1 pW/Hz$^{1/2}$. In addition to the noise measurements, we present the detection characteristics of the diode at a frequency of 2.5 THz. The measured THz laser response deviates from conventional theoretical prediction based on pure rectification. The reasons for the discrepancy will be discussed.

*On leave from VTT Information Technology, Espoo, Finland
†Contribution of U.S. Government, not subject to copyright in the U.S.

1. Introduction

At present, millimeter-wave imaging focal-plane arrays are based on microwave monolithic integrated circuits (MMICs), and their pixel count is limited by the complexity and cost of the components and assembly [1]. Moreover, the maximum operation frequency is limited by the availability of low noise amplifiers to $\lesssim$ 200 GHz. Bolometers can provide a low cost alternative and cover frequencies up to the infrared, albeit with lower sensitivity [2,3,4]. Schottky diodes, the workhorse technology for high frequency mixers, can also operate as direct detectors up to several THz, though they typically display an RC -limited rolloff above a few hundred GHz. Their great drawback has been $1/f$ noise, brought about by the DC bias required for sufficient curvature of the IV curve.

In this paper we present the $1/f$ noise performance of a recently introduced direct detector, the InAs/AlSb/GaSb heterostructure backward diode [5,6,7]. The principle of operation is similar to that of an Esaki tunnel diode in that the controlling current mechanism is interband tunneling between adjacent energetically offset semiconductor regions. For low Al concentrations, the Type II band gap lineup between InAs and AlGaSb, in which the conduction band minimum of InAs lies energetically below the valence band maximum of AlGaSb, creates a natural asymmetry in the current flow with bias direction. The special feature of these diodes is that they exhibit highly non-linear IV characteristics near zero bias. In the ideal case, the absence of DC bias means that the devices do not suffer from $1/f$ noise. These merits combined suggest that the quantum tunneling diodes can provide an attractive choice for large integrated millimeter-wave imaging arrays of the future.

2. Diode *IV* measurements

The InAs/AlSb/GaSb tunnel diode layers are deposited using molecular beam epitaxy on semi-insulating GaAs substrates, using an InAs buffer layer process, identical to that described in Ref. 8. The InAs layer is followed by a 200 nm thick GaSb anode, 20 nm of undoped AlGaSb, a 3.2 nm thick undoped AlSb tunnel barrier, a 50 nm thick InAs cathode, and finally an n^+ InAs contact layer.

Three nominally-identical diodes with a junction area of 2 μm $\times$ 2 μm were characterized by first measuring their DC IV characteristics, while monitoring the differential conductance dI/dV and its derivative d^2I/dV^2 using a lock-in technique. A 100 Hz excitation voltage $V_{\rm osc} = 10$ mV was superimposed on the DC bias voltage through a transformer coupled to the diode bias circuit. The resulting AC voltage across a series resistor, R_L, was connected through a low-noise preamplifier with a gain of 100 to the lock-in input. The measured IV and dI/dV are shown in Fig 1. a. At $V = 0$, the three diodes had a mean junction resistance of $\overline{R}_j =$ (14.4 ± 0.1 kΩ). The current responsivity (or curvature) of the diode is given by $\gamma = (d^2I/dV^2)/(dI/dV)$, and for the three measured diodes we obtained a mean curvature of -39.5 A/W at zero bias. A measured $\gamma(V)$ for one diode is shown in Fig. 1 b. The divergences at 43 mV and at 111 mV correspond to the points where

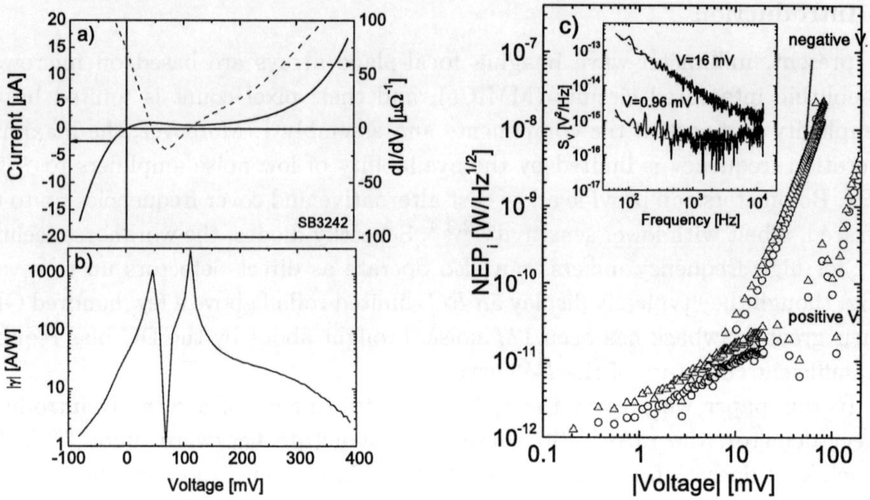

Fig. 1. a) The IV characteristics (solid line) of the diode with the dI/dV curve (dashed line) as a function of the bias voltage. b) The absolute value of current responsivity $\gamma(V)$. c) The matched (circles) and unmatched (triangles) NEP of a diode as a function of the absolute value of dc bias. The minimum NEP is obtained at $V = 0$ due to the rapid increase in $1/f$ noise which overcomes the increase in the responsivity at small positive bias. The inset shows noise spectra recorded at V=0.96 mV and V=16 mV. The dashed lines indicate the fits to the spectra.

dI/dV crosses zero.

3. Noise measurements

IV measurements, combined with noise characterization, determine the true figure of merit for a direct detector, the noise equivalent power (NEP). The diode was again biased in series with the 10 kΩ current sensing resistor, R_L, using a programmable battery power supply. A low-noise preamplifier with a gain of 100 and an input equivalent noise floor of 1.5 nV/$\sqrt{\text{Hz}}$ was used to amplify the voltage noise fluctuations across the current sensing resistor. The output of the preamplifier was connected to a spectrum analyzer, and noise spectra were recorded at different bias voltages in a frequency band from 62 Hz to 12.5 kHz. As the rectified RF power produces a small positive DC voltage across the device, we concentrate on the noise at positive bias. Two representative noise spectra are shown in the inset of Fig. 1 c. At $V = 0$ the noise corresponds to the Johnson noise of the junction resistance R_j. At finite values of V the $1/f$ and current shot noise appear. To quantify the $1/f$ noise, each spectrum was fit with a voltage noise power spectral density given by $S_{\text{fit}} = \alpha V^m/f^r + v_n^2$, where the white noise term v_n includes contributions from the Johnson noise and the shot noise. For the three diodes, the mean fitted values yielded $\alpha = 3.18 \cdot 10^{-6}$ Volts^{2-m}Hz^{r-1}, $m = 2.58$ and $r = 1.33$. Even though the

$1/f$ noise does not directly affect the NEP at zero bias, we note that it does have an effect on the dynamic range of the device.

The origin of the $1/f$ noise in our diodes is likely due to a similar scenario as that described in Ref. 9, according to which thermally excited traps states modulate the barrier height and thus its transmission coefficient. Whether the noise arises from the bulk of the device or is associated with surface states can be determined by measuring the $1/f$ noise properties of diodes with different junction area, and is a subject of future work.

In the following discussion of the NEP, a typical diode small signal circuit is used, with a resistance, R_s, in series with the parallel combination of R_j and C_j. The values for the circuit parameters are from earlier network analyzer and numerical modeling results with $R_s = 17.6\ \Omega$ and $C_j = 13.6$ fF. A noteworthy property of the tunnel diodes is that C_j is only weakly dependent of bias voltage [10]. The intrinsic NEP of the diode across a pre-detection bandwidth $\Delta\omega$ can be calculated from $\text{NEP}_i = \sqrt{S_V/\mathcal{R}_{V0}^2}$, where $\mathcal{R}_{V0} = R_j\gamma/2\{(1+R_s/R_j)[1+R_s/R_j + R_sR_jC_j^2(\Delta\omega^2/4+3\omega_c^2)/3]\}$ is the average voltage responsivity around a center frequency ω_c for the case of a matched source impedance (e.g. the antenna) and infinite load impedance [11]. Considering first the zero bandwidth case ($\Delta\omega = 0$), when operated at zero bias, and at $\omega_c/2\pi \gg (R_jR_sC_j^2)^{-1/2} \approx 23$ GHz, the voltage responsivity yields with the given junction parameters $\mathcal{R}_{V0} \approx 16\,000$ V/W [12], with the noise given by the Johnson noise of R_j yielding $\text{NEP}_i|_{\Delta\omega=0} \approx \sqrt{16k_BTR_j\omega_c^2C_j^2R_s/\gamma} \approx 1$ pW/$\sqrt{\text{Hz}}$ at 95 GHz. Taking into account reflection loss from impedance mismatch only, the extrinsic NEP is given by $\text{NEP}_e = \text{NEP}_i/\eta$ where $\eta = 1-|\Gamma|^2$ with Γ the reflection coefficient. For example, when connected to a 75 Ω load, the extrinsic NEP yields 4.7 pW/$\sqrt{\text{Hz}}$ at $\omega_c/2\pi = 95$ GHz. The best NEP at 100 Hz is obtained at $V=0$ as the rapid onset of the $1/f$ noise overwhelms the increase in the responsivity at small positive bias. The NEP values can be compared with those obtained with room temperature antenna coupled microbolometers, where the best demonstrated intrinsic NEP is about 10 pW/$\sqrt{\text{Hz}}$, but can only be obtained at a modulation frequency of 50 kHz [13]. It should be noted that due to their purely real impedance, bolometers can be easily matched and have an intrinsic NEP that is independent of operating frequency.

4. THz measurements

The ever-growing interest in sensitive detectors in the THz frequencies lead us to investigate whether rectification can be observed at these frequencies which lie well above the R_jC_j rolloff of the diodes. The THz source was a far-IR cavity, pumped by a 15 W CO_2 laser. The output power of the far-IR laser was approximately 8 mW at the 2.52 THz methanol line. A mylar beam splitter was placed in the far-IR beam to couple a fraction of the power to a pyroelectric detector for real-time power monitoring. After the beam splitter, an $F/1$ off-axis paraboloid with a focal length of 25.4 mm was used to focus the far-IR power onto the diode. The diode was mounted on a 50 μm thick mylar membrane spanning a 2.75 mm diameter hole in a printed circuit board. Aluminum wire bonds from the PCB were used for the electrical connections. It is likely that at least part of the bond wires and pads serve to couple the THz to the diode. The maximum observed signal across the diode was small, $\approx 2\ \mu$V. A large amount of power is not coupled as the diode active area is likely orders of magnitude smaller than the $\pi(1.22\lambda)^2$ area of the beam.

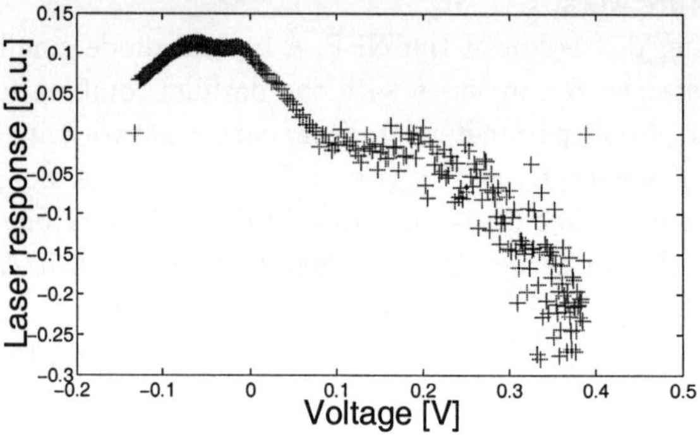

Fig. 2. Measured laser response at 2.5 THz.

The diode's THz response was also measured as a function of applied bias to determine whether or not the laser signal follows the responsivity curve determined from the $IV - dI/dV$ measurements. The measured laser response is shown in Fig. 2, where a signal is present at zero bias, a clear indication of rectification. However, there seems to be no agreement with γ (Fig. 1c). We are currently investigating the qualitative disagreement of the predicted and measured laser curves to see whether a combined bolometric and diode response might yield better fit.

5. Conclusions

We have measured the noise properties of zero bias quantum tunneling diodes and shown that these non-optimized devices have already reached a sensitivity that challenges other existing room temperature detectors in this wavelength range. The high sensitivity, low cost and reproducibility of the diodes make them an attractive newcomer to the millimeter and submillimeter-wave detector arena. The absence of $1/f$ noise and the resulting excellent narrow band NEP of the zero bias diodes can be used to its full potential in, for example, active detection where the signal is narrow band. For passive detection of thermal sources, the broadband matching poses a fundamental challenge which can be addressed by significant reduction in the junction resistance and capacitance in future devices. Rectification was observed at 2.5 THz, yet the bias dependence of the response did not agree with the DC measurements.

Acknowledgments

The authors would like to thank Aaron Miller for useful discussions.

References

1. L. Yujiri, H. Agravante, M. Biedenbender, G. Dow, M. Flannery, S. Fornaca, B. Hauss, R. Johnson, R. Kuroda, K. Jordan, P. Lee, D. Lo, B. Quon, A. Rowe, T. Samec, M. Shoucri, K. Yokoyama, and J. Yun, "Passive millimeter-wave camera," in *Passive Millimeter-Wave Imaging Technology* (R. M. Smith, ed.), vol. 3064, (Bellingham, Washington 98227-0010 USA), pp. 15–22, SPIE, SPIE, 1997.
2. T.-L. Hwang, S. Schwarz, and D. Rutledge, "Microbolometers for infrared detection," *Appl. Phys. Lett*, vol. 34, no. 11, pp. 773–776, 1979.
3. M. E. MacDonald and E. N. Grossman, "Niobium microbolometers for far-infrared detection," *IEEE Trans. Microwave Theory Tech.*, vol. 43, pp. 893–896, April 1995.
4. A. Luukanen and V.-P. Viitanen, "Terahertz imaging system based on antenna-coupled microbolometers," in *Proc. SPIE Vol. 3378, p. 34-44, Passive Millimeter-Wave Imaging Technology II, Roger M. Smith; Ed.*, pp. 34–44, Aug. 1998.
5. J. Schulman and D. Chow, "Sb-heterostructure interband backward diodes," *IEEE Electron Device Letters*, vol. 21, pp. 353–355, July 2000.
6. J. N. Schulman, D. H. Chow, E. T. Croke, C. W. Pobanz, H. L. Dunlap, and C. D. Haeussler, "Sb-heterostructure zero-bias diodes for direct detection beyond 100 GHz," in *Proc. SPIE Vol. 4111, p. 221-226, Terahertz and Gigahertz Electronics and Photonics II, R. Jennifer Hwu; Ke Wu; Eds.*, pp. 221–226, Dec. 2000.
7. J. Schulman, D. Chow, and D. Jang, "Ingaas zero bias backward diodes for millimeter wave direct detection," *IEEE Electron Device Letters*, vol. 22, pp. 200–202, May 2001.
8. R. G. Meyers, P. Fay, J. N. Schulman, S. Thomas, III, D. H. Chow, Y. K. Boegeman, J. Zinck, and P. Deelman, "Bias and temperature dependence of sb-based heterostructure millimeter wave detectors with improved sensitivity," *IEEE Electron Device Letters*, vol. 25, pp. 4–6, Jan 2004.
9. S. Ng and C. Surya, "Theory and experiments on flicker noise in $In_{0.53}Ga_{0.47}As/AlAs/InAs$ resonant tunneling diodes," *J. Appl. Phys.*, vol. 73, pp. 7504–7508, June 1993.
10. P. Fay, J. Schulman, S. Thomas III, D. Chow, Y. Boegeman, and K. Holabird, "Performance and modeling of antimonide-based heterostructure backward diodes for millimeter-wave detection," in *IEEE Lester Eastman Conference on High Performance Devices*, pp. 334–342, IEEE, Aug 2002.
11. I. Bahl and P. Bhartia, *Microwave solid state circuit design*, ch. 11, pp. 541–549. New York: John Wiley & Sons, 1988.
12. J. Schulman, V. Kolinko, M. Morgan, C. Martin, J. Lovberg, S. Thomas, III, J. Zinck, and Y. Boegeman, "W-band direct detection circuit performance with sb-heterostructure diodes," *IEEE Microwave and Wireless Component Letters*, vol. 14, pp. 316–318, July 2004.
13. A. Luukanen, *High performance microbolometers and microcalorimeters: from 300 K to 100 mK.* PhD thesis, University of Jyväskylä, Department of Physics, P.O. Box 35 (YFL) , FIN-40014 University of Jyväskylä, Finland, May 2003. http://www.phys.jyu.fi/theses/arttu_luukanen.pdf.

International Journal of High Speed Electronics and Systems
Vol. 14, No. 3 (2004) 670–675

World Scientific
www.worldscientific.com

TEMPERATURE DEPENDENCE OF TERAHERTZ EMISSION FROM SILICON DEVICES DOPED WITH BORON

R. T. TROEGER, T. N. ADAM†, S. K. RAY‡, P.-C. LV, S. KIM, AND J. KOLODZEY

Department of Electrical and Computer Engineering, University of Delaware 140 Evans Hall, Newark, DE 19716, USA

In this paper, we report on electrically pumped terahertz emitters based on silicon doped with boron acceptors. At cryogenic temperatures, three narrow spectral emission lines attributed to radiative transitions from p-like excited hydrogenic states to the s-like Γ^8 ground state associated with the boron dopants were observed centered around 8 THz. The spectral emission line center frequencies were in remarkable agreement with values reported from absorption measurements and theoretical calculations. The total time-resolved terahertz emission power was found to be up to 31 μW per device facet. We have solved the rate equations describing the populations in the hydrogenic dopant states involved in the emission mechanism and derived expressions for the current pumping and temperature dependence of the emitted terahertz power, yielding excellent agreement with the experimental data. These results suggest that silicon-based terahertz emitters may be fabricated without epitaxial quantum wells. The observed temperature dependence suggests that electric field assisted thermal escape of carriers from upper hydrogenic states may be responsible for lower output powers at higher temperatures.

1 Introduction

In the last few years, the terahertz range of the electromagnetic spectrum has garnered much interest for applications in such fields such as biomedical non-ionizing imaging, broadband communications, chemical sensing, and ranging.[1] Compact, reliable, and low-cost terahertz active devices are required for the commercialization of these techniques. Quantum cascade lasers based on III-V semiconductors operating in the terahertz range have been reported recently.[2] Silicon-germanium quantum cascades are attractive for CMOS-compatible terahertz emitters, however; operation has been achieved only at low output powers.[3] Lasing at wavelengths around 100 μm from Si-based resonant state lasers involving population inversion between strain-split dopant states formed in SiGe quantum wells whose central regions were delta-doped with boron has recently been reported.[4] Terahertz emission from donor doped Si has also been demonstrated under *optical* pumping.[5]

Recently, we have reported on electrically pumped terahertz emission at cryogenic temperatures from silicon devices doped with boron acceptors.[6] In this paper, we derive theoretical expressions for the pumping and temperature dependence of the emitted terahertz power. Excellent agreement with the experimental data has been achieved.

† T. N. Adam is now with IBM, 2070 Route 52, M/S AE-1, Hopewell Junction, NY 12533
‡ S. K. Ray is now with the Indian Institute of Technology Kharagpur, Kharagpur - 721302, INDIA

2 Device Fabrication

Emitter devices were fabricated from 380 μm thick silicon substrates with a boron doping concentration of ~ 0.5 to 1×10^{16} cm^{-3}. A standard lift-off procedure was used to pattern electron-beam evaporated Ti/Pt/Au layers (20/20/500 nm) into rectangles typically 120×190 μm^2 in size for use as topside electrical contacts and as mesa etch masks. Large area Ti/Pt/Au (20/20/200 nm) backside contacts were deposited without breaking vacuum. Devices were placed close (~ 30 μm) to the substrate cleaved edge to permit free-space radiation without obstruction. The substrates were then etched anisotropically to a depth of 17 μm, using cyclic multi-step processing in a PlasmaTherm 790 reactive ion etching (RIE) apparatus.[7] Substrates containing several device mesas were cleaved to a chip size of 2×2 mm^2, and mounted onto copper sample holders using conductive epoxy. Electrical connections were fabricated by wedge-bonding gold wires to the top-side metal contacts.

3 Electroluminescence Measurements

Electroluminescence measurements at cryogenic temperatures were carried out using a liquid helium cooled cryostat. Spectra were recorded using a ThermoNicolet Nexus870 Fourier Transform Infrared spectrometer (FTIR) operating in amplitude-modulation step-scan mode. A liquid helium cooled silicon bolometer in combination with a lock-in amplifier was used to detect the electroluminescence signal. Devices were electrically pumped by applying trains of pulsed negative bias with a train repetition frequency of 413 Hz to the top metal contacts. The number of pulses in each train and their duty cycles were chosen to prevent Joule heating in the sample. Pumping currents were measured using an inductive probe and oscilloscope.

4 Results and Discussion

A typical emission spectrum measured at a temperature of 4.2K is depicted in Fig. 1. Three closely spaced spectral lines attributed to radiative transitions from *p*-like excited states to the *s*-like ground state associated with the Γ_8 valence band were observed centered around 8 THz. The center frequencies of the emission lines were in remarkable agreement with those reported from far-infrared absorption measurements of lightly boron doped silicon[8] and from theoretical calculations,[9] as summarized in Table 1. We have achieved time-resolved peak powers of up to 31 μW per facet at 4.2 K, corresponding to a time averaged emission power of 384 nW, as determined using a recessed-cone-type blackbody radiator with a known emission spectrum and calibrated emitted power.

The following mechanism explains the features of the observed emission. At low temperatures, most carriers are frozen out in the ground states of the dopants, and in the absence of voltage bias, only a small number of free carriers exist in the device (~ 10^8 cm^{-3} for a doping concentration of ~ 10^{15} cm^{-3}).[10] The electric field necessary to accelerate the free carriers to gain enough energy to ionize the neutral impurities, evidenced by an abrupt onset voltage for current flow (approximately 15 V for the devices investigated here), is on the order of a few hundred V/cm.[10] Carriers excited to the valence band from acceptors may be captured under phonon emission into a number of upper level hydrogenic states associated with the ionized boron dopants, and subsequently cascade via multiple acoustic phonon emission to the *p*-like states involved in the radiative transition.[11] The inset of Fig. 1 schematically illustrates this mechanism and the involved energy states.

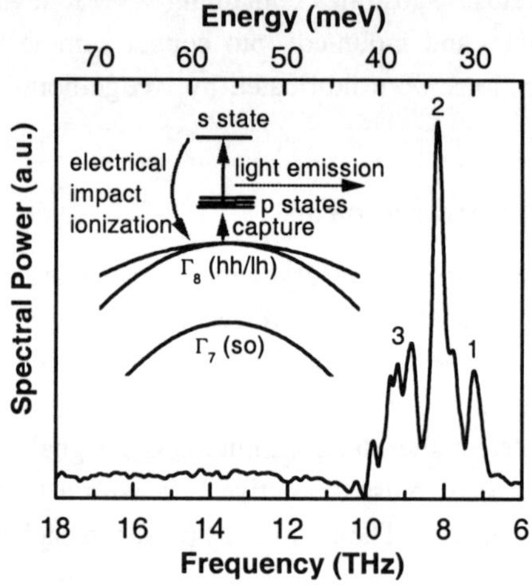

Fig. 1. Observed electroluminescence spectrum from boron-doped silicon devices. The spectral resolution used during electroluminescence measurements was 4 cm^{-1}. The spectral peaks have been labeled 1-3 for ease of comparison with data given in Table 1. *Inset:* Hole transitions underlying observed THz emission. Shown are the *s*- and *p*-like hydrogenic states associated with the boron acceptors, and the Si heavy-hole and light-hole (Γ8) and spin-orbit (Γ7) valence bands.

Line	This work	Reported absorption lines[8]	Theory[9]
1	29.58 meV	30.38 meV	29.39 meV
2	33.5 meV	34.53 meV	33.48 meV
3	38.0 meV	38.35 meV	37.67 +38.94 +39.04 meV

Table 1: Comparison of observed THz emission lines 1-3 as shown in Fig. 1 with experimentally determined absorption lines for boron-doped Si and theoretical values of acceptor level transition energies in boron doped silicon. Line 3 may contain contributions from several different radiative transitions.

The total emitted terahertz power is related to the population density n_P in the p-like initial states involved in radiative transitions to the s-like $\Gamma 8$ ground state via

$$P = V \sum_{lines} \eta_{ext} h f_0 \frac{\eta_{\text{int}} n_P}{\tau_{P \to S}}, \tag{1}$$

where V is the active device volume, $\tau_{P \to S}$ is the lifetime associated with transitions between the p-like and s-like hydrogenic acceptor states, with a fraction of radiative transitions η_{int}, f_0 is the center frequency of the spectral line, h is Planck's constant, and η_{ext} is the ratio of emitted-to-generated photons. The sum extends over all spectral lines. The change in n_i, the population density in the ith state involved in the emission mechanism, is described by the rate equation

$$\frac{\partial n_i}{\partial t} = \sum_{k \neq i} \frac{n_k}{\tau_{k \to i}} - n_i \cdot \sum_{k \neq i} \frac{1}{\tau_{i \to k}}, \tag{2}$$

where n_k is the volume density of carriers in the kth state, $\tau_{i \to k}$ are transition times between the ith and the kth state, and the summation extends over all states. The reciprocal lifetimes that appear in the formula are understood to represent sums over individual inverse lifetimes, to account for the various possible excitation and relaxation mechanisms, into and out of the ith state. In the steady state, a linear system of equations results, which must be solved for n_P under the constraint that the population densities in all states must add up to the boron acceptor doping concentration. The inverse lifetimes for impact excitation processes may be related to the current density J in the device as $\sigma J/q$,[12] where σ is an ionization cross section on the order of 10^{-13} cm^2,[11] and q is the elementary charge. The inverse lifetime associated with thermal excitation is $exp(-E_a/kT)/\tau_0$, where E_a is an activation energy dependent on the electric field in the device, k is Boltzmann's constant, T is the absolute temperature, and $1/\tau_0$ is an "attempt-to-escape" frequency on the order of 10^{13} s^{-1}.[13] In principle, carriers may be thermally excited to the valence band from any confined hydrogenic state; however, due to the ionization energy of the ground state and to the low measurement temperatures, thermal excitation is important only from upper hydrogenic states. The lifetimes associated with phonon-assisted and radiative relaxation were assumed to be constant with the temperature and electric field. Auger recombination and optical carrier excitation have been neglected.

From Eq. (1)-(2), in the small current limit, we obtain a power dependence of the form

$$P \approx \eta_{ext} VN \left(\alpha J - \left(\beta_1 + \beta_2 \exp\left(-E_a/kT \right) \right) J^2 \right), \tag{3}$$

where α, β_1, and β_2 are constants depending on lifetimes and capture cross sections. Fig. 2 (a) compares the emitted power versus current density with the theoretical expression. At 20K, we observe a deviation from linearity at high pumping currents as expected when the population of the p-like upper hydrogenic states is affected by field-assisted thermal escape of carriers to the valence band.

Under constant current pumping, we obtain an emission power versus temperature dependence of the form,

$$P = D_1 - D_2 \exp(-E_a/kT), \tag{4}$$

where D_1 and D_2 are parameters. Fig. 2 (b) compares the measured data with Eq. (4), yielding excellent agreement. The activation energy for thermal excitation of carriers to the valence band was extracted from the experimental data by fitting using Eq. (4), and was found to strongly decrease with increasing current density, which may be caused by electric field-dependent barrier lowering. The exact mechanism responsible for the observed current dependence of the activation energy is currently under investigation.

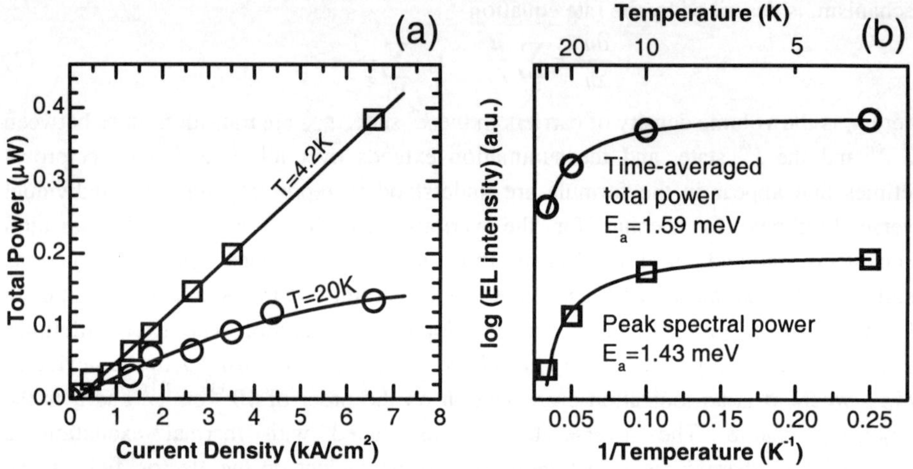

Fig. 2. *a)* Dependence of the total time-averaged emission power on the pumping current density at sample temperatures of 4.2K and 20K (symbols). The solid lines represent fits to the data using Eq. (3). *b)* Temperature dependence of the emitted THz intensity under a constant pumping current of 0.5A (symbols). The solid lines represent fits to the data using Eq. (4), yielding the electric field dependent activation energy E_a.

5 Conclusions

In summary, we have demonstrated the fabrication and characterization of simple boron-doped silicon devices that emitted THz radiation centered around 8 THz with tens of microwatts of total pulsed peak power using technology compatible with silicon integrated circuits, without need for epitaxial growth. We have solved the rate equations describing the populations in the hydrogenic dopant states involved in the emission mechanism and derived expressions for the current pumping and temperature dependence of the emitted terahertz power, yielding excellent agreement with the experimental data. The observed temperature dependence implies that field-assisted thermal escape of carriers from upper hydrogenic states may be responsible for lower output powers at higher temperatures.

Acknowledgements

The authors gratefully acknowledge R. Colombelli, C. Gmachl, K. Goossen, M. S. Kagan, E. Martinez, M. A. Odnoblyudov, G. Pomrenke, R. Soref, N. Sustersic, and I. N. Yassievich for fruitful discussions and suggestions. This work was supported by AFOSR contract F49620-03-1-0380, CRDF project RP2-2552-MO-02, DARPA contract MDA972-03-C-0102, and by the NSF under Award 033815.

References

1. P. H. Siegel, "Terahertz technology", *IEEE Trans. Microwave Theory Tech.*, **50** (2002) 910-928.
2. R. Köhler, A. Tredicucci, F. Beltram, H. E. Beere, E. H. Linfiled, A. G. Davies, D.A. Ritchie, R. C. Iotti, and F. Rossi, "Terahertz semiconductor-heterostructure laser", *Nature* **417** (2002) 156-159
3. G. Dehlinger, L. Diehl, U. Gennser, H. Sigg, J. Faist, K. Ensslin, D. Grützmacher, and E. Müller, "Intersubband electroluminescence from silicon-based quantum cascade structures", *Science* **290** (2000) 2277-2280
4. M. S. Kagan, I. V. Altukhov, E. G. Chirkova, V. P. Sinis, R. T. Troeger, S. K. Ray, and J. Kolodzey, "THz lasing of SiGe/Si quantum-well structures due to shallow acceptors", *Phys. Stat. Sol.* **B235** (2003) 135 –138
5. S. G. Pavlov, R. Kh. Zhukavin, E. E. Orlova, V. N. Shastin, A. V. Kirsanov, H.-W. Hübers, K. Auen, and H. Riemann, "Stimulated Emission from Donor Transitions in Silicon", *Phys Rev. Lett.* **84** (2000) 5220-5223
6. T. N. Adam, R. T. Troeger, S. K. Ray, P.-C. Lv, and J. Kolodzey, "Terahertz electroluminescence from boron-doped silicon devices", *Appl. Phys. Lett.* **83** (2003) 1713-1715
7. T. Adam, R. T. Troeger, S. Ray, U. Lehmann, S. Shi, D. Prather, and J. Kolodzey, "The design and fabrication of microdisk resonators for terahertz frequency operation", *Proc. IEEE Lester Eastman Conference on High Performance Devices*, Newark, DE, August 2002, pp. 402-408
8. A. Onton, P. Fisher, and A. K. Ramdas, "Spectroscopic investigation of the group-III acceptor states in silicon", *Phys. Rev.* **163** (1967) 686-703
9. R. Buczko and F. Bassani, "Shallow acceptor resonant states in Si and Ge", *Phys. Rev.* **B45** (1992) 5838-5847
10. W. Kaiser and G. H. Wheatley, "Hot electrons and carrier multiplication in silicon at low temperatures", *Phys. Rev. Lett.* **3** (1959) 334-336
11. M. Lax, "Cascade capture of electrons in solids", *Phys. Rev.* **119** (1960) 1502-1523
12. M. Markmann, A. Sticht, F. Bobe, G. Zandler, K. Brunner, G. Abstreiter, and E. Müller, "Efficient light emission at 1.54 μm from Er in Si excited by hot electron injection through thin suboxide layers", *J. Appl. Phys.* **91** (2002) 9764-9771
13. K. W. Böer, *Survey of Semiconductor Physics*, Van Nostrand Reinhold, New York, 1990, p. 941

International Journal of High Speed Electronics and Systems
Vol. 14, No. 3 (2004) 676–683

TWO-DIMENSIONAL ANALYTICAL MODELING AND SIMULATION OF RETROGRADE DOPED HMG MOSFET

R.S GUPTA, KIRTI GOEL AND MRIDULA GUPTA

Semiconductor Device Research Laboratory, Department of Electronic Science, University of Delhi, South Campus, Benito Juarez Road, New Delhi 110021, India

MANOJ SAXENA

Department of Physics & Electronics, Deen Dayal Upadhyaya College, University of Delhi, New Delhi110015,India

A novel device architecture, Retrograde doped HMG (**H**etero-**M**aterial **G**ate) MOSFET is presented which combines the advantages of an epitaxial layer in the channel region with the HMG MOSFET. This combination tackles the problems of short channel effects (SCEs), hot electron effects and gate transport efficiency in a single structure by reducing threshold voltage and modifying the electric field pattern and surface potential profile along the channel. The expression for the 2-D potential profile in the channel has been derived. The results of surface potential, electric field and threshold voltage have been plotted and compared with their corresponding simulated results.

1 Introduction

Scaling of MOSFETs towards the submicron range causes deviation of the long channel behavior due to reduction in effective channel length, 2D potential distribution and high electric field in the channel region. Thus the long channel 1-D analysis has to be replaced by the short channel 2-D analysis, where SCEs, hot electron effects and gate transport efficiency are of prime importance.

SCEs can be reduced by a retrograde channel profile by which resistance can be reduced and high breakdown voltage of the S/D junctions can be maintained.[1,2] The presence of an epitaxial-layer also reduces the threshold voltage[3] that apparently means reduction of applied bias to turn on the device. Epi-structure has fast switching speed since the low threshold voltage has the advantage of improving the speed of device[4].

Since the epi-structure is unable to improve the gate transport efficiency, two materials in the gate region have been used such that the work function of metal gate 1 (M1) is greater than metal gate 2 (M2) i.e. $\phi_{M1} > \phi_{M2}$. Hence threshold voltage $V_{T1} > V_{T2}$ which has the inherent advantage of improving the gate transport efficiency by modifying the electric field pattern such that the average electric field increases.[5,6]

In this paper, 2-D analytical model of novel device architecture retrograde doped HMG MOSFET, which combines all the advantages of epi-MOSFET, and HMG MOSFET is presented and results are validated by comparison with the simulated results obtained from the 2-D device simulator ATLAS.[7]

1.1. *Model Formulation*

The channel region under the gate has been divided into two parts since two materials are used in Retrograde doped HMG-MOSFET and the potential is denoted as $\psi_{11}(x,y)$ under metal gate M1 in the Epi-layer and as $\psi_{12}(x,y)$ under metal gate M2 in the Epi-layer. The potential profile for the channel region below the two gates is different due to difference in flat band voltages, which depends on metal work function of the two materials [6].

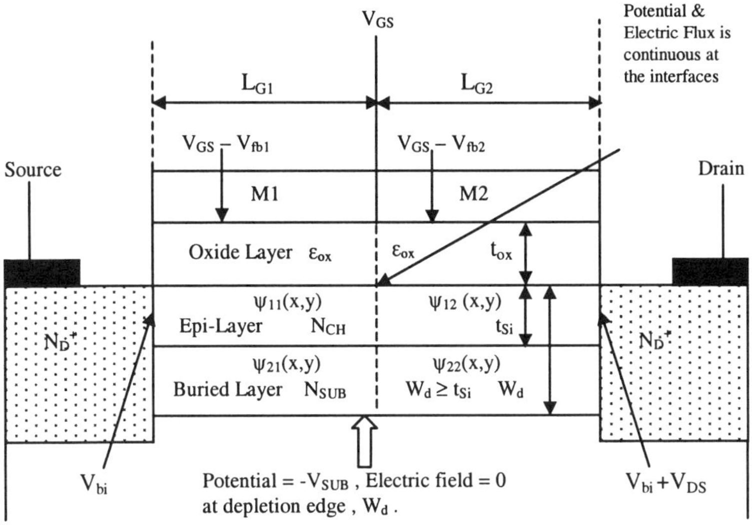

Figure 1 Schematic structure of Retrograde doped HMG MOSFET

The 2-D potential distribution along the channel can be obtained by solving 2-D Poisson's equation. It is presumed that the Epi-layer is fully depleted, so W_d (depletion width) must be larger than t_{Si} (Epi-layer thickness), or the analysis is valid when

$$t_{Si} < \sqrt{\frac{2\varepsilon_{Si}\left(2\phi_f + V_{SUB}\right)}{qN_{CH}}} \tag{1a}$$

where ε_{Si} is the permittivity of silicon, ϕ_f is thc fcrmi-level, q is the electronic charge, N_{CH} is the epi-layer doping concentration and V_{SUB} is the substrate to source voltage. The 2-D Poisson's equation for potential in the channel region is given below

$$\frac{d^2\psi_{11}(x,y)}{dx^2} + \frac{d^2\psi_{11}(x,y)}{dy^2} = \frac{qN_{CH}}{\varepsilon_{Si}} \quad 0 < x < L_{G1} \qquad 0 < y < t_{Si} \tag{1b}$$

$$\frac{d^2\psi_{12}(x,y)}{dx^2} + \frac{d^2\psi_{12}(x,y)}{dy^2} = \frac{qN_{CH}}{\varepsilon_{Si}} \quad L_{G1} < x < L_{G1} + L_{G2} \qquad 0 < y < t_{Si} \tag{1c}$$

$$\frac{d^2\psi_{21}(x,y)}{dx^2} + \frac{d^2\psi_{21}(x,y)}{dy^2} = \frac{qN_{SUB}}{\varepsilon_{Si}} \quad 0 < x < L_{G1} \qquad t_{Si} < y < t_{Si} + W_d \tag{1d}$$

$$\frac{d^2\psi_{22}(x,y)}{dx^2}+\frac{d^2\psi_{22}(x,y)}{dy^2}=\frac{qN_{SUB}}{\varepsilon_{Si}}\; L_{G1}<x<L_{G1}+L_{G2}\; t_{Si}<y<t_{Si}+W_d \quad (1e)$$

The potential profile is assumed to be a two-region polynomial potential distribution in the vertical direction [1] and it reduces to that of a uniform channel when t_{Si} is zero. The potential distribution is given by

$$\psi_{1i}(x,y)=\psi_{Si}(x)+C_{i1}(x)y+C_{i2}(x)y^2 \quad (2a)$$

$$\psi_{2i}(x,y)=\psi_{Si}(x)+\left(\frac{q(N_{SUB}-N_{CH})t_{Si}^{\;2}}{2\varepsilon_{Si}}\right)+C_{i3}(x)y+C_{i4}(x)y^2+C_{i5}(x)y^3 \quad (2b)$$

where i is region number, i =1 for region M1, i =2 for region M2, and are the surface potentials under M1 and M2 respectively. The Poisson's equation is solved separately under the two regions M1 and M2 and the coefficients in equation (2a)-(2b) are obtained using the boundary conditions as shown in Fig.1. The 2-D Potential distribution under M1 and M2 in the epi-layer is obtained as

$$\psi_{1i}(x,y)=\psi_{Si}(x)-\left(\frac{V'_{Gi}-\psi_{Si}(x)}{\beta t_{ox}}\right)y$$
$$+\left(\frac{(2W_d^{\;2}+\beta t_{ox}(3W_d-t_{Si}))(V'_{Gi}-\psi_{Si}(x))}{W_d^{\;2}(W_d+t_{Si})\beta t_{ox}}\right)y^2$$
$$-\left(\left(\frac{q(N_{SUB}-N_{CH})(W_d-t_{Si})^3}{2\varepsilon_{Si}W_d^{\;2}(W_d+t_{Si})}\right)+\left(\frac{(3W_d-t_{Si})(V_{SUB}+V'_{Gi})}{W_d^{\;2}(W_d+t_{Si})}\right)\right)y^2 \quad (3)$$

where, $V'_{GS}=V_{GS}-V_{fbi}$, V_{GS} is the gate to source voltage V_{fbi} is the flat band voltage of the metals and N_{SUB} is the doping concentration of the buried-layer and $\beta = C_{si}\, t_{si}/C_{ox}\, t_{ox}$ On substituting (2a) in (1a) and (1b) respectively we obtain,

$$\frac{d^2\psi_{Si}(x)}{dx^2}+\frac{V'_{Gi}-\psi_{Si}(x)}{\xi^2}=\frac{qN_{CH}}{\varepsilon_{Si}}+\frac{q(N_{SUB}-N_{CH})(W_d-t_{Si})^3}{\varepsilon_{Si}W_d^{\;2}(W_d+t_{Si})}+\frac{2(3W_d-t_{Si})(V'_{Gi}+V_{SUB})}{W_d^{\;2}(W_d+t_{Si})} \quad (4)$$

where ξ is characteristic length and is given as

$$\xi=\sqrt{\frac{\beta t_{ox}W_d^{\;2}(W_d+t_{Si})}{2(2W_d^{\;2}+\beta t_{ox}(3W_d-t_{Si}))}} \quad (5)$$

Introducing a new variable $\alpha_i(x)$ such that

$$\alpha_i(x)=\psi_{Si}(x)-V'_{Gi}\,a+v\xi^2 \quad (6)$$

solution of Eq.(4) can be expressed as

$$\psi_{Si}(x) = \alpha_i(x) + V'_{Gi}\, a - v\xi^2 \tag{7}$$

where

$$a = 1 - \frac{2(3W_d - t_{Si})}{W_d^{\ 2}(W_d + t_{Si})} \tag{8}$$

$$v = \frac{qN_{CH}}{\varepsilon_{Si}} + \frac{q(N_{SUB} - N_{CH})(W_d - t_{Si})^3}{\varepsilon_{Si} W_d^{\ 2}(W_d + t_{Si})} + \frac{2(3W_d - t_{Si})V_{SUB}}{W_d^{\ 2}(W_d + t_{Si})} \tag{9}$$

$\alpha_i\,(x)$ is found using the boundary conditions given in Fig.1 and is

$$\alpha_i(x) = \left(\alpha_{1i} \sinh\left(\left(\sum_{j=1}^{i} L_{Gj} - x \right) \Big/ \xi \right) + \alpha_{1i} \sinh\left(\left(x - \left(\sum_{j=1}^{i} L_{Gj} \right) + L_{Gi} \right) \Big/ \xi \right) \right) \Big/ \sinh(L_{Gi}) \tag{10}$$

Where,

$$\alpha_{11} = V_{bi} - V'_{G1}\, a + v\xi^2 \tag{11a}$$

$$\alpha_{22} = (V_{DS} + \alpha_{11}) - (V_{fb1} - V_{fb2})a \tag{11b}$$

$$\alpha_{12} = \frac{\alpha_{11} \sinh(L_{G2}/\xi) + \alpha_{22} \sinh(L_{G1}/\xi) + (V'_{G1} - V'_{G2})a \sinh(L_{G2}/\xi)\cosh(L_{G1}/\xi)}{\sinh(L_{G1}/\xi)\cosh(L_{G2}/\xi) + \sinh(L_{G2}/\xi)\cosh(L_{G1}/\xi)} \tag{11c}$$

$$\alpha_{21} = \frac{\alpha_{11} \sinh(L_{G2}/\xi) + \alpha_{22} \sinh(L_{G1}/\xi) + (V'_{G2} - V'_{G1})a \sinh(L_{G1}/\xi)\cosh(L_{G2}/\xi)}{\sinh(L_{G1}/\xi)\cosh(L_{G2}/\xi) + \sinh(L_{G2}/\xi)\cosh(L_{G1}/\xi)} \tag{11d}$$

Where, V_{bi} is the built in potential and V_{DS} is the source to drain voltage. As the electron transport velocity through the channel is related to the electric field pattern along the channel, the electric field component in the x-direction is given as

$$E_{1i}(x) = \left. \frac{d\psi_{1i}(x, y)}{dx} \right|_{y=0} = \frac{d\psi_{Si}(x)}{dx}$$

The position of minimum surface potential x_{min} lies under M1 and is approximated to be $x_{min} \approx (L_{G1}/2) - (\xi / 2)\, \ln(\alpha_{21} / \alpha_{11})$ if $L_{G1} >> \xi$. By substituting for switching, threshold voltage can be obtained.

1.2. *Result and Discussion*

In Figure 2, the variation of surface potential along the normalized channel length is shown for different metal work functions of metal M2. Comparison of four structures has been made a) Conventional Si-MOSFET, b) Epi-MOSFET, c) HMG MOSFET and d) Retrograde doped HMG MOSFET). Apart from Al (ϕ_M ~ 4.08 V) and Ru (ϕ_M ~ 4.71 V), recently Tantalum (Ta) based binary alloy system of Ta-Pt and Ta-Ti gives a wide range of continuous Metal Workfunction (ϕ_M) Modulation from 4.16 V to 5.05 V.

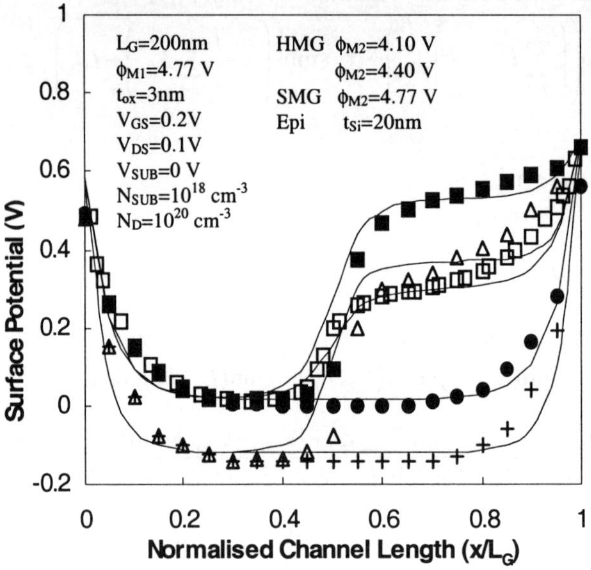

Figure 2. Surface potential variation with normalized channel position for different metal work functions of metal M2. ■■■ - is used for retrograde HMG-MOSFET (ϕ_{M2}=4.10V), □□□ - for retrograde HMG-MOSFET (ϕ_{M2}=4.40 V), ΔΔΔ - HMG MOSFET, ●●● - Epi-MOSFET and + + + - Uniform conventional Si-MOSFET. Line gives the analytical model for all cases.

In Figure 3, the variation of electric field along the normalized channel length is shown for all four structures. It is seen that the peak of *structure d* is lowest at the drain end compared to rest of the structures. This implies that reduction of hot carrier effect is the best in *structure d*.

In Figure 4 it is observed that there is a step change of potential along the channel at the interface of the two metal gates. This is due to the difference in metal work functions. The larger the difference in metal work functions, the more is the positive offset voltage and more is the screening of region M1 from variation of V_{DS}. The step change of potential shifts towards the source side on reduction of the gate length ratio (L_{G1}/L_{G2}), which results in enhancement of screening effect and thus small improvement of DIBL.

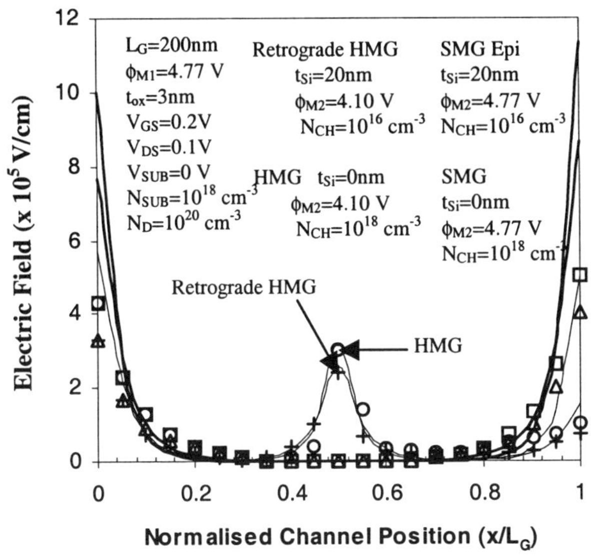

Figure 3. Electric field variation with normalized channel position. +++ - is used for Retrograde HMG-MOSFET, ○○○ - for HMG MOSFET, ΔΔΔ - for Epi-MOSFET and □□□ - for uniform conventional Si-MOSFET. Line gives the analytical model for all cases.

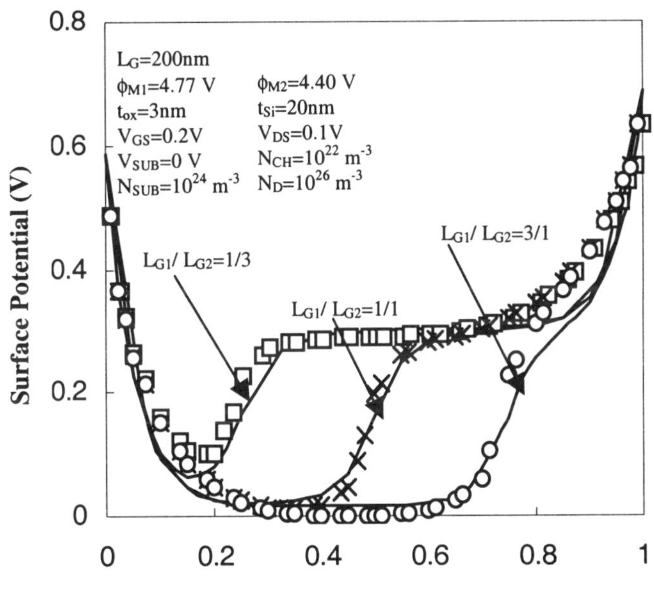

Figure 4. Surface potential variation with normalized channel position for different combinations of L_{G1} and L_{G2} of M1 and M2 respectively, keeping the sum constant (L_{G1}+ L_{G2}= L_G). Square is used for (L_{G1}/ L_{G2}=1:3), cross for (L_{G1}/ L_{G2}=1:1), circle for (L_{G1}/ L_{G2}=3:1). Line gives the analytical model for all cases.

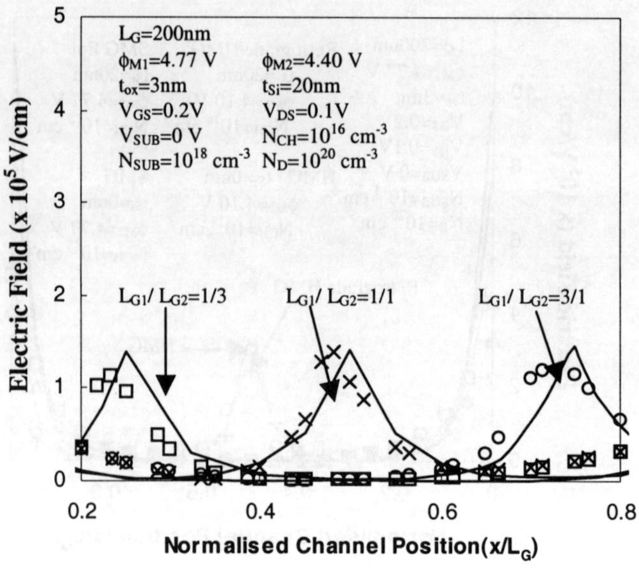

Figure 5. Electric field variation with normalized channel position for different combinations of L_{G1} and L_{G2} of M1 and M2 respectively, keeping the sum constant (L_{G1}+ L_{G2}= L_G). □□□ - is used for (L_{G1}/ L_{G2}=1/3), xxx - for (L_{G1}/ L_{G2}=1/1), ○○○ - for (L_{G1}/ L_{G2}=3/1). Line gives the analytical model for all cases.

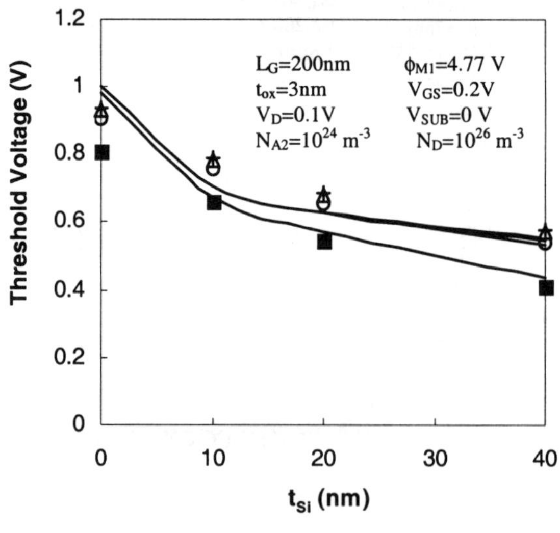

Figure 6. Threshold Voltage variation with epi-layer thickness (t_{Si}). Square is used for L_{G1}/L_{G2}=0.25/0.75, circle for L_{G1}/L_{G2}=0.5/0.5, triangle for L_{G1}/L_{G2}=0.75/0.25 and plus for L_G=L_{G2}.,L_{G2}=0. Line gives the analytical model for all cases.

In Figure 5 the variation of electric field along the normalized channel position for different gate length ratios is shown. Corresponding to shifting of step change of potential on reduction of L_{G1}/L_{G2} ratio, electric field also shifts. This is because the point of maximum barrier lies at the interface of the two metal gates. This peak causes more uniformity of the electric field in the channel resulting in higher carrier drift velocity and device speed. So decreasing L_{G1}/L_{G2} ratio increases the carrier transport efficiency. Since on decreasing L_{G1}/L_{G2} ratio, the position of peak electric field at the interface shifts towards the source end.

In Figure 6 threshold voltage variation with Epi-layer thickness for different L_{G1}/L_{G2} is shown. The threshold voltage of *structure b* and *d* is lower than that of *structure a* and *c* respectively due to presence of low concentration region epilayer. Thus for high value of epilayer thickness (t_{Si}), threshold voltage is lower. It is seen that *structure d* has the highest positive offset, which results in highest screening of region M1 from variation of $V_{DS,}$ that results in more reduction in DIBL.

1.3. *Conclusions*

The Retrograde HMG MOSFET has been compared with three other structures and is found to have superiority above them with respect to SCEs, hot carrier effects and carrier transport efficiency.

Acknowledgments

The authors are grateful to the University Grants Commission (UGC) and Defence Research and Development Organization (DRDO), Ministry of Defence, Government of India, for the necessary financial assistance.

References

1. K. Suzuki, *IEEE Trans. Electron Dev.*, **47**(12), 2372, (2000).
2. N. D. Arora, "MOSFET Models for VLSI Circuit Simulation-Theory and Practice," Berlin, Germany: Springer-Verlag Wien New York, 1993.
3. B. Jarvis, Jacobs, D. Antoniadis, *IEEE Trans. Electron Dev.*, **42**(5), 870, (1995)
4. K. C. Jang, J. S. Jhung, H. K. Jung and K. Taniguchi, *Physica B*, **314**, 391, (2002)
5. W. Long, H. Ou, J.-M. Kuo and K. K. Chin, *IEEE Trans. Electron Dev.*, **46**(5), 865, (1999)
6. M. Saxena, S. Haldar, M. Gupta and R. S. Gupta, *IEEE Trans. Electron Dev.*, **49**(11), 1928, (2002).
7. ATLAS: 2-D Device Simulator, SILVACO International, 2000.

International Journal of High Speed Electronics and Systems
Vol. 14, No. 3 (2004) 684–689

ELECTRICAL EFFECTS OF DNA MOLECULES ON SILICON FIELD EFFECT TRANSISTOR*

G. XUAN[†], J. KOLODZEY AND V. KAPOOR
Electrical and Computer Engineering Dept.
University of Delaware, Newark, DE 19716 USA

G. GONYE
Dept. of Pathology, Anatomy and Cell Biology
Thomas Jefferson University, Philadelphia, PA 19107 USA

Much research has been devoted to the field of DNA detection for biotechnology and medical diagnostics. Conventionally, this has involved lab-scale large instruments such as fluorescent microscope, with DNA binding and tagging. Recently, however, the possibility of label-free, rapid, sensitive and miniaturized DNA detection electronically has attracted increasing interest in the field. To investigate the feasibility of DNA detection by the semiconductor field effect without tagging or binding agents, we studied the effects of DNA molecules directly on the gate oxide of field effect transistors (FET). DNA solution was deposited onto the gate region of conventional silicon FET devices without gate metal, and current-voltage measurements were performed. Our work showed that the presence of DNA molecules in an aqueous solution on the gate decreased channel current I_{ds} and increased threshold voltage. With increasing concentration in the solution, the threshold voltage increased monotonically. We observed that the channel current was reduced systematically during the hybridization of complementary DNA strands. The mechanism that induced the threshold voltage shift is attributed to the negative charges the DNA molecules carry and to changes in work function (Fermi Energy). Our results showed the possibility that field effect transistors may provide a vehicle for a label-free, miniaturized and semiconductor-based DNA detector.

1 Introduction

Currently the popular methods for detecting DNA include fluorescence, enzymatic or radioactive labeling methods. [1] However, these measurement techniques present some inconvenience due to the requirement for relatively large volumes of DNA solutions, the need for expensive and cumbersome supporting systems, and the extra procedure of labeling and separation of DNA probes. Recently, alternative ways have been investigated to detect DNA molecules and among them, the surface field effect approach receives much attention for its simplicity in device design, compatibility with planar CMOS technology and without the need to label the DNA. [2, 3] In such approaches, DNA probes were generally immobilized onto the gate oxide by silanization or similar binding agent. This is compatible with current DNA printing technology, but it prevents the subsequent re-use of the biochip for other measurements. In this paper, the effects of DNA on transistor properties without the binding agent are investigated.

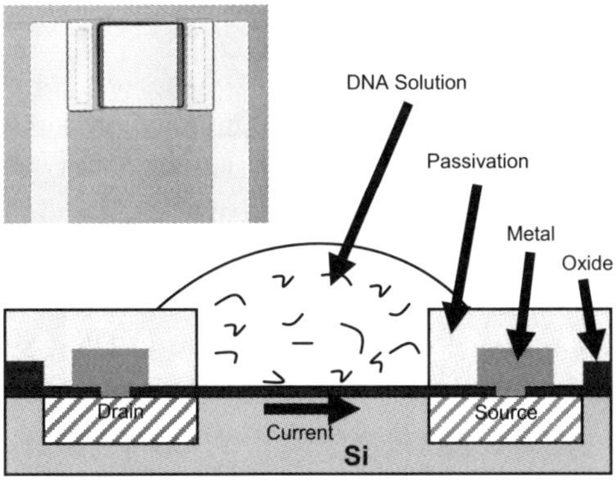

Figure 1. Top shows the microscopic image of a single device, with large square gate region in the center. Bottom shows cross-section scheme of the FET device with DNA solution on gate oxide.

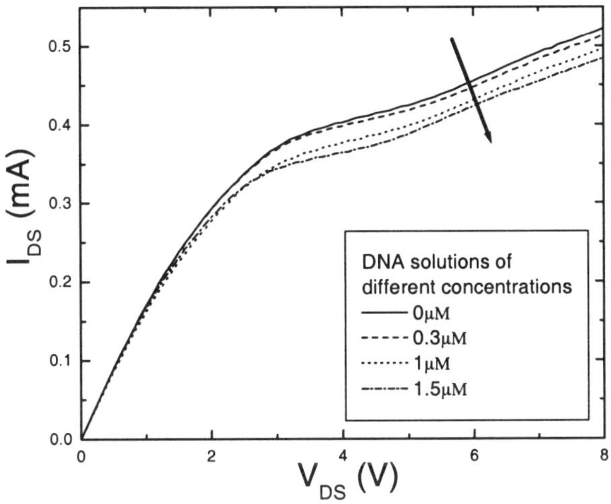

Figure 2. I-V characteristics of a W/L=200/200μm FET versus DNA solutions at concentrations of 0, 0.3, 1 and 1.5μM. The arrow shows the concentration dependent trend.

The idea of using field effect transistors for sensor purposes was first raised by Bergveld in the early 1970s in the study of ion-sensitive FET (ISFET) devices [4], which later gained commercial success as PH sensors. Since the mid-90s, much attention has been given to DNA sensing using FET-based devices [5]. The gate region of an FET is sensitive to the electrode material composition and electrical charge. These effects are produced by differences in work function, Fermi energy and electrostatic potential. For this paper, the effect of DNA solutions on FET channel current was studied.

2 Experimental

The FETs were fabricated using standard NMOS processing but with no gate metal. After fabrication, the sample was passivated with 2μm thick photoresist, leaving only the gate oxide uncovered. No further treatment of the gate oxide was performed other than rinsing with DI water. Figure 1 shows both the cross-section and the microscopic plan view of the FET. The gate dimensions on the 2 types of samples were width/length: 200μm/200μm and 200μm/40μm. Several experiments were performed to characterize the effects on the FET drain-source current in response to DNA solutions of different concentrations and hybridization states. All device measurements were done at room temperatures.

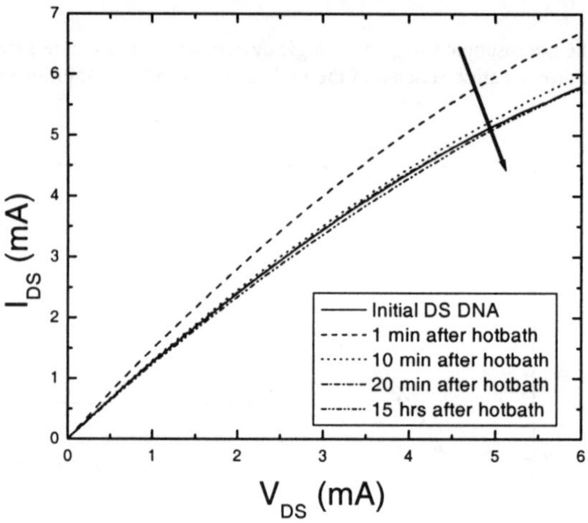

Figure 3. I-V characteristics of a W/L=200/40μm FET versus thermal re-hybridization. The curves shows the FET response at 1, 10, 20 minutes and 15 hours after the vial was taken out from the hot bath. The arrow shows the time dependent trend.

First we studied the FET current response to DNA solutions of different concentrations placed on the gate. Sonicated salmon sperm double-stranded DNA (DS DNA) of lengths between 5,000 and 20,000 bases long was diluted to four different concentrations with DI water: 0, 0.3, 1 and 1.5 μM. In a series of measurements a volume of 0.5 μL of each of these DNA solutions was deposited onto the exposed gate oxide region using a micropipette. Prior to each measurement, we followed a procedure of rinsing the FET with running DI water and drying by nitrogen. A range of voltages V_{DS} was then applied across the drain and source electrodes, and the channel current I_{DS} was recorded. Figure 2 shows the I-V characteristics of a W/L=200/200μm FET in response to DNA solutions of different concentrations with increasing concentration, the FET current decreased, which can be modeled as a more algebraically positive threshold voltage V_T, which has well known contributions from gate material work function and charge state.

The effect on FET I-V characteristics from different DNA hybridization states was also studied. A volume of 0.5μL of 1.52μM sonicated salmon sperm DS DNA was deposited on the gate region of a W/L=200/40μm FET which was measured as a baseline reference, followed by thorough rinsing of the FET as described previously. Then the small vial containing the DNA sample was put into a 95 °C hot bath for 15 minutes with the intention to melt the DNA into single-stranded DNA (SS DNA), followed by quenching in an ice bath for 30 seconds with the intention to begin the process of re-hybridization. The vial was then placed at room temperature. In a time periodic sequence of experiments, 0.5μL of DNA from this vial was deposited on the FET gate and the I-V measurement was repeated. Figure 3 shows the I-V characteristics of the FET with the gate covered by the DNA solutions initially, and at 1, 10, 20 minutes, and15 hours after removing the vial from the hot bath. One minute after the melted DNA was quenched, the current is significantly higher than for the initial baseline. At times 10, 20 and 15 hours the current decreases toward the baseline. This difference in current may be due to differences in the degree of DNA hybridization that occurred after melting. The time constant of 10 min. is in good agreement with the hybridization time in similar setups [3].

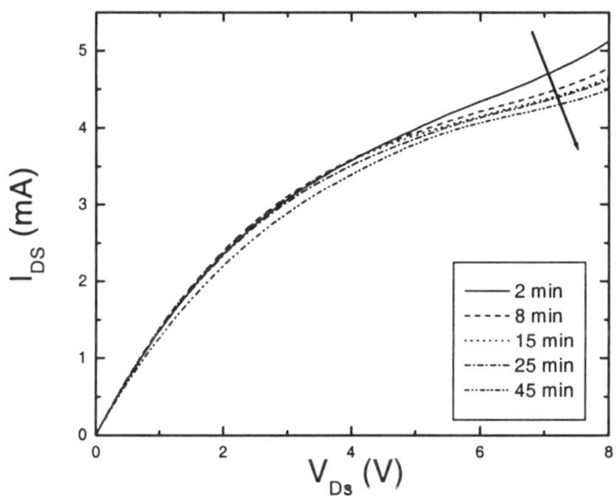

Figure 4. I-V characteristics of a W/L=200/40μm FET versus time after complementary oligonucleotides were mixed together. The arrow shows the time-dependent trend.

Another experiment was conducted to study the effect on FET characteristics from the mixture of two complementary oligonucleotides (60 bases long, 20μM). The vials containing these DNA solutions were placed in an ice bath for 20 minutes to stabilize them before the experiment. Equal amounts of these two oligonucleotide solutions were mixed into a new vial at room temperature. At the times indicated, 0.5uL of solution was drawn from this mixture and deposited on the gate region of a W/L=200/40um FET followed by I-V measurements. The times the measurements were performed were 2, 8, 15, 25, and 45 minutes after the mixing of the two strands. Figure 4 shows the trend of hybridization process on the transistor characteristics - with increasing time, the drain current decreased.

In order to check the hybridization state of the DNA solutions in the afore-mentioned experiments, fluorescence microscopy on these samples was performed using an ABI Prizm 7000 Sequence Detection system (ABI Applied BioPhysics). For both the

DS DNA and the mixture from complementary DNA solutions the measured melting temperature, which determines the hybridization state of DNA, was around 82 °C. The result confirmed that after the 95 °C heat treatment, both solutions comprised mainly of complementary denatured DNA molecules, and that after the solutions were quenched in ice and then brought to room temperature, the denatured DNA molecules began re-hybridizing back to double-stranded form.

3 Discussion

From the FET response versus DNA concentration study it was shown that the presence of DNA molecules directly on gate oxide reduced channel current proportionally to DNA concentration in the solution. DNA molecules are known to carry strong net negative charges, and hence the reduction in channel current might be attributed to the negative charges the DNA molecules introduced into the solution. However, it could also be attributed to the work function changes in the solution with different DNA concentrations.

In both the second and third experiments, the FET channel current was reduced over time as the complementary single-stranded DNA pairs were hybridizing. Thus the shifting I-V curve versus time reflected the hybridization process in the DNA solution. And it is also shown that DNA in hybridized states reduced channel current more than denatured states.

4 Summary

FET devices without gate metal were fabricated to investigate the effect of DNA solutions on FET I-V characteristics. DNA solutions of different concentrations and different states of hybridization were found to have reproducible changes on the FET channel current. These results suggest that FET devices have the potentials for DNA monitoring in bio-electronic applications.

Acknowledgements

The authors are pleased to thank Pengcheng Lv, Ralph T. Troeger and Thomas Adam for useful discussion. This research work was supported by NSF SGER award ECS0129535.

References

1. J. Dicesare; B. Grossman; E. Katz; E. Picozza; R. Ragusa; T. Woudenberg, *BioTechniques* **15** 152 (1993)
2. F. Pouthas, C.Gentil, D. Cote, and U. Bockelmann, *Appl. Phys. Lett.* **84** 1594 (2004)
3. F. Uslu, S.Ingebrandt, D. Mayer, S. Bocker-Meffert, M.Odenthal, A. Offenhausser, *Biosens. Bioelectron.* **19** 1723 (2004)
4. P. Bergveld, *IEEE Trans. Biomed. Eng.***19** 342 (1972)
5. P. Bergveld, *Sens. Actuators A* **56** 65 (1996)

International Journal of High Speed Electronics and Systems
Vol. 14, No. 3 (2004) 690–695

ANALYSIS OF OPERATIONAL TRANSCONDUCTANCE AMPLIFIER FOR APPLICATION IN GHz FREQUENCY RANGE

A. GHORI and P. GHOSH

Department of Electrical Engineering and Computer Science
Syracuse University, Syracuse, NY 13244, USA

Operational Transconductance Amplifier (OTA) is an excellent current mode device suited very well for VLSI implementation. In this contribution we report realization of OTA using Silicon-On-Insulator (SOI) structure based MOSFETs and compared them to OTA designed with bulk MOSFET. SOI based OTA outperformed bulk MOSFET OTA giving close to 10 GHz improvement in high frequency f_T. A band-pass filter was implemented with SOI based OTA with a center frequency of 7 GHz and a bandwidth of 480 kHz.

1 Introduction

Operational Transconductance Amplifier (OTA) is an excellent active current mode device with wide band capability. Its simple, regular, modular, and programmable structure suits very well with the VLSI implementation. OTA is considered to be a potential candidate for the design of high speed integrated analog circuits [1-3]. Silicon-on-Insulator devices are inherently better for high frequency applications than their bulk counterparts because of the reduced junction capacitance. Figure 1 shows a schematic diagram of an SOI device. The buried oxide layer isolates the active silicon region from the substrate, resulting in lower junction capacitance between the source/drain and the bulk. Therefore, it seems logical to use SOI based transistors for high frequency OTA design.

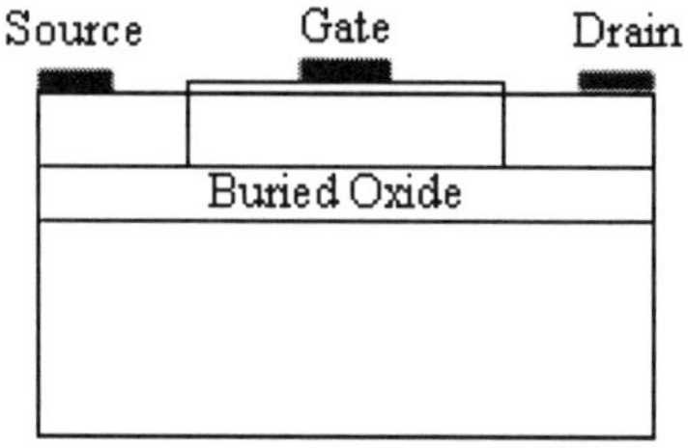

Figure 1. A silicon-on-insulator MOSFET

We start with the design and analysis of a SOI structure based MOSFET and compares its frequency response with a comparable bulk MOSFET. The model for the bulk MOSFET was derived from a foundry process of 0.25 microns [6] while the SOI model used was an example netlist from the University of Berkeley SOI modeling group [7]. Output characteristics of the bulk MOSFET and SOI devices were simulated using AIM-SPICE 4. The SOI MOSFET used for analysis had 0.25 μm long channel over a 10 nm active region and a 30 nm thick buried oxide layer. The doping concentration of the active silicon region was 3.37×10^{17} cm^{-3} acceptor atoms. A 0.25 μm bulk MOSFET was also analyzed for comparison purposes. The boost in high frequency performance of SOI device compared to bulk MOSFET is evident in their frequency responses shown in fig. 2.

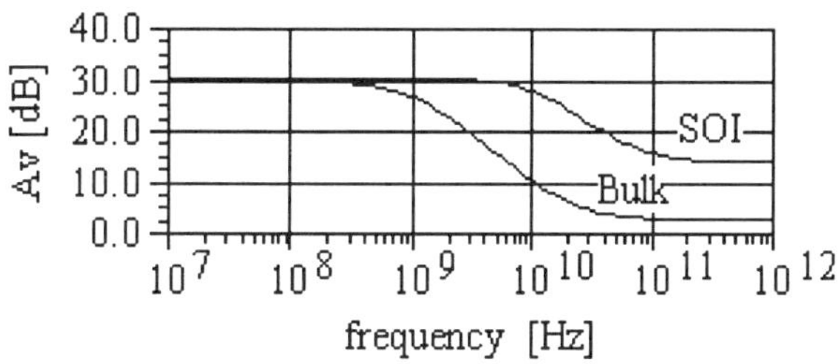

Figure 2. Gain of SOI vs. Bulk Device

2 SOI vs Bulk OTA Performance

A basic single-stage OTA design was realized with SOI devices and compared with bulk MOSFET based circuits. Figure 3 depicts the general OTA topology that was used for this work. The OTA circuits were analyzed with a resistive load at the output terminal.

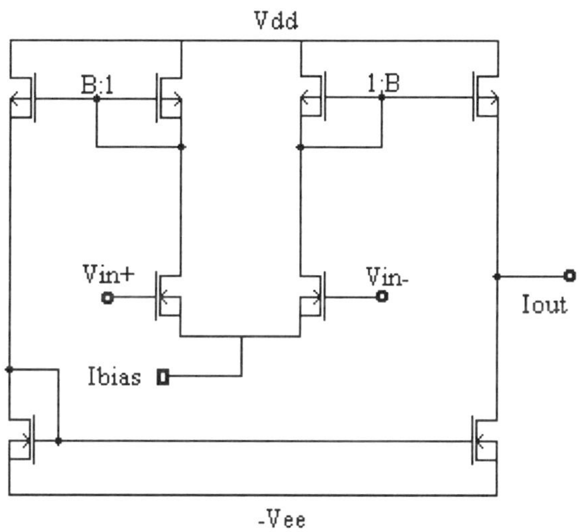

Figure 3. Basic OTA Design

2.1 Frequency Behavior

For performance simulation of the OTAs, identical circuit parameters were maintained for both SOI and bulk MOSFET based structures. The current mirror ratio (B) was set at 20 and a load resistor of 100kΩ was placed on the output terminal. Results shows that the SOI based OTA has a transit frequency (f_T) of around 12.7 GHz with a phase margin of 30° while the bulk MOSFET OTA has a transit frequency of around 3 GHz and a phase margin of 64°. The gain-frequency behaviors and phase margins of a specific SOI and a bulk MOSFET based OTAs are shown in figs. 4 and 5, respectively.

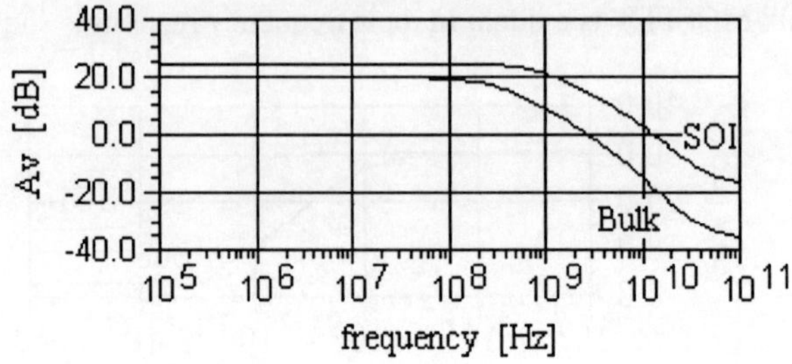

Figure 4. Gain plot of SOI and Bulk OTA

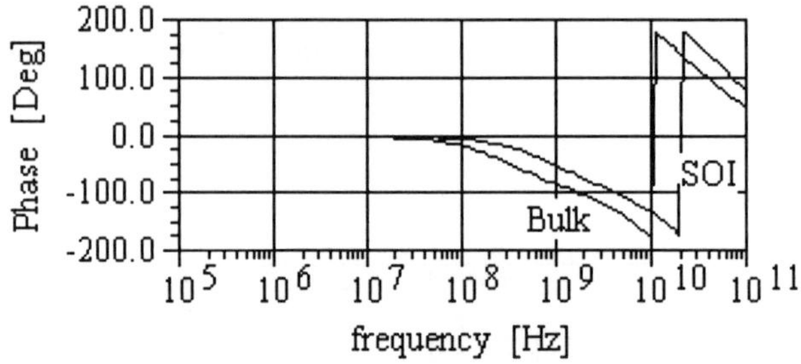

Figure 5. Phase plot of SOI and Bulk OTA

2.2 Sinusoidal Property

Designed OTA circuits preserve the sinusoidal characteristic of the transmitted signal as depicted in figs. 6 and 7. Figures show the input voltage waveform and the output current waveform of OTA at 5 GHz with a 100 kΩ load resistor and when the bias current is 1mA.

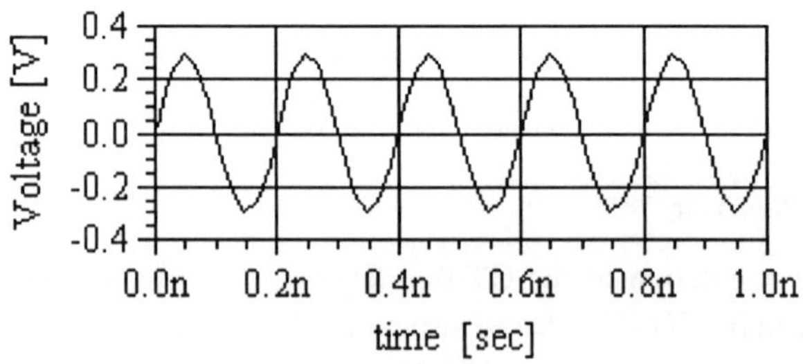

Figure 6. Differential Input Voltage to OTA

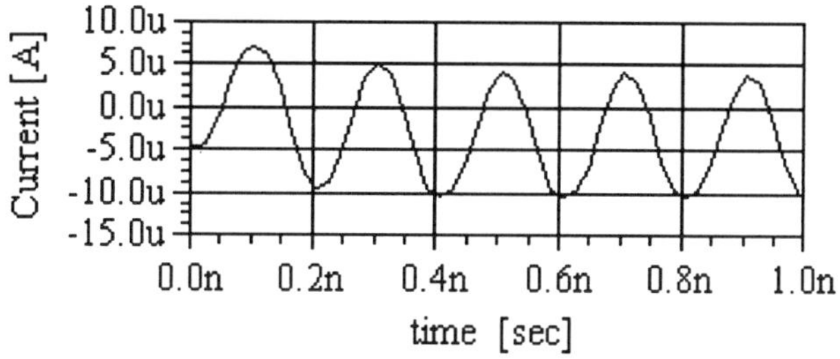

Figure 7. OTA Output Current

2.3 Switching Characteristics and Power Dissipation

To estimate the switching characteristics of OTAs, the pulse responses of the circuits were considered. For an input pulse with a rise-time of 1 ns, the SOI based OTA showed a rise-time of 1.08 ns while for the bulk design the rise-time turned out to be 1.55 ns. Next, the static power dissipation in different circuits was calculated and results indicate comparable power dissipation were 85 mW for the SOI based design against 80 mW for the bulk version.

2.4 Effects of Transistor Channel length

Changing the gate length of a MOSFET device affects, among other parameters, the effective capacitance of the individual transistor, thereby changing the frequency response of the device. To evaluate the effects of shorter channel length on the overall circuit performance we have analyzed OTAs designed with different channel length SOI MOSFETs. SOI transistors of gate lengths 250 nm, 100 nm, 75 nm and 50 nm were tried keeping current mirror ratio and the load resistor the same. To keep the W/L ratio of each transistor, the widths of the transistors were scaled proportionally. All other circuit parameters were kept the same. Bias currents ranged from 1 mA for the 250 nm device to 0.25 mA for the 50 nm design. Results (Table 1) demonstrated that reduction in the gate-length of the transistor raised the transit frequency of the OTA and improved the phase margin at the expense of the circuit gain.

Table 1. Effects of SOI MOSFET gate length shortening on OTA circuit performance

Gate Length (nm)	Power Dissipation (mW)	A_{vo}	f_T (GHz)	Phase Margin °
250	85	24.7	12.7	30
100	43	20.3	21.4	46.3
75	42	20.2	25.2	48.4
50	22	20.1	27.4	55

3 Band Pass Filter

To demonstrate the applicability of SOI MOSFET based OTA in the design of high frequency analog circuits, a band-pass filter was realized using two cascaded OTAs [5] as shown in fig. 8.

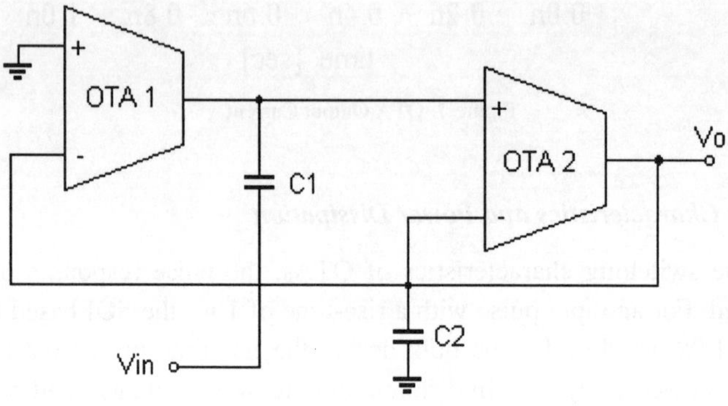

Figure 8. Band Pass Filter

Results clearly indicate high frequency band pass filter characteristics of the designed cascaded circuit. Representative frequency characteristics of SOI transistor based band pass filter along with a bulk MOSFET based filter are shown in fig. 9. The designed SOI MOSFET based filter has center frequency of 7 GHz with a bandwidth of 480 kHz as compared to bulk MOSFET band pass filter which has a center frequency of 1.12 GHz with a bandwidth of 460 kHz. These results clearly establish GHz frequency application potential of SOI MOSFET based OTA circuits and filters.

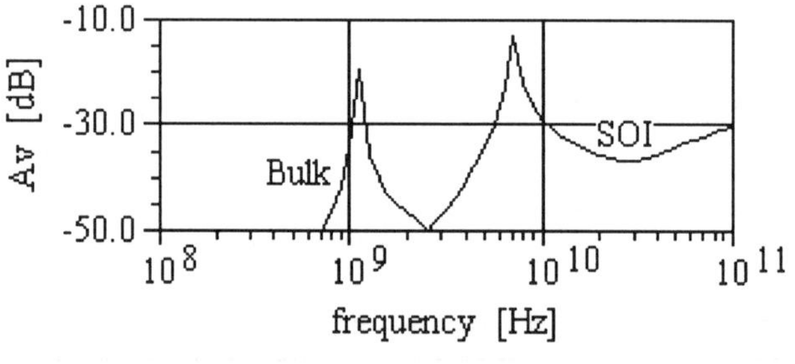

Figure 9. SOI vs. Bulk OTA Based Band Pass Filter Frequency Plot

4 Conclusion

SOI MOSFET based OTA circuit responses were compared with OTA designed with bulk MOSFET. The design began with the showing of higher frequency capacity of SOI MOSFETs. The higher frequency response of the single-stage OTAs with SOI MOSFETs establishes its application potential in GHz range while maintaining the sinusoidal characteristic of the input signal. The transition frequency (f_T) of the SOI based OTA is around 12.7 GHZ with acceptable power dissipation whereas that of bulk MOSFET based OTA is around 3 GHz. Results also indicate that both f_T and phase margin value depends on the transistor channel length. Reduction in the MOSFET channel length increases frequency response with lower power dissipation but at the same time the circuit gain is reduced. Simulation results also establish the reality of designing GHz frequency band-pass filter using SOI based OTA as a building block.

References

1. S. Jambugalingam, O. Rodrigues, A. Bangle, A. Venkatachalam, J. Mody, and P.Ghosh, "Characteristic of Operational Amplifier Circuit in the Microwave Frequency Range", INCEMIC, 2003
2. B. N. Ray, P. Palchoudhuri, P. K. Nandi and P. Ghosh, "Development of Operational Transconductance Amplifier based Bandpass Filters", The IASTED Conference, Applied Modeling and Simulation", 2002
3. J. P. Eggermont, D. Flandre, R. Gillon, J. P. Collinge, "A 1-GHz Operational Amplifier in SOI Technology", Proceedings of the International SOI Conference, 1995
4. AIM-SPICE student version 3.8f, AIM-SOFTWARE, 1993.
5. Geiger, R., Sanchez-Sinencio, E., "Active Filter Design Using Operational Transconductance Amplifiers: A Tutorial", IEEE Circuits and Devices Magazine, 1985
6. TSMC SPICE BSIM3 Version 3.1 Parameters, **http://www.mosis.org**, January, 2004
7. FD NMOS/PMOS, BSIMSOI Version 2.0 example netlist, University of California – Berkeley, **http://www-device.eecs.berkeley.edu/~bsimsoi** April, 1999

International Journal of High Speed Electronics and Systems
Vol. 14, No. 3 (2004) 696–701

World Scientific
www.worldscientific.com

LIFETIME OF NONEQUILIBRIUM CARRIERS IN AlGaN EPILAYERS WITH HIGH Al MOLAR FRACTION

J. MICKEVICIUS, R. ALEKSIEJUNAS, AND M. S. SHUR
Department of ECE and CIE, Rensselaer Polytechnic Institute, Troy, U.S.A.

J. P. ZHANG, Q. FAREED AND R. GASKA
Sensor Electronic Technology, Inc., Columbia, SC 29209, U.S.A.

G. TAMULAITIS
MTMI, Vilnius University, Vilnius, Lithuania

AlGaN epilayers with different aluminum content have been studied by photoluminescence (PL) and light-induced transient grating (LITG) techniques. The epilayers were grown by conventional metal organic chemical vapor deposition (MOCVD) and by migration enhanced MOCVD (MEMOCVDTM). The carrier lifetime and PL intensity in epilayers containing the same amount of aluminum are shown to be inversely proportional to the dislocation density, which was determined using etch pit technique. The carrier diffusion length is shown to be close to the average distance between two first-neighbor dislocations. Enhancement of the lifetime and PL intensity in epilayers grown on sapphire by MEMOCVDTM is demonstrated.

1 Introduction

Deep ultraviolet (UV) detectors and emitters might find various applications in biology chemistry, and medicine as well as in detection of hazardous agents and in solid state lighting. The development of light emitting diodes (LED) and laser diodes (LD) operating in deep UV requires using AlGaN with a high Al molar fraction. However, it still encounters serious problems due to structural defects in high-Al-content AlGaN epilayers: achieving low dislocation densities remains a challenge. Recently introduced low-defect AlN/AlGaN superlattice buffers, deposited on sapphire by using Migration Enhanced Metal Organic Chemical Vapor Deposition (MEMOCVD^{TM}), act as effective filters decreasing density of threading dislocations.[1]

In this report, we compare AlGaN epilayers with different Al content in terms of their photoluminescence (PL) intensity and nonequilibrium carrier lifetimes. The carrier diffusion length was determined by LITG and compared to the average distance between two first-neighbor dislocations, which was roughly estimated by using etch pit technique. AlGaN epilayers grown by MEMOCVD^{TM} technique were compared with the epilayers grown by conventional MOCVD and are shown to have stronger PL intensity and longer lifetimes.

2 Experiment

The AlGaN layers were deposited using conventional low pressure metalorganic chemical vapor deposition (MOCVD) and migration enhanced MOCVD (MEMOCVD™). Trimethyl aluminum, triethyl gallium, and ammonia were used as precursors for Al, Ga, and N, respectively. The samples were nominally of the same thickness of ~2.5 μm. Samples in Set A were grown on basal plane sapphire by the conventional MOCVD with Al content of 17, 26, 33, 43, and 50% in samples A1 to A5, respectively. High-temperature AlGaN layers were deposited on the low-temperature 200 Å-thick AlN buffers. Samples in Set B contain 26% (B1) and 65% (B2) of aluminum and were deposited with the insertion of defects-filtering AlN/AlGaN superlattices by MEMOCVD™ technique. MEMOCVD™ is an improved version of Pulsed Atomic Layer Epitaxy (PALE), which deposits quaternary $Al_xIn_yGa_{1-x-y}N$ layers by repeats of a unit cell grown by sequential metal organic precursor pulses of Al, In, Ga, and NH_3.[2] While in PALE the duration of each pulse in the unit cell is fixed, and the NH_3 pulse always follows each metalorganic pulse, the duration and waveforms of precursor pulses in MEMOCVD™ are optimized, and the pulses might overlap. MEMOCVD™ combines a fairly high growth rate for buffer layers with reduced growth temperature (by more than 150°C) and improved quality for active layers. Introduction of MEMOCVD™ enabled us to achieve a better mobility of pre-cursor species on the surface and, thus, better atomic incorporation and improved surface coverage.

Luminescence of the epilayers was measured under quasi-steady-state excitation using the 5th harmonic radiation (213 nm) of a YAG:Nd laser with the pulse width of 4 ns. The spectra were recorded using a UV-enhanced intensified CCD and a TRIAX 550 spectrometer. The unit HOLO-2 of *EKSPLA Ltd.*, which exploits the light induced grating technique, was used to measure the carrier lifetime and diffusion coefficient. Two beams of the 4th harmonic (266 nm) of YAG:Nd mode-locked laser radiation (pulse duration 20 ps) overlapping at the sample surface created a periodic pattern of carrier density by invoking band-to-band excitation of the AlGaN epilayers, while delayed pulses of the 2nd harmonic (532 nm) probed the decay of the light-induced grating.

3 Results and discussion

The time evolution of the diffraction efficiency for several samples is shown in Figure 1 (a), where the decay curves are normalized to peak value for comparison. The LITG decay in all samples is exponential, thus we suppose that it is caused only by two mechanisms: the carrier recombination and diffusion. Therefore, determination of the LITG decay time, τ_G, at different grating spacing, Λ, enables extraction of the lifetime, τ_R, and the coefficient of ambipolar diffusion D_a using the relationship:[3]

$$\frac{1}{\tau_G} = \frac{1}{\tau_R} + \frac{4\pi^2 D_a}{\Lambda^2}. \tag{1}$$

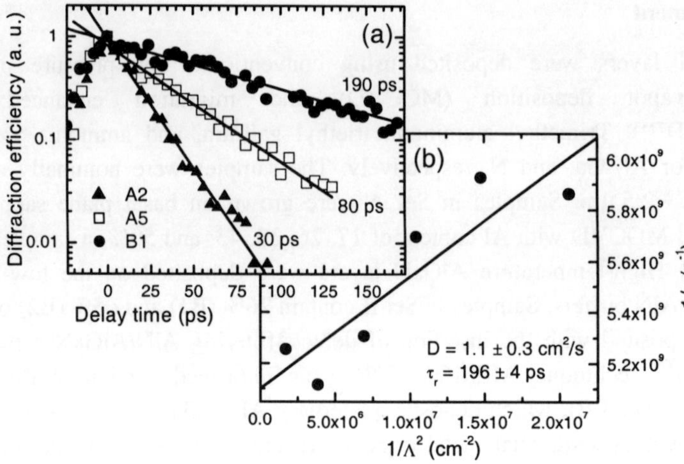

Figure 1. Decay kinetics of the light-induced transient gratings (period 10.2 μm) in AlGaN samples with different Al content (points) and the best fit to experimental points (lines); grating decay times are indicated (a), and decay time versus grating spacing measured in the sample B1; diffusion coefficient and carrier lifetime are indicated (b).

The decay rate τ_G at a large grating spacing Λ, when diffusion can be neglected, provides a good estimate for the carrier lifetime τ_R. A plot of the grating decay time versus the grating spacing according to Eq. 1 is presented in Figure 1(b) for sample B1 taken from Reference.[4] The spacing dependence of the decay time in other samples was too weak was for such a plot with sufficient accuracy. Thus, the diffusion coefficient was reliably estimated for sample B1. Since the sample was highly doped n-type, the photoexcited holes are minority carriers, and the ambipolar diffusion coefficient, D_a, should be equal to the hole diffusion coefficient, D_p. Determination of both lifetime and diffusion coefficient enables an estimation of the diffusion length, $L_D = \sqrt{D\tau_R}$. For the values of $D = 1.1 \pm 0.3$ cm^{2}s^{-1} and τ_R = 196 ± 4 ps measured in the sample B1, the diffusion length equals 0.15 ± 0.03 μm. This value is close to the average distance between two first-neighbour dislocations, r_{dd} = 0.13 μm, which was roughly estimated using the dislocation density of $2 \cdot 10^9$ cm^{-2} obtained by the etch pit technique and a simple relation between the dislocation density and interdislocation distance $N_{dd} = 1/\pi r_{dd}^2$. Thus, it is reasonable to assume that the carrier lifetime and diffusion is dislocation-limited and the carrier lifetime in our samples is limited predominantly by non-radiative recombination at dislocations.

Figure 2 shows the PL spectra for the samples containing the same amount of Al, but grown by different techniques: sample B1 is grown on low-defect AlN/AlGaN superlattice buffer, while sample A1 is grown without such buffer. The superlattice buffer acts as a filter decreasing the dislocation density from 10^{10} cm^{-2} (in sample A2) down to 2 10^9 cm^{-2} (in B1). This decrease in dislocation density results in the increase of PL intensity by a factor of 6, which is close to the ratio of the lifetimes (6.5), and similar to the ratio of the dislocation densities (5). This is reasonable, since the radiative emission proceeds at a constant and much lower rate than the nonradiative recombination, which limits the lifetime. Thus, the PL intensity in AlGaN epilayers containing the same Al content is in direct proportion to the carrier lifetime.

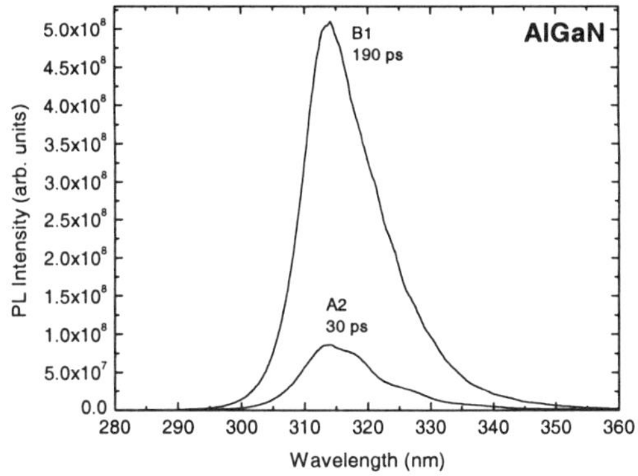

Figure 2. PL spectra of AlGaN samples with the same Al content, but grown by conventional MOCVD and migration enhanced MOCVD (MEMOCVD™).

Spectra of the AlGaN epilayers with similar lifetimes but different aluminum contents are shown in Fig. 3. It is clearly seen that the PL intensity decreases with increasing Al content. Since the carrier lifetime is limited by nonradiative recombination, this behavior can be explained in terms of decreasing rate of the radiative recombination.

A strong correlation of the photoluminescence intensity with the dislocation density is in agreement with the results of our comparative study of photoluminescence in AlGaN layers grown on SiC and bulk AlN substrates.[5] The PL spectra of identical multiple quantum well structures deposited on conventional SiC substrate and on bulk AlN substrate are presented in Fig. 4 (the samples were excited by excimer laser radiation at 193 nm). Introduction of bulk AlN substrate also showed an increase in the PL intensity by a factor of 28. This PL enhancement we interpreted as caused by the decrease in the dislocation density of the AlGaN layers grown on bulk AlN compared to those grown on SiC substrates.

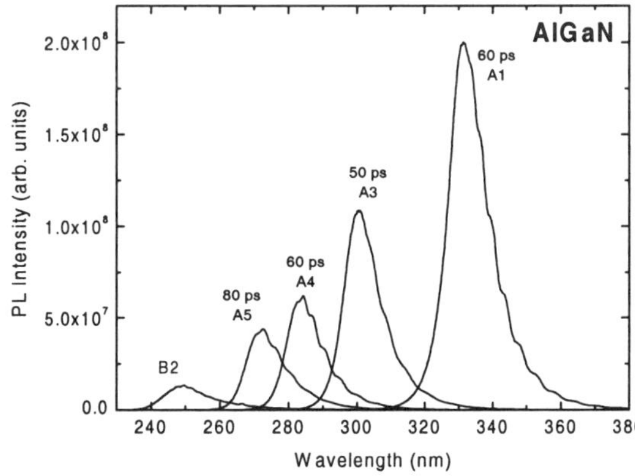

Figure 3. PL spectra of the AlGaN epilayers with different Al content. Carrier lifetimes determined by LITG are indicated.

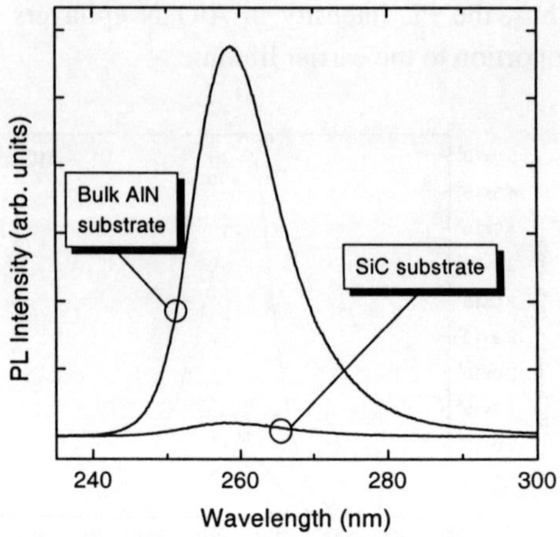

Fig. 4. Photoluminescence spectra of identical multiple quantum well structures containing three 5 nm-wide Al0.5Ga0.5N wells separated by 20 nm-wide AlN deposited on conventional SiC substrate and on bulk AlN substrate.

4 Conclusions

In conclusion, the study of lifetime, diffusion length, and PL intensity in AlGaN layers with different Al content and dislocation densities show that the lifetime of nonequilibrium carriers in these epilayers is primarily limited by the carrier diffusion to the dislocations, where the nonradiative recombination occurs. In AlGaN epilayers with the same Al content but different dislocation densities the radiative recombination proceeds at the same rate. The radiative decay rate decreases with increasing Al content. We demonstrate that the MEMOCVD^{TM} growth technique enables an increase of the PL intensity for the AlGaN epilayers grown on sapphire substrates by at least up to a factor of 6, compared to the epilayers containing the same fraction of Al but grown by conventional MOCVD.

Acknowledgments

One of the authors (G. T.) acknowledges the Lithuanian Ministry of Education and Science for his Fellowship. The work at RPI was partially supported by the National Science Foundation (Program Director, Dr. V. Rao.)

References

1. H. M. Wang, J. P. Zhang, C. Q. Chen, Q. Fareed, J. W. Yang, M. Asif Khan, Appl. Phys. Lett. 81, 604 (2002).
2. J. Zhang, E. Kuokstis, Q. Fareed, H. Wang, J. Yang, G. Simin, M. Asif Khan, R. Gaska, and M. Shur, Appl. Phys. Lett. **79**, 925 (2001).
3. Miller, Transient Grating Studies of Carrier Diffusion and Mobility in Semiconductors, in: E. Garmire, A. Kost (Eds.), Nonlinear Optics in

Semiconductors II (Semiconductors and Semimetals, vol. 59), Acad. Press., 1999, p. 287.

4. G. Tamulaitis, J. Mickevicius, R. Aleksiejunas, M. S. Shur, J. P. Zhang, Q. Fareed, and R. Gaska, "Dislocation-limited Lifetime of Nonequilibrium Carriers in AlGaN Epilayers", to be published in Proceedings of the 27th International Conference on the Physics of Semiconductors, July 26-30, 2004, Flagstaff, Arizona.
5. R. Gaska, C. Chen, J. Yang, E. Kuokstis, A. Khan, G. Tamulaitis, I. Yilmaz, M. S. Shur, J. C. Rojo, L. J. Schowalter, Appl. Phys. Lett. **81**, 4658 (2002).

International Journal of High Speed Electronics and Systems
Vol. 14, No. 3 (2004) 702–707

NOISE CHARACTERISTICS OF 340 nm AND 280 nm GaN-BASED LIGHT EMITTING DIODES

SHAYLA SAWYER, SERGEY L. RUMYANTSEV, NEZIH PALA, AND MICHAEL S. SHUR

Department of Electrical, Computer, and Systems Engineering Center for Broadband Data Transport Science and Technology CII 9017, Rensselaer Polytechnic Institute, Troy, NY 12180-3590, USA

YURIY BILENKO AND REMIS GASKA

Sensor Electronic Technology, Inc. 1195 Atlas Road Columbia, SC 29209, USA

PAVEL V. KOSTERIN AND BRIAN M. SALZBERG

Departments of Neuroscience and Physiology, University of Pennsylvania School of Medicine, Philadelphia, PA 19104, USA

Low frequency fluctuations in light intensity of 340 nm and 280 nm GaN-based light emitting diodes (LEDs) are compared with noise properties of other commercially available UV and visible wavelength LEDs and halogen lamps. At low frequencies, LEDs can exhibit lower levels of noise than halogen lamps. An LED noise quality factor β is estimated for the UV LEDs.

1 Introduction

Deep ultraviolet (UV) light sources are now employed in multiple biological and physical experiments, as well as for water and air purification. Protein molecules strongly absorb in the deep ultraviolet (200-300nm) range and exhibit fluorescence at longer wavelengths. The detection of fluorescence from protein molecules is an effective method to discover *in situ* the presence of small amount of proteins, which may include hazardous biological pathogens[1]. Mercury and xenon lamps, solid state or gas lasers, and even stable tungsten-halogen lamps that are used in such detection systems are large, costly, heavy, and difficult to use[1]. Novel 280nm and 340 nm GaN-based light emitting diodes (LEDs) might substitute for existing UV light sources in many of these applications.

For biological experiments, the noise properties of the light sources are very important. The detection system, including the excitation source, must exhibit high sensitivity; i.e. low noise and high stability over tens of minutes.

To the best of our knowledge, there are no noise studies of light intensity fluctuations for UV LEDs.

2 Experimental Details

Table I summarizes the electrical and optical parameters of the light sources under investigation. To measure the light intensity fluctuations from LEDs and halogen lamps, we used the UV enhanced Si photodiode UV-100L from UDT Sensors, Inc. The photodiode was biased by a low noise battery using a 10 - kΩ load resistor. The voltage fluctuations across the resistor were amplified by a Signal Recovery (model 5184) low noise amplifier and analyzed using an SR 770 Network Analyzer. The LEDs were biased by a low noise battery with a 100 Ω resistor connected in series. The load resistance was at least 10 times higher than the LED differential resistance at high current. The halogen lamps were powered using a high capacity, low noise car battery.

Table 1. Electrical and optical parameters of the light sources under investigation

Light source	Peak wavelength	Maximum forward current	Forward voltage	Luminous intensity (mcd), radiant flux (W), or luminous flux (lm)
SET UVTOP^{tm}	280nm	50mA	5-7V	0.1mW
SET UVTOP^{tm}	340nm	50mA	5-6.5V	1mW
NICHIA NSHU550A	375nm	25mA	3.5-4V	2 mW
NICHIA NSPE510S	505nm	30mA	3.5-4V	3200 mcd
Roithner Lasertechnik, LED740-66-60	740nm	750mA	>9V	>1W
Halogen lamp Gilway L7389A		4A	12V	900 lm
Halogen lamp OSRAM HLX 64625		8.3A	12V	3600 lm

3 Results and Discussion

Fig.1 shows an example of as measured voltage fluctuations on the 10 kΩ resistor connected in series with photodiode illuminated with 280 nm LED (the background noise measured in the dark is also shown). At low frequencies, $f<$ (10^2-10^4) Hz, the noise spectra of all LEDs depend on frequency as $1/f^{\gamma}$ with $\gamma = 1$-2.

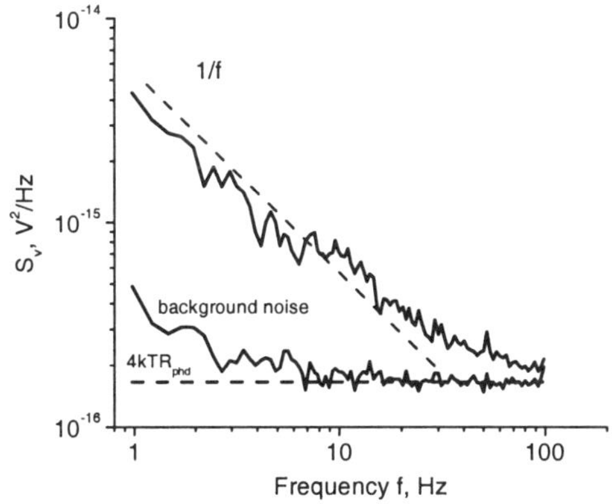

Fig.1. Noise spectrum of the voltage fluctuations S_v on the 10 kΩ resistor in series with the photodiode for 280 nm LED. LED current I_{LED}=50mA, photodiode current I_{phd}=0.3μA. The background noise measured in the dark is also shown.

Because of low photodiode sensitivity at 280nm, the photodiode current was small while measuring UV LEDs, and thermal noise dominated at high frequencies (see Fig.1). For 340 nm LEDs, visible wavelength LEDs and halogen lamps, the classical shot noise dominated at high frequencies.

Figure 2 shows the LED noise spectra, S_I, obtained after subtraction of the background noise measured in the dark (also shown in the figure). The LED bias current was the maximum current specified by the manufacturer. As seen, at low frequencies, the $1/f^{\gamma}$ noise with γ=1-2 is dominant for all these LEDs. The photodiode current, I_{phd}=250μA, was adjusted to be the same for all these light sources by varying the amount of light reaching the photodiode. In this case the spectral noise density S_I of the $1/f$-like noise was always proportional to the square of the photodiode current $S_I \sim I_{phd}^2$. Since photodiode current of 250μA was not achievable for the 280 nm UV LEDs, the noise spectra for these LEDs in Fig.2 were normalized assuming $S_I \sim I_{phd}^2$.

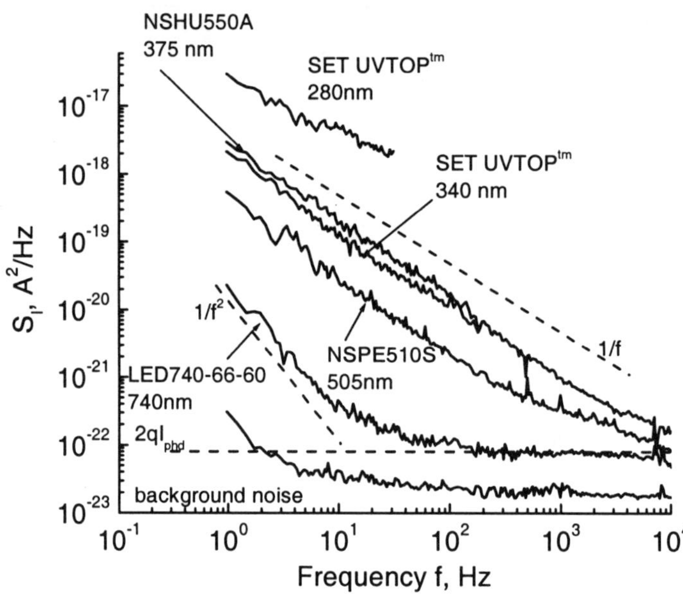

Fig. 2. Noise spectra of light intensity fluctuations for different LEDs. The background noise measured in darkness ("dark noise") and the level of the shot noise S_I=$2qI_{phd}$ are also shown.

Fig. 3 compares noise spectra for UV LEDs and two different halogen lamps. The halogen lamps demonstrated unstable behavior. One of the possible reasons for the unstable behavior of the halogen lamps might be the fluctuation of thermal sink due to the fluctuation of the air flow around the lamp. However, we found that housing of the halogen lamp into boxes of different dimensions does not affect the noise. Cooling of the lamp with fan drastically increases the low frequency noise. The noise amplitude varied from measurement to measurement by about one order of magnitude at low frequency (each measurement session took approximately 5 min).

In Fig. 3, the hatched areas show the range of this variation for the two halogen lamps under study. The noise spectra of the halogen lamps are close to a $1/f^4$ law. This rather unusual spectral shape is probably a signature of nonstationary behavior of the halogen lamps. As seen in Fig. 3 at low frequency the noise amplitude for UV LEDs is smaller than that for halogen lamps.

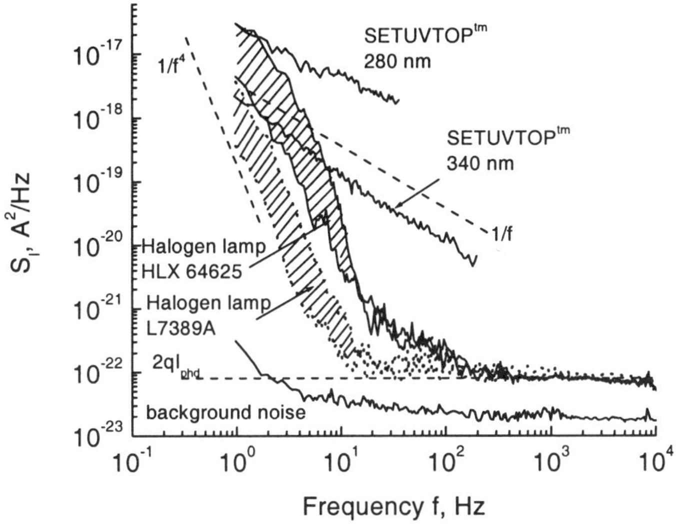

Fig.3. Comparison of the noise spectra of the light intensity fluctuations for two Halogen lamps (hatched areas) and UV LEDs. Background noise, and shot noise are shown for comparison. I_{phd}=250μA.

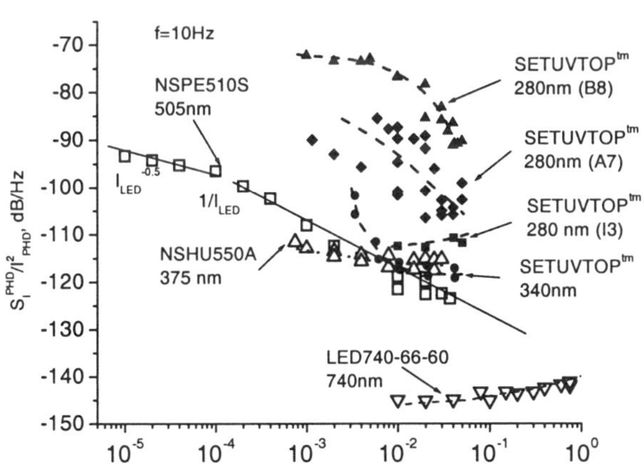

Fig.4. Dependence of relative noise spectra S_I/I_{phd}^2 on LED current, I_{LED}, for different LEDs.

Figure 4 shows S_I/I_{phd}^2 at 10 Hz as a function of the LED current for all LEDs studied. In ref. 3, we introduced the LED noise quality factor of LEDs:

$$\beta = \frac{S_I}{I_{phd}^2} fn \frac{\tau}{q} I_{LED}, \qquad (1)$$

where f is the frequency, n is the number of chips connected in series, τ is the radiation life-time, and q is the electronic charge. The lower the value of β, the better is the LED noise quality. This parameter is similar to the Hooge parameter[2] used as a figure of merit for the $1/f$ noise in semiconductor materials and devices. For LEDs, the product $(n\tau/q)I_{LED}$ represents the total number of charge carriers, N_{light}, involved in the light emission process. Parameter N_{light} plays the same role as the total number of carriers, N, in the expression for the Hooge parameter:

$$\alpha = \frac{S_I}{I^2} fN \qquad (2)$$

Fig. 5 shows the dependence of β on the LED current, I_{LED} (symbols are the same as in Fig.4). For a crude estimate, we assumed $\tau/q \sim 10^{10}$ A^{-1} for all devices. As seen, our 340nm and 280 nm LEDs demonstrate the quality factor β of the same order of magnitude as 375nm and 505nm LEDs produced by Nichia. In a relatively narrow range of currents, the noise level of the 280nm LEDs and quality factor β approach the same level as those for longer wavelength LEDs. The values of β for LED740-66-60 are considerably smaller. These LEDs demonstrate the lowest noise among all the LEDs that we measured[3].

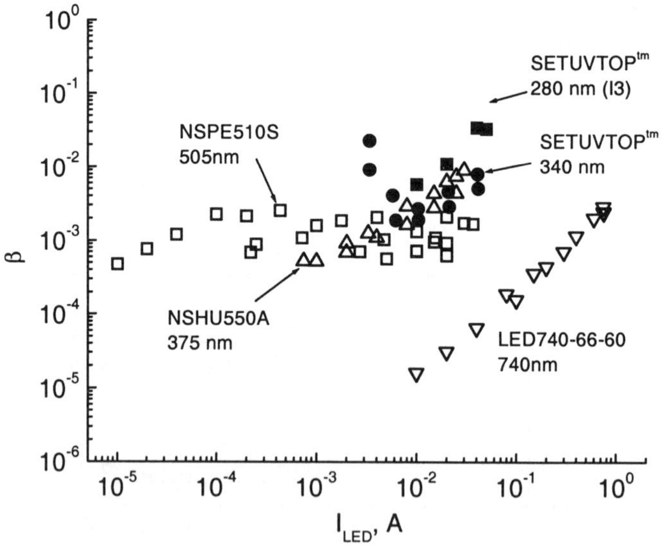

Fig.5. Quality factor β as function of the LED current, I_{LED}, for different LEDs.

4 Conclusions

At low frequencies $f<(10^2\text{-}10^4)$ Hz, the noise spectra of LEDs depend on frequency as $1/f^{\gamma}$ with $\gamma = 1\text{-}2$. This $1/f$-like noise dependence on current is different for different LED types. Halogen lamps exhibit markedly unstable noise behavior. At frequencies below 10 Hz, UV light emitting diodes can provide lower levels of noise than halogen lamps. UV LEDs exhibit a quality factor β of the same order of magnitude as visible wavelength LEDs. These results show that 280 nm and 340 nm UVTOP™ LEDs are suitable for studying steady state and time-varying UV fluorescence of biological materials.

References

1. Field Analytical Chemistry & Technology, **3** (4-5) (1999), Special Issue: Chemical and Biological Aerosol Detection and Identification with Field Analytical Instrumentation. Issue Edited by A. Peter Snyder
2. F. N. Hooge, T.G.M. Kleinpenning, and L.K.J. Vandamme, Experimental studies on 1/f noise, Rep. Progr. Phys. **44**, 479-532 (1981)
3. S.L. Rumyantsev M.S. Shur, Yu. Bilenko, P.V. Kosterin, and B.M. Salzberg, Low frequency noise and long-term stability of non-coherent light sources, Journal of Applied Physics, **96**, 966-969 (2004)

International Journal of High Speed Electronics and Systems
Vol. 14, No. 3 (2004) 708–713

World Scientific
www.worldscientific.com

JUNCTION-TEMPERATURE MEASUREMENTS IN GaN UV LIGHT-EMITTING DIODES USING THE DIODE FORWARD VOLTAGE

Y. XI
Future Chips Constellation
Department of Physics, Applied Physics, and Astronomy
Rensselaer Polytechnic Institute, Troy, New York 12180, USA
xiy@rpi.edu

E. F. SCHUBERT*
Future Chips Constellation
Department of Physics, Applied Physics, and Astronomy
Department of Electrical, Computer, and Systems Engineering
Rensselaer Polytechnic Institute, Troy, New York 12180, USA
EFSchubert@rpi.edu

A theoretical model for the dependence of the diode forward voltage (V_f) on junction temperature (T) is developed. A new expression for dV_f/dT is derived that takes into account all relevant contributions to the temperature dependence of the forward voltage including the intrinsic carrier concentration, the bandgap energy, and the effective density of states. Experimental results on the junction temperature of GaN UV LEDs are presented. Excellent agreement between the theoretical and experimental temperature coefficient of the forward voltage (dV_f/dT) is found. The experimentally found linear dependence of the junction temperature on forward current is explained by a thermal conduction model. A thermal resistivity of 342.2 K/W is found for the UV LED

III–V nitride semiconductors have a direct bandgap and thus are very suitable for solid-state ultraviolet (UV) light sources. Such UV sources have a wide variety of applications, including UV-induced fluorescence, lighting, displays, spectrofluometry, photo-catalytic processes, and high resolution optics. Double heterostructure GaInN/AlGaN UV LEDs emitting at 371 nm with external quantum efficiency of 7.5% and output powers of 5 mW have been demonstrated[1]. The junction temperature is a critical parameter and affects internal efficiency, maximum output power, reliability, and other parameters. Several groups have reported measurements of the junction temperature of laser diodes (LDs) using micro Raman spectroscopy[2], threshold voltage[3], thermal resistance[4], photothermal reflectance microscopy (PRM)[5], electroluminescence[6] and photoluminescence[7].

In this paper, the diode forward voltage is employed to measure the junction temperature of GaN UV LEDs emitting at 375 nm. The theoretical foundation of the temperature dependence of the forward voltage is analyzed in detail. The experimental and theoretical results are compared and discussed.

From the well-known Shockley equation, for a forward voltage $V_f >> kT/e$, we obtain

* Future Chips Constellation Rensselaer Polytechnic Institute, Troy, New York 12180, USA

$$\frac{dV_f}{dT} = \frac{d}{dT}\left[\frac{n_{ideal}\, kT}{e} \ln\left(\frac{J_f}{J_s}\right)\right] \tag{1}$$

where J_s is the saturation current density, n_{ideal} is the diode-ideality factor, and the other symbols have their usual meaning. J_s can be expressed as

$$J_s = e\left[\sqrt{\frac{D_n}{\tau_n}}\frac{n_i^2}{N_D} + \sqrt{\frac{D_p}{\tau_p}}\frac{n_i^2}{N_A}\right] \tag{2}$$

where D_n and D_p are diffusion constants of electrons and holes, respectively, and τ_n and τ_p are the minority carrier lifetimes of electrons and holes, respectively. Both, the diffusion constants and the lifetimes are temperature dependent. For phonon scattering, the diffusion constants decrease with temperature according to a $T^{-1/2}$ dependence. The carrier lifetimes can decrease (non-radiative recombination) or increase (radiative recombination) with temperature. Dopants with concentration N_D and N_A are assumed to be fully ionized so that the free carrier concentration has no temperature dependence.

The intrinsic carrier concentration, n_i, which is strongly temperature-dependent, is given by

$$n_i = \sqrt{N_C N_V}\ \exp\left(-\frac{E_g}{2kT}\right) \tag{3}$$

where N_C and N_V are effective densities of states at the conduction band and valence band edge, respectively. The effective densities of states are given by[8]

$$N_C = 2\left(\frac{2\pi m_{de} kT}{h^2}\right)^{\frac{3}{2}} M_c \propto T^{\frac{3}{2}} \tag{4}$$

$$N_V = 2\left(\frac{2\pi m_{dh} kT}{h^2}\right)^{\frac{3}{2}} \propto T^{\frac{3}{2}} \tag{5}$$

where m_{de} and m_{dh} are the density-of-state effective mass for electrons and holes respectively. M_c is the number of equivalent minima in the conduction band.

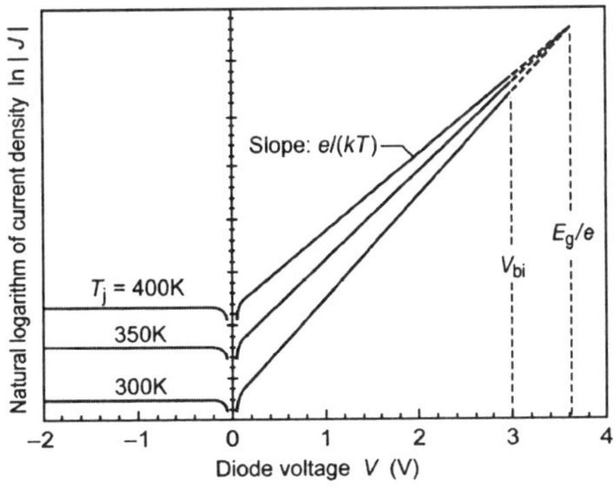

FIG. 1. Schematic dependence of $\ln|J|$ on the voltage V of a GaN p-n junction for junction temperatures T_j = 300 K (low), 350 K (middle) and 400 K (high).

FIG. 1 shows the schematic dependence of $\ln|J|$ on the voltage V for different junction temperatures, as inferred from the Shockley equation. Inspection of FIG. 1 shows that, for a constant current density, the junction voltage decreases with increasing temperature.

By substituting Eqs. (2), (3), (4) and (5) into Eq. (1), the temperature dependence of n_i, E_g, N_C, and N_V are taken into account. Executing the derivative yields

$$\frac{dV_f}{dT} = \frac{eV_f - E_g}{eT} + \frac{1}{e}\frac{dE_g}{dT} - \frac{3k}{e}. \tag{6}$$

This equation gives the fundamental temperature dependence of the forward voltage. The first summand on the right-hand side of the equation is due to intrinsic carrier concentration. The second summand is due to the temperature dependence of the bandgap energy. The third summand, $3k/e$, is due to the temperature dependence of N_C and N_V. Note that the second summand was not included in earlier derivations[9]. Inclusion of the temperature dependences of diffusion constants and lifetimes would only yield a minor contribution to the temperature coefficient and we therefore neglect these contributions. Light-emitting diodes are typically operated with their forward voltage close to the built-in voltage, i.e. $V_f \approx V_{bi}$. For non-degenerate doping concentrations, we can write

$$eV_f - E_g \approx kT\ln\left(\frac{N_D N_A}{n_i^2}\right) - kT\ln\left(\frac{N_C N_V}{n_i^2}\right) = kT\ln\left(\frac{N_D N_A}{N_C N_V}\right). \tag{7}$$

Furthermore, the bandgap energy can be expressed as

$$E_g = E_0 - \frac{\alpha T^2}{\beta + T} \tag{8}$$

where α and β are the Varshni parameters. For GaN[8], α = 0.77 meV/K^2, β = 600 K. Substituting Eqs. (7) and (8) into Eq. (6) yields

$$\frac{dV_f}{dT} \approx \frac{k}{e}\ln\left(\frac{N_D N_A}{N_C N_V}\right) - \frac{\alpha T(T+2\beta)}{e(T+\beta)^2} - \frac{3k}{e}. \tag{9}$$

This equation is a very useful expression for the temperature coefficient of the forward voltage. Note that this expression is a *lower limit* for the magnitude of dV_f/dT, because the *junction voltage* is *less* than the *built-in voltage* in all practical cases. For GaN with $N_D = N_A = 2 \times 10^{16}$ cm^{-3}, one obtains $dV_f/dT = -1.76$ mV/K. For a Si p-n junction with the same doping concentration, one obtains $dV_f/dT = -1.74$ mV/K, in good agreement with Millman and Halkias[9].

The diode forward voltage can be used to assess the junction temperature of p-n junction diodes. The forward-voltage method consists of two series of measurements, a calibration measurement and the actual junction-temperature measurement. In the calibration measurement, a pulsed forward current (with duty cycle 0.1 %) drives the LED sample located in a temperature-controlled oven. The small duty cycle ensures that the junction temperature is equal to the ambient temperature. An oscilloscope is used to measure the forward voltage V_f of the LED sample at different oven temperatures. The calibration measurement establishes the relation between the forward voltage and the junction temperature.

Experimental results reveal that the experimental V_f versus T relation is very close to linear and can be fitted by the equation

$$V_f = A + BT_o \tag{10}$$

where T_o is the oven temperature, A and B are fitting parameters. Since the duty circle of the pulsed current is 0.1%, the heat generated by the pulsed current can be neglected and the junction temperature can, with very good accuracy, be assumed to be equal to the oven temperature. The same process is repeated for different (pulsed) current levels.

After the calibration measurement, a DC forward current is applied to the sample and the DC forward voltage values are measured. Using the calibration measurement, the junction temperatures for different DC currents are given by:

$$T_j = \frac{(V_f - A)}{B}. \tag{11}$$

To demonstrate the viability of the method, the junction temperature of GaN UV LEDs (Nichia Corp. with peak wavelength of 375 nm) was determined. Equipment for the experiments included a HP 214B pulse generator, a Wisconsin oven with thermal sensor, a Tektronix TDS 3054B four channel color digital oscilloscope, an Ando AQ-6315A optical spectrum analyzer, an Agilent E3649A dual output DC power, and HP 34401A multimeter to measure the DC forward voltage.

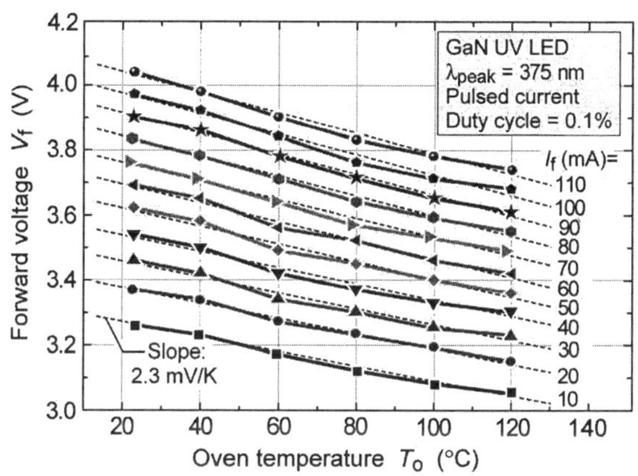

FIG. 2. Experimental forward voltage versus oven temperature for different pulsed injection currents. Also shown is a linear fit for the experimental data (dashed line).

During the calibration measurement, the pulsed current increased from 10 mA to 110 mA in 10 mA increments. The relation of the forward voltage vs. junction temperature is shown in FIG. 2. The dashed lines are linear fits to the experimental data. From FIG. 2, the temperature coefficient of the forward voltage at low currents is –2.3 mV/K, slightly larger in magnitude than the theoretical result. The small difference between the experimental temperature coefficient and the theoretical result of Eq. (9) is fully consistent with Eq. (9) giving the lower limit to dV_f/dT. The difference could also be due to a decrease in p-type GaN layer resistivity, caused by more complete accepter activation at elevated temperature.

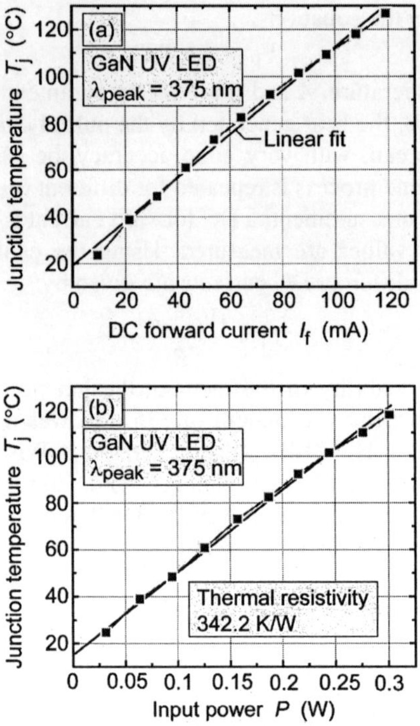

FIG. 3 Junction temperature as a function of (a) DC forward current and (b) input power (IfVf) for a GaN UV LED.

FIG. 3 (a) shows the junction temperature vs. the DC forward current. The T_j-vs.-I_f curve is approximately linear. As the DC forward current increases from 10 mA to 110 mA, the junction temperature increases from 23 to 126 °C. FIG. 3 (b) shows the junction temperature as a function of the diode power, $I_f V_f$. Inspection of the figure reveals that this dependence is also approximately linear.

We next explain the linear relation between the junction temperature and the DC forward current. According to the thermodynamics, heat transfers from a high-temperature object to a low-temperature object if they contact each other. It is well known that the *conduction rate* P_{cond} (heat energy transferred per unit time) is proportional to the temperature difference between the two objects

$$\Delta T \propto P_{cond} \,. \tag{12}$$

Under steady-state conditions, the diode electric input power is (i) converted to light, (ii) consumed by unavoidable non-radiative recombination, and (iii) by the series resistance. Heat will be generated due to non-radiative recombination and the series resistance. Thus the electrical power P has two parts: one part, P_l, is converted to light. The other part, P_d, the dissipated power, is converted to heat. In the steady state, the dissipated power is equal to the conduction rate. The total power is thus given by:

$$P = P_d + P_l = I_f V_f = I_f (V_j + I_f R_s) \tag{13}$$

where R_s is the series resistance, and V_j is the junction voltage. P_l can be expressed as

$$P_l = \eta I_f (V_f - I_f R_s) \tag{14}$$

where η is the external quantum efficiency. In GaN UV LEDs, η typically ranges between 0.01 and 0.1. From Eqs. (13) and (14), we can get

$$P_d = P_{cond} = P - P_l = (1-\eta)I_f V_f + \eta I_f^2 R_s . \quad (15)$$

Since I_f is very small in our case, less than 120 mA, and $\eta << 1$, $\eta I_f^2 R_s$ can be neglected. Substituting Eq. (15) into Eq. (12) yields

$$\Delta T = T_j - T_a \propto I_f V_f \quad (16)$$

where T_a is the ambient temperature. Because the changes of V_f with current are much smaller than V_f (i. e. $\Delta V_f << V_f$), Eq. (16) indicates that the junction temperature depends linearly on the forward current. This is indeed found in our experiments, as shown in FIG. 3 (b). The thermal resistivity of the UV LED, 342.2 K/W is obtained from the slope of the linear fit of FIG. 3(b).

In conclusion, a new theoretical expression of the temperature coefficient of the diode forward voltage is developed. The three main contributions to the temperature coefficient are (i) the intrinsic carrier concentration, (ii) the bandgap energy, and (iii) the effective density of states. Experimental results are presented for GaN UV LEDs. For I_f = 10 mA, the experimental temperature coefficient is dV_f/dT = –2.3 mV/K. The junction temperature increases from 23 °C to 126 °C when the DC forward current increases from 10 mA to 110 mA. The theoretical temperature coefficient, which is shown to be a lower limit in magnitude, is –1.7 mV/K, close to the experimental coefficient. A thermal conduction model is proposed to explain the linear relation between the junction temperature and the forward current in UV LEDs. The linear experimental dependence of the junction temperature on power is in excellent agreement with the model. . A thermal resistivity of 342.2 K/W is obtained for the UV LED.

Acknowledgments

Support through the DARPA SUVOS program is gratefully acknowledged.

References

1. T. Mukai, D. Morita and S. Nakamura, "High-power UV InGaN/AlGaN double heterostructure LEDs", *J. Crystal Growth*, **189** (1998) 778 -781.
2. S. Todoroki, M. Sawai and K. Aiki, "Temperature distribution along the striped active region in high-power GaAlAs visible lasers", *J. Appl. Phys.* **58**, (1985) 1124 -1128.
3. H. I. Abdelkader, et al., "Temperature rise and thermal rise-time measurements of a semiconductor laser diode", *Rev. Sci. Instrum.* **63**, (1992) 2004 - 2007.
4. S. Murata and H. Nakada, "Adding a heat bypass improves the thermal characteristics of a 50 pm spaced a-beam laser diode array", *J. Appl. Phys.* **72**, (1992) 2514 -2516.
5. P. W. Epperlein, "Reflectance modulation-a novel approach to laser mirror characterization" *in Proceedings of 17th International Symposium of Gallium Arsenide and Related Compounds, IOP Conference Series ~ IOP*, London, **112**, (1990) 633 - 638.
6. P. W. Epperlein and G. L. Bona, "Influence of the vertical structure on the mirror facet temperatures of visible GalnP quantum well lasers", *Appl. Phys. Lett.* **62**, (1993) 3074.
7. D. C. Hall, et al., "Technique for lateral temperature profiling in optoelectronic devices using a photoluminescence microprobe", *Appl. Phys. Lett.* **61**, (1992) 384 - 386.
8. http://www.ioffe.rssi.ru/SVA/NSM/Semicond/.
9. J. Millman and C. Halkias, *Integrated Electronics* (McGraw-Hill, New York, 1972), pp. 752 -754.

International Journal of High Speed Electronics and Systems
Vol. 14, No. 3 (2004) 714–719

World Scientific
www.worldscientific.com

High Speed 0.9 μm Lateral p-i-n Photodetectors Fabricated in a Standard Commercial GaAs VLSI Process

WOJCIECH P. GIZIEWICZ* and CLIFTON G. FONSTAD Jr.†
Department of Electrical Engineering and Computer Science,
Massachusetts Institute of Technology,
77 Massachusetts Ave., Cambridge, MA, 02139, USA
**wpgiziew@mit.edu, †fonstad@mit.edu*

SHEILA PRASAD§
Electrical and Computer Engineering Department, Northeastern University
360 Huntington Ave., Boston, MA, 02115, USA
§prasad@ece.neu.edu

Lateral detectors fabricated in an unmodified commercial GaAs VLSI process are presented. The detectors exhibit previously observed characteristic sensitivity and capacitance properties. High speed measurements indicate a bandwidth in excess of 4 GHz.

1. Introduction

The detector of choice for many applications has traditionally been the metal-semiconductor-metal (MSM) photodetector because it is an intrinsically simple device that is planar, has good responsivity, and has low parasitic capacitances. It would seem that the MSM detector is an obvious candidate for integration with GaAs integrated circuits (ICs) using MESFETs. However, in practice it is necessary to make substantial additions to commercial integrated circuit processes to realize high performance MSM photodetectors. These changes are complicated and costly to implement, and there remains a great need for an integrated photodetector that can be fabricated without making changes to commercial GaAs MESFET IC processes.

Drawing on a reported structure for a silicon-on-insulator process[1], it was previously demonstrated[2] that high performance lateral p-i-n photodiodes can be formed using MESFET source and drain regions implanted into relatively light p-type device "tubs" without making any modifications to the standard IC process. A structure for device simulation was subsequently developed and matched with experimental results[3].

This paper will present the performance characteristics of a subsequent generation of lateral p-i-n diode photodetectors fabricated in the standard Vitesse Semiconductor Corporation[4] H-GaAs-V IC process, demonstrating superior performance to those of previous reports.

2. Lateral Diode Structure

The lateral device structure used in the current work is shown in Figure 1. The pattern is tiled to create a large area device with a width of 75 μm. This lateral structure can be fabricated in the standard Vitesse H-GaAs-IV and H-GaAs-V processes in which MESFETs are formed in lightly-doped p-type wells created by ion implantation into a nominally semi-insulating GaAs wafer. The H-GaAs-IV and V processes provide for isolation implants around the p-wells, and p+ ohmic contacts to these wells.

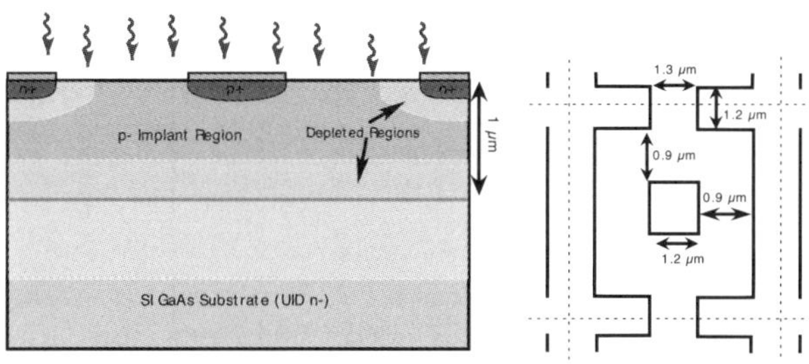

Fig. 1. Schematic of the side and top views of the lateral p-i-n device structure, including ion implant and ohmic contact structures but excluding interconnect metal. The width of the i region is defined photolithographically as the separation between the p and n implant regions.

In the device pictured in Figure 1, the p-well forms the i-region and the p+ and n implants are used to form the p and n-regions, respectively. The width of the i-region is determined photolithographically by the separation between the two later implants, 0.9 μm in the present study. Referring to Figure 1, the n implant is patterned around the p+ implant in a mirrored "C" shape. The intent of this layout is to achieve a compromise between the proportions of device area covered by metal contacts, and the exposed area between the n and p+ regions. The same geometry was used in the previously reported studies [2,3].

3. DC Characteristics

3.1. *I-V Behaviour*

The DC electrical characteristics were measured using a HP 4145B semiconductor parameter analyzer. They are summarized in Table 1. The devices exhibit low dark current at bias points of interest and a high breakdown voltage. An ideality factor greater than 1 is expected in a p-i-n device.

Table 1. Static electrical characteristics.

Dark Current (VBias = 10 V)	< 90 pA
Dark Current (VBias = 5 V)	< 30 pA
Breakdown voltage	> 30 V
Ideality Factor	1.9

3.2. *Capacitance*

For a one-dimensional p-n junction with uniform doping on either side of the junction, a plot of C^{-2} vs. V yields a straight line whose slope is dependent solely on the material parameters of the device (doping, semiconductor properties). C-V and C^{-2}-V curves are shown in Figure 2.

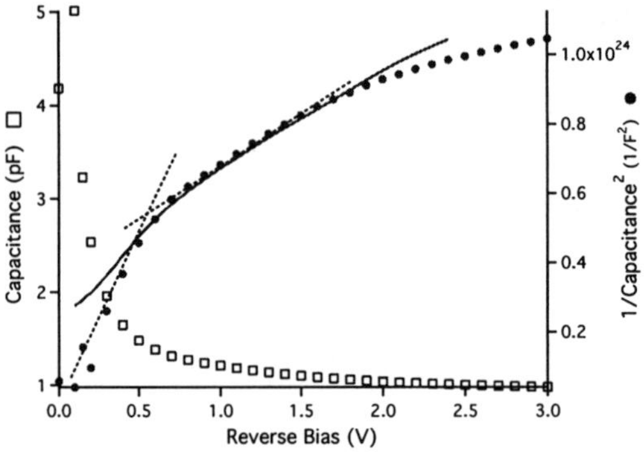

Fig. 2. Capacitance (squares) and Capacitance^{-2} (dots) as a function of reverse bias. The solid line overlapping the dots represents simulated Capacitance^{-2} values.

The capacitance data indicate that when connected to a 50 Ω load, and assuming that performance is RC-limited, the detectors should be capable of performance over 1 GHz at very low reverse bias. The C^{-2} vs. V data contain two linear regions of particular interest that are indicated by dashed lines in Figure 2. The intersection of the extrapolations of those linear regions occurs at $V_{rev} \approx 0.6$ V. Simulations suggest that there is a change of shape in the depletion region within the device at that reverse bias. It is at this level that the region under the n-type implant becomes fully depleted. In previous devices, this is also a performance transition point where both sensitivity and speed behavior changed[3].

3.3. *DC Sensitivity*

The photocurrent was measured as a function of incident light power at a constant reverse bias. A Thorlabs V3-780-TO-DA VCSEL with emission at nominally 780 nm was used as the light source, free-space coupled to detectors under measurement.

The detectors exhibit a linear intensity vs. photocurrent behaviour (from 0.5 μW to 1 mW of incident light power).

The photocurrent was also measured as a function of reverse bias at varying light intensities, as shown in Figure 3. At low reverse bias, the photocurrent rises substantially with increasing bias. At higher bias, the photocurrent vs. reverse bias curve flattens significantly. At low light levels, the initial rise is steepest and the subsequent plateau is flattest. Extrapolating the linear regions of the curve at lower and high reverse bias results in an intersection at approximately 0.6 V, the same point observed in the CV data. Sensitivity reaches a value of 0.17 A/W for a reverse bias of 5 V. Taking into account the metal fill-factor of each detector cell, the quantum efficiency of the devices is over 60% at that operating point.

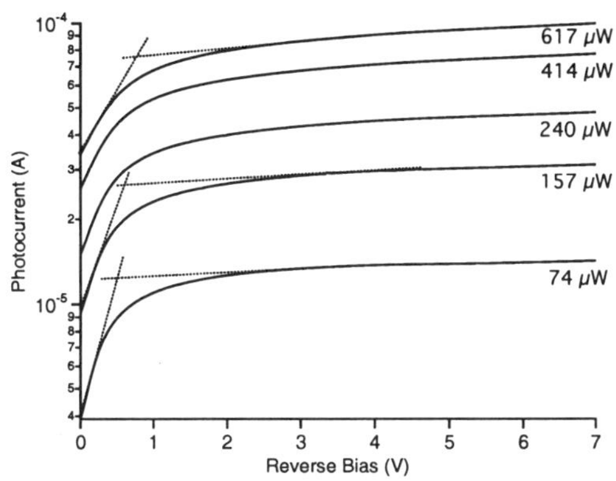

Fig. 3. Photocurrent as a function of reverse bias and incident light power. Sensitivity is strongly dependent on reverse bias at low bias levels and reaches a plateau at higher reverse bias, indicating a quantum efficiency of over 60%.

4. Transient Characteristics

4.1. *AC Source*

The VCSEL and detector were biased using HP 6633A voltage/current sources. The VCSEL was modulated with an HP 8341B signal source in the frequency range 100 MHz to 5 GHz, and the small signal photocurrent was measured using a Minicircuits ZJL-6G amplifier and a HP 8510B network analyzer.

The results of these measurements are shown in Figure 4 for reverse biases varying from 0 V to 7 V. The data was fitted to a $(1 + (\frac{f}{f_{3dB}})^c)^{-1/2}$ dependence (a single-pole solution for c = 2) that is represented by the thick lines in the figure. This relationship provides a fit to the experimental data allowing a convenient definition of the 3dB frequency, but does not explicitly take into account the physical processes responsible for the behaviour, except for the RC (single pole) limit.

The 3dB bandwidth exceeds 2 GHz at 5 V reverse bias, and is 4.5 GHz at 7 V reverse bias which is the RC limit of the device and 50 Ω load. Because the capacitance is proportional to device area, smaller devices will have correspondingly larger bandwidths.

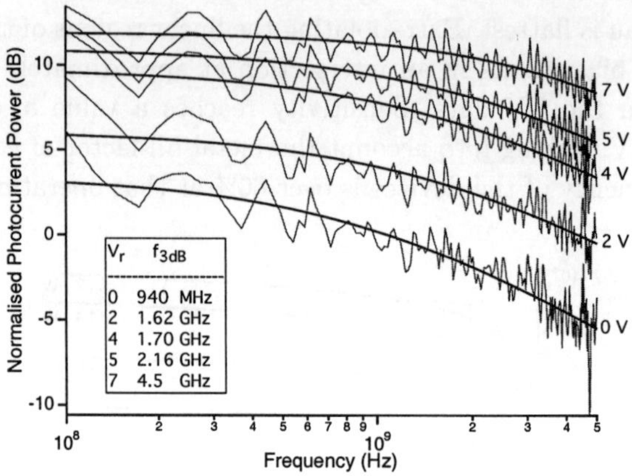

Fig. 4. AC photocurrent response as a function of frequency at different reverse bias levels. The fitting parameter "c" increases with reverse bias (going from a value of 1.2 at 0 V to 1.9 at 7V)

4.2. *Pulse Response*

In the device structure schematic (Figure 1), it is apparent that there are two p-n junctions in the device: 1) the n implant to the p- well, and 2) the p- well to the semi-insulating (physically lightly n-) substrate. The substrate is a hanging node in normal operation, but may be a factor in device performance because carriers in the p-well may diffuse towards the substrate-well junction. A pulse response measurement was carried in an attempt to extract the effect of the bottom junction from the overall AC performance. The device was illuminated by a Ti-Sapphire (830 nm) mode-locked laser producing picosecond pulses with a repetition rate of 81 MHz. The detected signal was measured using a Tektronix 11801C sampling oscilloscope. Characteristic results of this measurement are shown in Figure 5.

The response is approximately second-order, due to bond wire inductance. The width of the first peak suggests a wire inductance of 8 nH which is consistent with the diameter and length of wire used. The second peak roll-off changes with reverse bias and is explained by a combination of RC and carrier diffusion time constants. In the 5 V case, the second peak FWHM of 0.45 ns agrees with the measured 3dB bandwidth of over 2 GHz. Significantly, the pulse response decays to level of noise quickly, therefore the bottom junction does not appear to affect device speed.

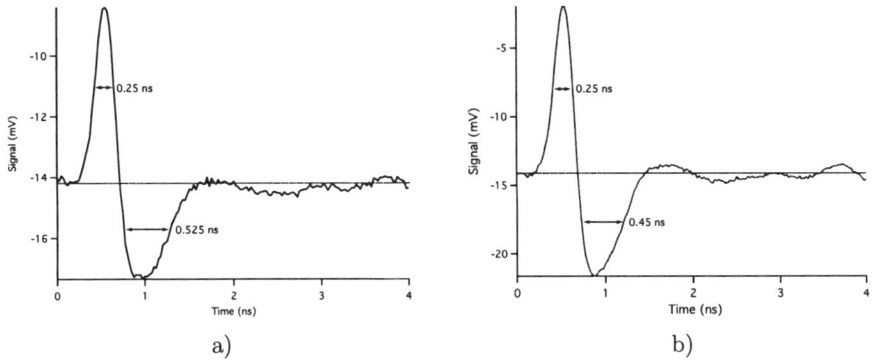

Fig. 5. Detector pulse response at a) 1 V and b) 5 V reverse bias.

5. Conclusion

High performance lateral p-i-n detectors have been fabricated in an unmodified commercial GaAs VLSI process. The detectors have low dark current and high breakdown voltage. Both the DC photocurrent data and the CV data exhibit a change in behaviour at a reverse bias of approximately 0.5 V. The AC and pulse response performance suggest that the devices are suitable for applications such as polymer fibre communication systems or gigabit Ethernet over fibre. Compared to other GaAs-based photodetectors, the current work demonstrates sensitivity and speed characteristics similar to those demonstrated for MSM diodes, but with a simpler fabrication process. Vertical heterojunction pin diodes may be designed to offer both higher sensitivity and speed, but are not suitable for incorporation into a standard VLSI process.

Acknowledgements

Special thanks to Juliet Gopinath, Peter Rakich, and Jason Sickler for help with the pulse measurement. This work was supported by the National Science Foundation.

References

1. T. Yoshida, Y. Ohtomo, and M. Shimaya, "A Novel p-i-n Photodetector Fabricated on SIMOX for 1 GHz 2V CMOS OEICs", *1998 IEDM Technical Digest*, IEEE, Piscataway, NJ, 1998, pp. 29-32.
2. H.K.H Choy, M. Kircherer, B. Ruedlinger, K. Sundarajan, J. Ahadian, J. Mikkelson, S. Prasad, and C.G. Fonstad Jr., "High Performance Lateral p-i-n Photodetectors Realized in a Standard Commercial GaAs IC Process", *Proceedings of SPIE*, **4746**, 2001, pp. 797-803.
3. W.P. Giziewicz, H.K.H. Choy, C.G. Fonstad, Jr., and S. Prasad, "Measurement and Modelling of High Performance Lateral p-i-n Photodetectors", *Proceedings of SPIE*, **5445**, 2004, pp. 114-119.
4. Vitesse Semiconductor Corporation, 741 Calle Plano, Camarillo, CA 93012-8543 USA.

International Journal of High Speed Electronics and Systems
Vol. 14, No. 3 (2004) 720–725

World Scientific
www.worldscientific.com

SELF-GUIDING IN LOW-INDEX-CONTRAST PLANAR PHOTONIC CRYSTALS

CAIHUA CHEN, ZHAOLIN LU, SHOUYUAN SHI, AND DENNIS W. PRATHER

Electrical and Computer Engineering Department, University of Delaware, Newark, Delaware, 19711, USA

This paper presents a new lateral confinement mechanism based on the self-collimation effect in planar photonic crystals(PhCs). In this mechanism, planar PhCs with approximately flat equi-frequency contours (EFCs) are utilized for laterally confining the light and total internal reflection (TIR) is used for the vertical confinement. Since this mechanism relies on dispersion engineering, no photonic band gap nor defect modes are required. Using this approach, it is possible to engineer flat EFCs completely below the light cone using relatively low-index-contrast material systems. Therefore, in this work we first engineer a low-index-contrast planar PhC with approximately flat EFCs in the first band, which is always below the light cone. Then, we investigate the lateral divergence of self-guided beams in this structure using the plane wave method (PWM). It is found that light with a narrow band frequency in the first band can be very well self-confined laterally.

1 Introduction

Photonic crystals (PhCs) [1] are periodic dielectric structures created in analogy with semiconductor materials. These structures have attracted tremendous attention since first proposed by Yablonovitch [2] and John [3] due to their great potential in a variety of applications, particularly in photonic integrated circuit. Planar PhCs are one type of such structures, which are normally made by periodically patterning air holes on material slabs. This type of PhCs have been of particular interest due to their convenience of integration, their ease of fabrication, and their compatibility with microelectronic circuits technology.

Planar PhC has two-dimensional periodicity, which enables engineering of in-plane photonic band gap. Within such a planar PhC, a line-defect can be introduced to create a waveguide, for which the in-plane band gap provides the in-plane light confinement. However, due to lack of periodicity in the vertical dimension, planar PhC line-defect waveguide relies on the total internal reflection (TIR) for vertical confinement. Therefore, in this type of waveguides, a light cone, which is defined by the TIR condition, is overlapped with otherwise truly guided waveguide modes by the in-plane band gap.[4] As such, defect modes guided by the in-plane band gap can be further divided into two groups, those outside the light cone and those inside the light cone. Former modes satisfy the vertical TIR condition and are thus confined in the waveguide vertically as well, while latter modes don't and thus have no light confinement in the vertical direction. Therefore, only former modes are truly guided modes and can propagate along the waveguide with no propagation loss, latter modes exhibit significant propagation loss and can't propagate far. Unfortunately, because in-plane band gap of planar PhC opens at least above the first band, line-defect modes normally consist of both types of modes. When a wave is launched into the waveguide, mostly both types of modes will be excited. As a result, a planar PhC line-defect waveguide is usually lossy.[5-7] Moreover, a planar PhC line-defect waveguide requires an in-plane band gap to provide the lateral confinement, which is typically achieved through the use of a wide band gap. For this reason, most PhC defect-based waveguides are realized using high-index-contrast structures, which greatly limits the material options for planar PhC devices.

Recently, a new type of PhC waveguides has been proposed.[8] Instead of the in-plane band gap, this type of PhC waveguides utilizes a unique mechanism, the self-collimation phenomenon, for its lateral confinement. This is achieved by engineering the anisotropic property of PhC such that its equi-frequency contours (EFCs) are approximately flat within certain angular and frequency regions. Since energy propagates along the direction perpendicular to the EFCs, such a PhC is capable of self-collimating and thus self-guiding light. Compared to PhC waveguides based on PBG, this light-guiding mechanism offers many advantages. For instance, engineering unique EFC shape within low bands allows light guiding completely below the light cone, which could lead to no loss propagation. In addition, this unique mechanism does not principally require high-index-contrast material systems. Low-index-contrast material systems, which are desired for more efficient coupling (due to the fact that the widely used fiber systems are low-index-contrast) and more diversified devices, can be also used. In this paper we investigate light guiding in planar PhCs consisting of low-index-contrast material systems.

2 Self-Collimation in Low-Index-Contrast Planar Photonic Crystals

In this section, an example will be used to demonstrate the self-collimation in low-index-contrast material system. In this example, a suspended SiN slab patterned with a square array of air holes, as illustrated in Fig. 1, is considered. The diameter of air holes is $0.7a$ and the thickness of the SiN slab is $1.2a$, where a is the lattice constant. The refractive index of SiN is 1.98. Figure 2(a) shows the dispersion surfaces of the first band of this structure and the light cone, which are obtained by using the three-dimensional (3-D) iterative plane wave method (PWM).[9] This figure indicates that the first band is completely below the light cone, which implies that all modes in the first band are capable of propagating through the slab without any radiation loss. Figure 2(b) is the corresponding EFCs of the first band. From this figure, one can see that EFCs with frequencies around $a/\lambda=0.36$ are approximately flat along the Γ-M direction. Since the relation between the group velocity v_g and the dispersion function $\omega(k)$ can be expressed as:

$$\mathbf{v}_g = \nabla_k \omega(k), \tag{1}$$

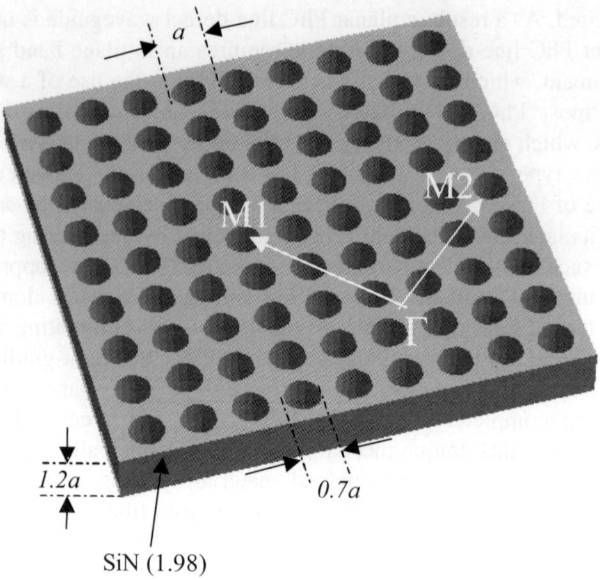

Figure 1. A low-index-contrast planar photonic crystal for studying the "self-collimation" phenomenon and the self-guiding application.

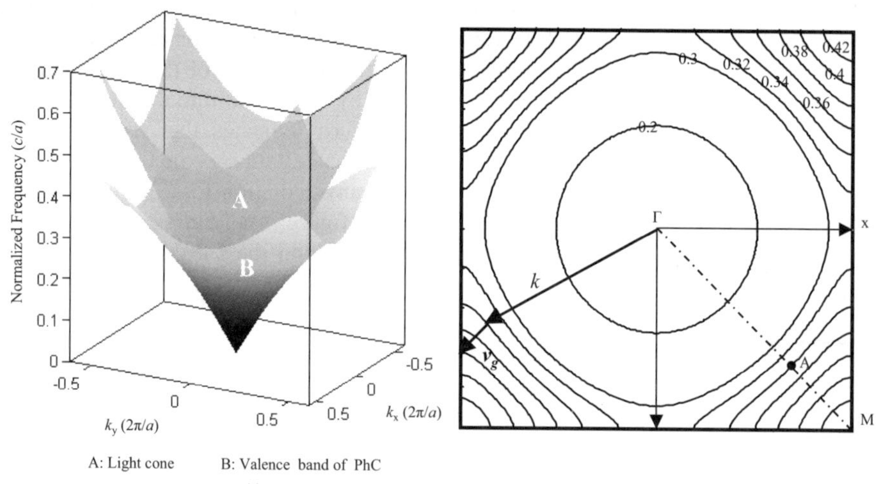

Figure 2. Dispersion relationship of the planar PhC shown in Fig. 1. (a) dispersion surfaces, (b) equi-frequency contours of the first band.

the group velocity, v_g, or the direction of light propagation coincides with the direction of the steepest ascent of the dispersion surface, and is perpendicular to the EFC, as indicated in Fig. 2(b). Therefore, when one varies the wavevector k over one approximately flat part along the EFC, energy propagation angles within the PhC maintain within a narrow range. This phenomenon is called "self-collimation" and was initially observed and experimentally demonstrated by Kosaka, et al. [10] in high-index-contrast material systems. To visualize the self-collimation in this structure, a dipole source with a wavelength of 1.5 μm is launched in the center of the central air hole of the structure and the three-dimensional (3-D) finite-difference time-domain (FDTD) technique [11] is implemented to simulate its propagation in the structure. The lattice constant is 550 nm, the diameter of air holes is 385 nm, and the thickness of the SiN slab is 660 nm. The FDTD resolution used in this simulation is *dx=dy=27.5 nm* (*a/20*) and *dz=66 nm* so that the SiN slab has exactly 10 cells in the vertical direction. On both top and bottom of the slab, perfectly matched layers (PMLs) are put twice the slab thickness away from the slab in order to reduce the effect of the evanescent waves from the slab on the calculation accuracy. Figure 3 shows the obtained steady-state field distribution of H_z component at the central plane of the slab.

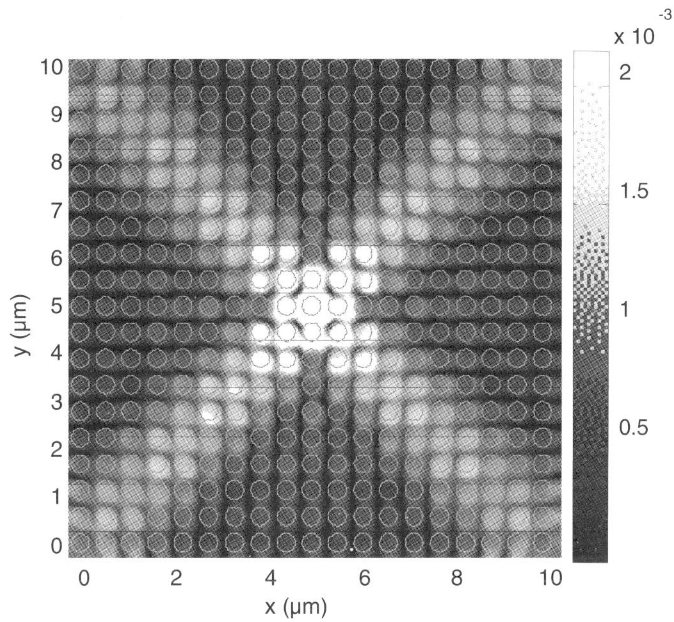

Figure 3. The steady-state field distribution of Hz component obtained using the 3D FDTD method when a dipole source is launched in the center of the structure shown in Fig. 1.

From this figure, one can see that indeed most of the energy goes along the Γ-M1 and Γ-M2 directions and is laterally confined within a small aperture in the slab plane. This lateral light confinement provided by the approximately flat EFC can be used to guide light in this low-index-contrast structure. In the following sections, we will discuss the possibility and performance of a non-channel waveguide based on this lateral confinement.

3 Self-Guiding in Low-Index-Contrast Planar Photonic Crystal

From the previous discussion, it is clear that the lateral light confinement provided by approximately flat EFCs and the vertical light confinement due to the total internal reflection (TIR) combine to allow non-channel self light guiding in the above-mentioned low-index-contrat planar PhC. To further investigate this self light guiding mechanism, we will look into the lateral beam divergence caused by the rounded corners of the EFCs.

From Fig. 2, one can see that although flat EFCs can be found outside the light cone in the first band, none of the EFC is absolutely flat, which implies that the self-collimated beam is divergent. To show its divergence properties, we plot the angular range of the incident beam from free space, within which the corresponding energy propagation angle of the beam in the planar PhC is less than 2 degree, versus the normalized frequency (*c*/*a*) in Fig. 4. In this case, the interface between the free space and the planar PhC is along one Γ-M direction and the surface normal is along the other Γ-M direction. All angles, the incident angle and the energy propagation angle in the planar PhC, are relative to the surface normal, as shown in Fig. 6. This figure shows that if a plane wave with a frequency within a/λ = 0.363~0.369 is incident on the PhC, waves with angular spectrum less than 15 degree will propagate in the PhC with energy propagation angles less than 2 degree.

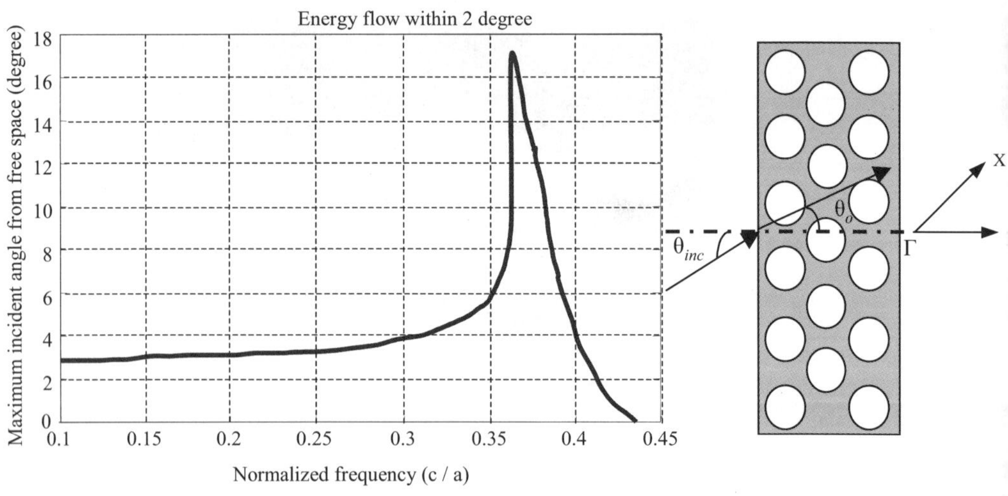

Figure 4. The divergence properties of the first band.

4 Conclusion

In conclusion, in this paper the self-collimation was demonstrated in a low-index-contrast planar photonic crystal. Light guiding, which is based on the lateral confinement provided by the self-collimation and the vertical TIR mechanism, was then investigated. The lateral divergence of the self-guided beam was also studied by varying the incident angle and frequency of the incident plane wave and calculating the corresponding energy propagation angle in the PhC. For the example studied in this paper, no-radiation-loss self-guiding can be observed in the first band. The bandwidth with energy propagation angle less than 2 degree is a/λ from 0.363 to 0.369 if the incident Gaussian beam has an angular spectrum with 15°.

References

1. J. D. Joannopoulos, R. D. Meade, and J. N. Winn, *Photonic Crystals: Molding the Flow of Light* (Princeton University Press, Princeton,N.J., 1995).
2. E. Yablonovitch, *Phys. Rev. Lett.* **58,** 2059-2062 (1987).
3. S. John, *Phys. Rev. Lett.* **58,** 2486-2489 (1987).
4. S. G. Johnson, S. Fan, P. R. Villeneuve, and J. D. Joannopoulos, *Phys. Rev. B* **60,** 5751-5758 (1999).
5. L. C. Andreani, *Physica Status Solidi B-Basic Research* **234,** 139-146 (2002).
6. Y. Desieres, T. Benyattou, R. Orobtchouk, A. Morand, P. Benech, C. Grillet, C. Seassal, X. Letartre, P. Rojo-Romeo, and P. Viktorovitch, *Journal of Applied Physics* **92,** 2227-2234 (2002).
7. L. C. Andreani and M. Agio, *Applied Physics Letters* **82,** 2011-2013 (2003).
8. D. W. Prather, S. Shi, D. Pustai, C. Chen, S. Venkataraman, A. Sharkawy, G. schneider, and J. Murakowski, *Opt. Lett.* **29,** 50-52 (2004).
9. S. G. Johnson and J. D. Joannopoulos, *Opt. Express* **8,** 173-180 (2001).
10. H. Kosaka, T. Kawashima, A. Tomita, M. Notomi, T. Tamamura, T. Sato, and S. Kawakami, *Appl. Phys. Lett.* **74,** 1212-1214 (1999).
11. A. Taflove and S. C. Hagness, *Computational Electromagnetics: The Finite-Difference Time-Domain Method* (Artech House, Boston, 2000).

International Journal of High Speed Electronics and Systems
Vol. 14, No. 3 (2004) 726–731

World Scientific
www.worldscientific.com

OMNI-DIRECTIONAL REFLECTOR USING A LOW REFRACTIVE INDEX MATERIAL

J.-Q. XI[1,2], MANAS OJHA[3], WOOJIN CHO[3], TH. GESSMANN[1,4], E. F. SCHUBERT[1,2,4], J. L. PLAWSKY[3] AND W. N. GILL[3]

1. Future Chip Constellation
2. Department of Physics, Applied Physics, & Astronomy
3. Department of Chemical and Biological Engineering
4. Department of Electrical, Computer, and Systems Engineering
110 8th Street, Rensselaer Polytechnic Institute
Troy, New York 12180, USA

Triple-layer omni-directional reflectors (ODRs) consisting of a semiconductor, a transparent quarter-wavelength dielectric layer and metal layer have high reflectivities at all angles of incidence. In this paper, triple-layer ODRs are demonstrated that incorporate nanoporous SiO_2, a novel low-refractive-index (low-n) material with refractive indices $n << 1.46$ as well as dense SiO_2 ($n = 1.46$). GaP and Ag serve as the semiconductor and metal layer materials, respectively. An angle-integrated transverse electric (TE) mode reflectivity of $R_{avg}|_{TE} = 99.9$ % and transverse magnetic (TM) mode reflectivity $R_{avg}|_{TM} = 98.9$ % are calculated for the triple-layer ODRs employing nanoporous SiO_2. Reflectivity measurements, including the angular dependence of R, are presented. Novel hybrid ODRs consisting of semiconductor, a several micron thick low-n dielectric material layer, a distributed Bragg reflector (DBR) and metal layer have outstanding reflectivities for all incident angles. GaP and Ag serve as the semiconductor and metal layer, respectively. Nanoporous SiO_2 is used as the low-n material. TiO_2 and dense SiO_2 serve as the DBR materials. The angle-intergrated reflectivities of the TE and TM modes are calculated to be larger than 99.9 % for the hybrid ODRs. The results indicate the great potential of the ODRs for light-emitting diodes with high light extraction efficiency.

Introduction

III-V nitride LEDs are becoming increasingly important for visible-spectrum [1] and UV [2] emitters. Different techniques have been employed to improve the light extraction efficiency [3] including flip chip packaging [4], interdigitated contacts, and the employment of triple-layer omni-directional reflectors (ODRs) [5, 6]. ODRs [5, 6] possess a much higher angle-integrated reflectivity than conventional metal and distributed Bragg reflectors (DBRs). It is well known that the refractive index contrast in optical components (in particular multi-layer filters and reflectors) is an important figure of merit. This creates a strong demand for novel low-refractive index (low-n) materials, i.e. materials with unprecedented low value of the refractive index, such as $n_{low} = 1.10$.

In this work, triple-layer ODRs and hybrid ODRs are studied. Triple-layer ODRs consist of a semiconductor, a quarter-wavelength dielectric layer, and a metal layer. Triple-layer ODR can be significantly improved by using nanoporous SiO_2 instead of dense SiO_2 as the dielectric layer. Nanoporous SiO_2 (also called mesoporous SiO_2) has a typical pore size ranging between 2 and 8 nm. It is a novel low-n material with refractive indices << 1.46 [7]. The high transparency of nanoporous SiO_2 in the visible and near-UV spectrum and its favorable mechanical and thermal characteristics [8, 9, 10] make it an ideal candidate for numerous low-n applications. The angular- and wavelength-dependent reflectivity of triple-layer ODRs with nanoporous SiO_2 and dense SiO_2 is measured and compared to calculations. Hybrid ODRs consist of a semiconductor, a several micron thick low-n material layer, a DBR, and a metal layer. The hybrid ODR can have outstanding performance with a low-n material with refractive index, $n_{low} = 1.1$. A calculation of its angular- and wavelength-dependent reflectivity is presented.

Triple-layer ODRs

The schematic of a triple-layer ODR consisting of a semiconductor, a quarter-wavelength dielectric layer, and a metal layer is shown in Fig. 1(a). At normal incidence, the reflectivity of the triple-layer ODR is given by [11],

$$R_{normal} = \left[\left|\frac{(n_{semi} - n_{die})(n_{die} + n_{metal} + i\kappa_{metal}) + (n_{semi} + n_{die})(n_{die} - n_{metal} - i\kappa_{metal})\exp(2\,i\,\beta)}{(n_{semi} + n_{die})(n_{die} + n_{metal} + i\kappa_{metal}) + (n_{semi} - n_{die})(n_{die} - n_{metal} - i\kappa_{metal})\exp(2\,i\,\beta)}\right|\right]^2, \quad (1)$$

where h is the thickness of the dielectric layer, $\beta = (2\pi/\lambda)n_{die}h$, and n_{semi}, n_{die}, n_{metal}, κ_{metal} are the refractive index of the semiconductor, the dielectric, the metal, and the metal extinction coefficient, respectively. The thickness h is chosen such that constructive interference at normal incidence is obtained. The optimal value of h, h_{opt}, in Eq. (1) can be derived from [11],

$$2\beta + \varphi_{normal} = \frac{4\pi}{\lambda} n_{die}\, h_{opt} + \varphi_{normal} = 2\pi \quad (2)$$

$$\tan\varphi_{normal} = \frac{2\,\kappa_{metal}\, n_{die}}{n_{metal}^{\;2} + \kappa_{metal}^{\;2} - n_{die}^{\;2}} \quad (3)$$

where φ_{normal} is the phase change at the interface of metal/dielectric layer at normal incidence. Since in general $\pi < \varphi_{normal} < (3/2)\pi$, the optimal thickness h_{opt} is smaller than the quarter-wavelength thickness $\lambda/(4n_{die})$. Silver is chosen as the metal and a double-side polished 300 μm thick optical GaP wafer as the semiconductor. At a wavelength $\lambda = 632.8$ nm, Ag and GaP have the following parameters: $n_{Ag} = 0.135$, $\kappa_{Ag} = 3.987$, $n_{GaP} = 3.31$, and $\kappa_{GaP} \approx 10^{-5}$ is negligibly small [12].

The angle-averaged reflectivity of the ODR, an important figure of merit for light-emitting diodes, can be calculated from

$$R_{avg} = \frac{\int_0^{\pi/2} R(\theta)\sin(\theta)\,d\theta}{\int_0^{\pi/2} \sin(\theta)\,d\theta} = \int_0^{\pi/2} R(\theta)\sin(\theta)\,d\theta\,. \quad (4)$$

The angular dependent reflectivity of the triple-layer ODR, $R(\theta)$, was measured using a HeNe Laser with the samples mounted on a goniometer. The detector is an ANDO AQ2140 optical multimeter. The precision of the measured reflectivity values is better than ± 0.15 %.

Due to high refractive index of the ambient GaP, the setup shown in Fig. 1(a) can measure the ODR reflectivity up to a maximum incident angle of $\theta = 17.6°$. However, as shown in Fig. 1(b), a high refractive index prism ($n = 1.778$) and a high index fluid ($n = 1.662$) are used to couple light into GaP at larger angles of incidence. This extends the maximum attainable angle θ to 29.5°. The high-n fluid between the prism and the

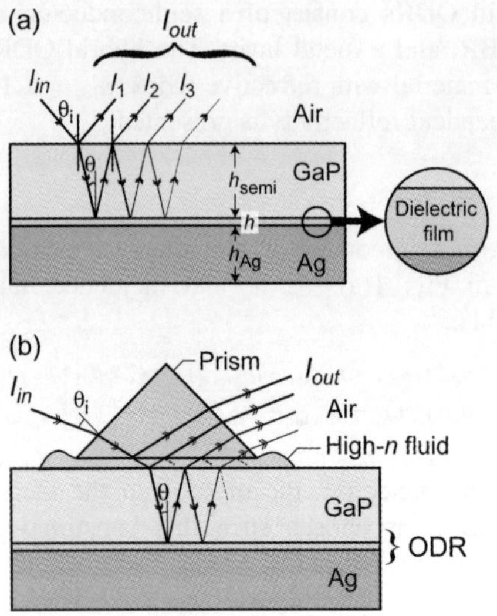

Fig. 1: (a). Schematic of a triple-layer ODR consisting of GaP, a transparent low-*n* dielectric layer, and silver. (b) Setup using a prism and high-index fluid to couple light into GaP at angles up to $\theta = 27.3°$.

ODRs with dense SiO_2 were fabricated by plasma-enhanced chemical vapor deposition (PECVD) on GaP. Nanoporous SiO_2 was fabricated by a sol gel process described in [7] with some modifications to satisfy the optical thickness requirements. GaP was treated in oxygen plasma to enhance the adhesion between the substrate and sol. The sol-gel solution was spun on the plasma treated GaP. Surface modification of nanoporous SiO_2 was then performed using Hexamethyl disilazane (HMDS), which does not affect the adhesion of the nanoporous SiO_2 film to the substrate. A 500 nm thick Ag layer was deposited by electron-beam evaporation. The refractive indices at $\lambda = 632.8$ nm and the thicknesses of the nanoporous SiO_2 and dense SiO_2 were determined by ellipsometer using Cauchy's formula with $n_{low} = 1.23$, $n_{SiO2} = 1.457$ [12], $h_{low} = 105$ nm ($h_{opt} = 104$ nm from Eqs. (2) and (3)), and $h_{SiO2} = 89$ nm ($h_{opt} = 84$ nm from Eqs. (2) and (3)) respectively.

GaP has a thickness of several tens of microns. The dependence of the reflectivity on the wavelength is measured with a Jasco V-570 spectrophotometer.

The angular dependent reflectivity $R(\theta)$ of the ODR can be obtained by correcting the measured result, I_{out}/I_{in}, for losses caused by Fresnel reflections at the GaP/air interface (Fig. 1(a)), and GaP/high-*n*-fluid/prism/air interface (Fig. 1(b)). Due to beam divergence at large angles of incidence, the measured TE and TM reflectivity are limited to 25.5° and 27.3° respectively. Multiple reflection events inside the GaP and high-*n* fluid layer are taken into account. Small absorption effects in the GaP, high-*n* fluid, and the prism are neglected, making the experimental reflectivities a lower limit of the actual ODR reflectivity.

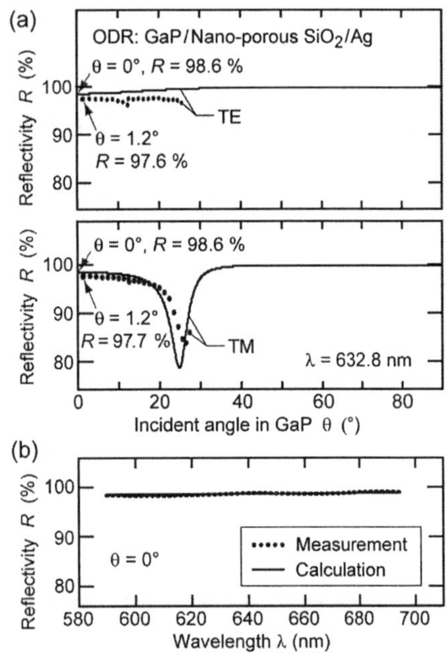

Fig. 2: Calculated (solid line) and measured (dotted line) reflectivity (a) versus angle of incidence and (b) wavelength for a triple-layer ODR with nanoporous SiO_2.

The measured reflectivity $R(\theta)$ of the ODR using nanoporous SiO_2 as dielectric layer is shown in Fig. 2(a). At near-normal incidence ($\theta = 1.2°$), a TE reflectivity of 97.6 % and TM reflectivity of 97.7 % are measured. Within the measured range, the TE reflectivity does not show a strong angular dependence. However, the TM reflectivity exhibits a Brewster-like dip near 27°. Fig. 2(b) reveals the broad spectral range of the high-reflectivity band of the ODR.

The calculated TE reflectivity increases with angle with the lowest reflectivity, 98.6 %, at normal incidence. The calculated TM reflectivity has a Brewster-like dip with a minimum value of 80.0 % at $\theta = 25.1°$. From Eq. (4), the ODR employing nanoporous SiO_2 has an outstanding angle-averaged reflectivity $R_{avg}|_{TE} = 99.9$ % for the TE polarization and $R_{avg}|_{TM} = 98.9$ % for the TM polarization.

The measured reflectivity $R(\theta)$ of the ODR with dense SiO_2 as dielectric material layer is shown in Fig. 3(a). The measured curves are qualitatively similar to the measurements shown in Fig. 2(a). However, the quantitative reflectivity values are lower. The near-normal incidence TE and TM reflectivity are 97.3 % and 97.4 %, respectively. This result can be attributed to the higher refractive index of dense SiO_2 compared with nanoporous SiO_2.

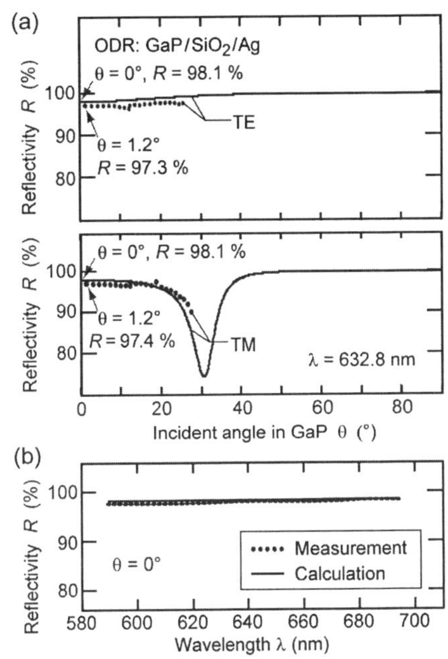

Fig. 3: Calculated (solid line) and measured (dotted line) reflectivity (a) versus angle of incidence and (b) wavelength for a triple-layer ODR with dense SiO_2.

The calculated TE reflectivity has the lowest reflectivity, 98.1 %, at normal incidence. The calculated TM reflectivity has a Brewster-like dip with a minimum value of 74.2 % at $\theta = 30.4°$. An angle-averaged reflectivity $R_{avg}|_{TE} = 99.8$ % for the TE polarization and $R_{avg}|_{TM} = 97.8$ % for the TM polarization can be calculated from Eq. (4).

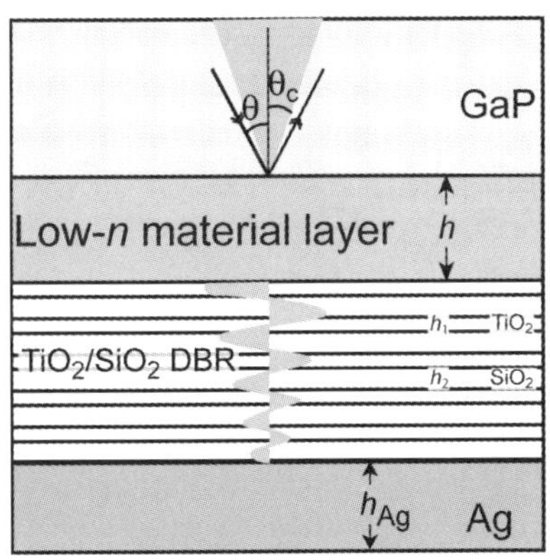

Fig 4. Schematic of a hybrid ODR consisting of GaP, a transparent low-n dielectric layer, a DBR and silver. The thickness of the low-n material is $h = 2.25 \cdot \lambda / n_{low}$. The DBR consists of 5 pairs of TiO_2 and SiO_2 layers.

Hybrid ODRs

Hybrid ODRs consist of semiconductor, thick low-n material layer, DBR and metal layer as shown in Fig 4. Although external ODRs have been demonstrated [13,14], internal ODRs have not been reported yet. The hybrid ODR proposed here has very high reflectivity inside the semiconductor ambient at any angle of incidence. In this work, a calculation of the angular dependent reflectivity is presented. GaP is chosen as the semiconductor ambient. Ag is chosen as the metal substrate. Nanoporous silica can be selected as the low-*n* material. Its refractive index can be as low as $n_{low} = 1.1$ [7]. To obtain the total internal reflection (TIR) at the interface GaP/low-*n* material, the thickness of the low-*n* material, *h*, has to be thicker than the penetration depth of the evanescent field into the low-*n* material, generally $h > 2\lambda/n_{low}$. To maximize the reflectivity at normal incidence, $h = 2.25\lambda/n_{low}$ is chosen as the thickness of the low-*n* material. TiO_2 and SiO_2 are chosen as the materials for the DBR due to their transparence in the visible part of the spectrum and high index contrast. The refractive index of SiO_2 is $n_2 = 1.457$ [12]. The refractive index of TiO_2 is $n_1 = 2.4$ [15].The thicknesses of TiO_2 and SiO_2 are $h_1 = \lambda/(4\,n_1 \cos(\sin^{-1}(n_{low}/n_1))) \approx \lambda/(4\,n_1\,0.912)$ and $h_2 = \lambda/(4\,n_2 \cos(\sin^{-1}(n_{low}/n_2))) \approx \lambda/(4\,n_2\,0.738)$, respectively. Inside the DBR, the maximum light beam angle corresponds to grazing incidence in the low-*n* material. In this case, the TM polarization has the smallest reflectance coefficient at the interface of TiO_2/SiO_2. The selected thicknesses result in constructive interference at the maximum light angle inside the DBR [3]. Hence, this design requires only 5 pairs to have the omnidirectionality.

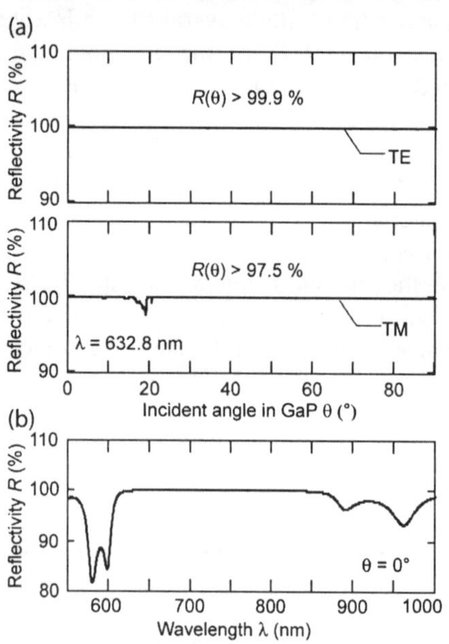

Fig. 5. Calculated reflectivity (a) versus angle of incidence and (b) wavelength for a hybrid ODR.

With the wavelength $\lambda = 632.8$ nm, at the incident angle, $\theta > \theta_c = 19.4\,°$, the reflectivity, $R(\theta) = 1$ due to TIR at the GaP/low-*n* material interface. At an incident angle, $\theta \leq \theta_c = 19.4\,°$, the reflectivity, $R(\theta) < 1$, which is dominantly contributed by the DBR and the Ag layer. The calculated reflectivity and its angular dependence of the hybrid ODR is shown in Fig. 5. The TE reflectivity of the hybrid ODR is $R > 99.9$ % at any incident angle. The TM reflectivity of the hybrid ODR has a small dip at an incident angle near $\theta = 19°$. The corresponding minimum reflectivity is $R \approx 97.5$ %. In any other range of incident angles, the reflectivity is $R > 99.5$ %. The angle-averaged reflectivity is $R_{avg} > 99.9$ % for both TE and TM polarizations, indicating that hybrid ODRs have outstanding reflection properties.

In conclusion, a triple-layer omni-directional reflector and a hybrid omni-directional reflector have been reported. The Triple-layer ODR consists of a semiconductor, a low-refractive-index dielectric layer, and a metal. Either nanoporous SiO_2 or dense SiO_2 are used as dielectric material. The angular dependent reflectivities of these ODRs were measured at 632.8 nm for TE and TM polarization and compared to calculations. For both polarizations the measured reflectivities agree very well with the calculation. It was found that the angle-averaged reflectivity for TE-polarized light $R_{avg}|_{TE}$ = 99.9 % and for TM-polarized light $R_{avg}|_{TM}$ = 98.9 % for the ODR employing nanoporous SiO_2 exceed the corresponding values $R_{avg}|_{TE}$ = 99.8 % and $R_{avg}|_{TM}$ = 97.8 % for the ODR with dense SiO_2 significantly. This can be directly attributed to the lower refractive index of the nanoporous SiO_2 as compared to dense SiO_2. Novel hybrid ODR consists of semiconductor, low-refractive-index dielectric layer, DBR and a metal. Nanoporous SiO_2 can be a good candidate for the dielectric layer. TiO_2 and SiO_2 are selected as the DBR material. Ag is the metal layer. A calculation of the angular dependent reflectivity has been reported. From the calculation, the angle-averaged reflectivity is R_{avg} > 99.9 % for both TE and TM polarization.

The authors gratefully acknowledge support by the National Science Foundation, the Defense Research Projects agency, and the Army Research Office.

References

[1] S. Nakamura, et al., "High-brightness InGaN blue, green, and yellow light-emitting diodes with quantum well structures", *Jpn. J. Appl. Phys.* **34**, L797 (1995)

[2] T. G. Zhu, et al., "AlGaN-GaN UV lighting-emitting-diode grown on SiC by metal-organic chemical vapor deposition", *IEEE Journal on selected topics on quantum electronics* **8**, 298 (2002)

[3] E. F. Schubert, *Light Emitting Diodes*, (Cambridge University Press, 2003)

[4] J. J. Wierer, et al., "High-power AlGaInN flip-chip light-emitting diodes", *Appl. Phys. Lett.* **78**, 3379 (2001)

[5] Th. Gessmann, et al., "GaInN light-emitting diodes with omni directional reflectors", *Proc. SPIE* **4996**, 139 (2003)

[6] J. K. Kim, et al., "GaInN light-emitting diodes with RuO_2/SiO_2/Ag omni-directional reflector", *Appl. Phys. Lett.* **84,** 4508 (2004).

[7] S. V. Nitta, et al., "Surface modified spin-on xerogel films as interlayer dielectrics", *J. Vac. Sci. Technol. B* **17**, 205 (1999)

[8] A. Jain, et al., "Porous silica materials as low-k dielectrics for electronic and optical interconnects, *Thin Solid Films* **398-399**, 513 (2001)

[9] A. Jain, et al., "Effects of processing history on the modulus of silica xerogel films", *J. Appl. Phys.* **90(11)**, 5832 (2001)

[10] A. Jain, et al., "Processing dependent thermal conductivity of nano-porous silica xerogel films", *J. Appl. Phys.* **91(5)**, 3275 (2002)

[11] M. Born and E. Wolf, *Principles of Optics,* (Pergamon Press, 6th Edition, 1980)

[12] E. D. Palik, *Handbook of Optical Constants of Solids,* (Academic Press, 1985). Value used comes from linear interpolation of data in this handbook.

[13] Y. Fink, et al., "A Dielectric Omnidirectional Reflector", Science 282, 1679 (1998)

[14] M. Deopura, et al., "Dielectric omnidirectional visible reflector", Optics Letter 26, 1197 (2001)

[15] H. K. Paulker, et al., "Refractive indices of TiO2 films produced by reactive evaporation of various titanium-oxygen phases", Appl. Opt. 15(12), 2986 (1976)

International Journal of High Speed Electronics and Systems
Vol. 14, No. 3 (2004) 732–737

World Scientific
www.worldscientific.com

MBE-Grown AlGaN/GaN HEMTs on SiC

SIDDHARTH RAJAN,* ARPAN CHAKRABORTY AND UMESH K. MISHRA
Electrical and Computer Engineering Department, University of California, Santa Barbara, Santa Barbara CA 93106, USA
CHRISTIANE POBLENZ, PATRICK WALTEREIT AND JAMES S. SPECK
Materials Department, University of California, Santa Barbara, Santa Barbara CA 93106, USA

We report on the development of AlGaN/GaN high-electron mobility transistors (HEMTs) grown on SiC using plasma-assisted molecular beam epitaxy (MBE). In this work, we show that performance comparable to state-of-the-art AlGaN/GaN HEMTs can be achieved using MBE-grown material. Buffer leakage was an important limiting factor for these devices. The use of either carbon-doped buffers, or low Al/N ratio in the nucleation layer growth were effective in reducing buffer leakage. Studies varying the thickness and concentration of the carbon doping were carried out to determine the effect of different carbon doping profiles on the insulating and dispersive properties of buffers. On devices without field plates, at 4 GHz an output power density of 12 W/mm was obtained with a power-added efficiency (PAE) of 46 % and gain of 14 dB. 15.6 W/mm with PAE of 56 % was obtained from these devices after field-plating. Two-tone linearity measurements of these devices were also carried out. At a C/I3 level of 30 dBc, the devices measured had an output power of 1.9 W/mm with a PAE of 53 %. The effect of the Al/N ratio in the AlN nucleation layer on buffer leakage was studied. N-rich conditions yielded highly insulating GaN buffers without carbon doping. At 4 GHz, devices without field plates delivered 4.8 W/mm with a PAE of 62 %. At a higher drain bias (50 V), 8.1 W/mm with a PAE of 38 % was achieved.

1. Introduction

We report on our work on AlGaN/GaN HEMTs grown by plasma-assisted MBE on 4H-SiC substrates. MBE HEMTs have recently demonstrated excellent power performance at microwave frequencies[1][2][3][4][5]. In-situ diagnostic techniques, a carbon-free and hydrogen-free growth environment, low source consumption, good interfaces and low point defect density are some of the advantages of MBE growth over MOCVD growth. One of the main issues in AlGaN/GaN HEMTs is obtaining an insulating, non-dispersive buffer. MBE-grown undoped buffers can be conductive unless grown under certain specific conditions.

To reduce the buffer leakage in these devices, we employed two methods. In the first method carbon doping was used to make the buffer insulating. Carbon, however, causes dispersion that degrades device performance.

It was found that moving the 2DEG away from the carbon removed dispersion. The

*e-mail: srajan@ece.ucsb.edu

insulating and non-dispersive buffers helped achieve excellent microwave power and linearity performance.

We also explored insulating buffers without the use of carbon doping. AlN nucleation layer growth in N-rich conditions resulted in a GaN buffer with reduced buffer conductivity. Using optimized growth conditions, uniformly insulating buffers were achieved. The devices grown on these buffers demonstrated high power and effi ciency.

2. Carbon-doped buffers

The devices in this work were grown on a Varian Gen II MBE system with an rf-plasma nitrogen source. Carbon was introduced using a CBr_4 gas delivery system[6], [7]. All HEMT structures were grown on semi-insulating 4H-SiC substrates from Cree, Inc.

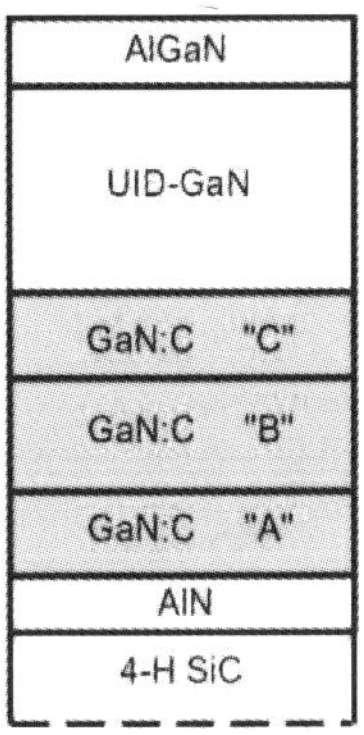

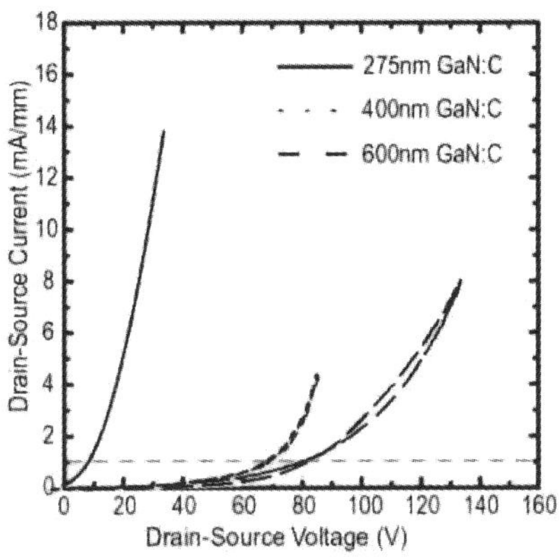

Fig. 1. (a) Epitaxial structure of 1st generation C-doped devices. The thickness of the C-doped layer was 400 nm. (b) Leakage between isolation patterns. The patterns were 3.4 μm apart.

In Fig. 1(a) we show the layer structure of a fi rst-generation C-doped device. A two-step buffer for dislocation reduction was employed [8], [9], [10]. In Fig. 1(b) the leakage from a C-doped samples with a varying C-doped layer thickness is shown. A tapered doping profi le was used with carbon doping concentration of $6 \times 10^{17} cm^{-3}$, $4 \times 10^{17} cm^{-3}$, $2 \times 10^{17} cm^{-3}$ in regions A, B, and C, respectively. Leakage between isolation patterns on these structures was studied. The ohmic pads were separated by 3.4 μm. As shown in the fi gure, it was seen that a higher carbon concentration does increase the insulating property of the buffer. For the doping concentration shown, 400 nm of C-doped GaN was suffi cient for an insulating buffer. The distance between the C-doped layer and the 2DEG was 300 nm.

DC and small-signal characterization of these devices was carried out after SiN passivation. A thin SiN layer was grown under the gate before processing to reduce gate leakage. Charge and mobility in these devices were similar to carbon-free samples. A saturated current > 1 A/mm was obtained with a current gain cutoff frequency (f_T) of 20 GHz and unity power gain frequency (f_{MAX}) of 44 GHz. The device had a gate length of 0.7 μm and were 150 μm wide. Large-signal load-pull measurements were carried out on the devices. At 4 GHz, a maximum output power of 4 W/mm with a power-added effi ciency (PAE) of 18 % was obtained at a drain bias of 35 V. These values were lower than expected, and large-signal RF-IV measurements were then conducted to investigate the problem. The large signal RF-IVs were measured using a microwave transition analyzer (MTA). As shown in Fig. 2 (a) the large-signal RF-IV showed signifi cant knee walkout. Comparable devices without carbon in the buffer did not exhibit this knee walkout after passivation, and this suggested that the carbon was responsible for the large-signal dispersion.

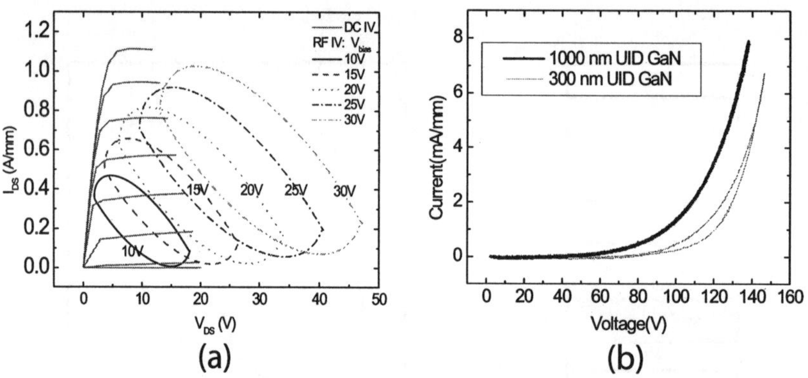

Fig. 2. (a) RF-IV curves at 4 GHz superimposed on the DC IV plot. Significant knee walkout was observed. (b) Leakage between isolation patterns for structures with 300 nm and 1 μm undoped GaN between the 2DEG and the C-doped region

To remove the effect of carbon, the distance between the carbon-doped layer and the 2DEG was increased from 300 nm to 1 μm. However, it was necessary to ensure that the buffer leakage did not increase due to the thicker unintentionally doped (UID) layer. In Fig. 2 (b), the leakage current in the buffer for different thicknesses of the undoped layer between the carbon and the 2DEG is shown. For a UID layer of up to 1 μm, there was no signifi cant increase in the buffer leakage. Devices with this thick UID GaN layer were then tested and RF-IV's on these devices showed that the RF dispersion was reduced. Excellent power and effi ciency performance was obtained from these devices. In Fig. 3(a), 4 GHz power data of a device biased at 30 V is shown. The device delivered 5 W/mm with a high PAE of 66 %. At 70 V, the same device delivered 12 W/mm with a PAE of 46 %. To the authors' knowledge, this is the highest reported power from a SiN-passivated device without fi eld plates on either MBE-grown or MOCVD-grown material.

The devices were then fi eld-plated [11] by depositing Ni/Au gates above the passivation layer. The fi eld-plate length was 0.7 μm and the overlap with the gate was 0.3 μm. The fi eld-plated devices showed an increase in output power and effi ciency. Fig. 3 (b) shows

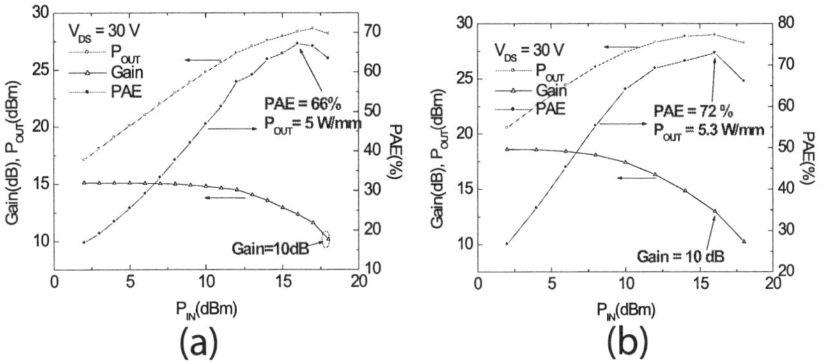

Fig. 3. 4 GHz Power characteristics of MBE-grown AlGaN-GaN HEMTs (a) without field-plate and (b) with field plate.

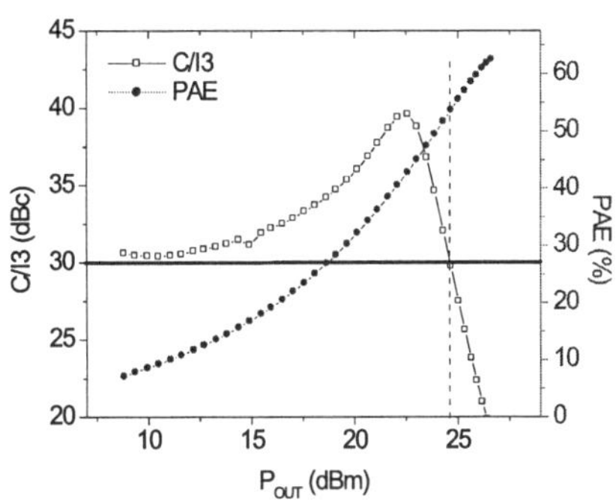

Fig. 4. Two-tone measurement results at 4 GHz. An output power of 1.9 W/mm with a PAE of 53 % was obtained at 4 GHz at a C/I3 level of 30 dBc. The drain bias was 30 V.

power data for fi eld-plated devices with 4 GHz cw input. The devices delivered an output power of 5.3 W/mm with a PAE of 72 %. At 70 V, the output power was 15.6 W/mm with a PAE of 55 %. This is currently the highest reported power value for MBE-grown AlGaN/GaN devices.

Two-tone measurements on these devices showed that they have excellent linearity characteristics. Results from two-tone measurements with the carrier to intermodulation product ratio (C/I3) maintained at greater than 30 dBc are shown in Fig. 4. An output power of 1.9 W/mm with a PAE of 53 % was achieved at a drain bias of 30 V. Measurements for a minimum C/I3 ratio of 45 dBc were also carried out. For a C/I3 greater than 45 dBc, an output power of 1 W/mm with a PAE of 37 % was achieved.

3. Undoped Buffers

MBE-grown buffers are not insulating under all growth conditions. We studied the influence of the nucleation layer growth conditions on the buffer layer conductivity. These growth studies revealed that the presence of Al droplets signifi cantly increased the leakage in the buffer. In Fig. 5(a) we show the current through isolation patterns. GaN buffers with the AlN nucleation layer grown in N-rich conditions were found to be most insulating. More detailed measurements showed that the wafer was uniformly resistive and that the isolation was as good or better than the C-doped samples.

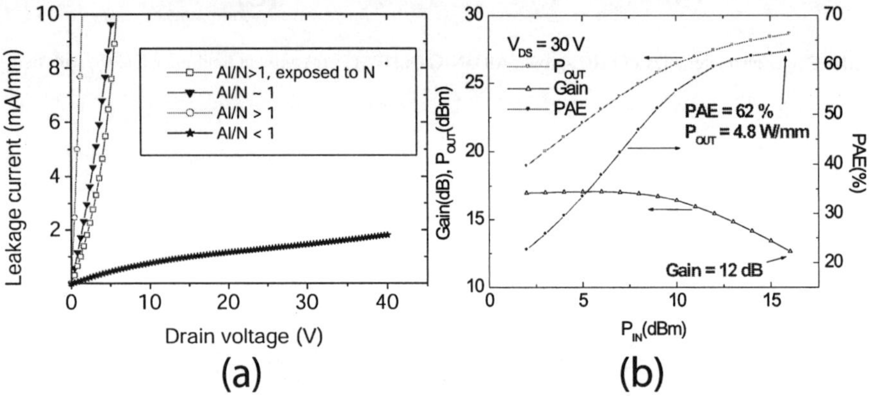

Fig. 5. (a) Comparison of buffer leakage between samples grown under different nucleation conditions. (b) 4 GHz power characteristics of undoped buffer devices.

Devices measured on these buffers had DC and small-signal properties similar to C-doped HEMTs. However, due the absence of a SiN insulating layer under the gate, the gate leakage in these devices was slightly higher than the C-doped devices discussed earlier in the work. In Fig.5(b) we show 4 GHz power measurement data for one of these devices. The device delivered 4.8 W/mm with a PAE of 62 % at a drain bias of 30 V. At a higher drain bias of 50 V, the transistor delivered 8.1 W/mm with a PAE of 38 %. These numbers indicate that undoped MBE buffers can be employed for high performance AlGaN/GaN transistors.

4. Conclusion

Carbon doping was shown to be a reliable and reproducible method to achieve highly insulating and non-dispersive buffers for AlGaN/GaN HEMTs. Excellent power and linearity characteristics were demonstrated in devices with C-doped buffers. Growth and electrical studies of the nucleation layer reveal that employing N-rich conditions during growth of the nucleation layer led to highly insulation buffers. Devices on these buffers had excellent DC and microwave properties.

Acknowledgments

We would like to acknowledge fi nancial support from DARPA (Program Manager Edgar Martinez) managed by ONR (Contract Monitor Harry Dietrich) and from the AFOSR (Program Manager Gerald Witt).

References

1. J. S. Moon, M. Micovic, P. Janke, P. Hashimoto, W.-S. Wong, R. D. Widman, L. McCray, A. Kurdoghlian, and C. Nguyen, "GaN/AlGaN HEMTs operating at 20GHz with a continuous-wave power density >6W/mm," *Electronics Letters*, vol. 37, p. 528, 2001.
2. D. S. Katzer and S. C. Binari and D. F. Storm and J. A. Roussos and B. V. Shanabrook and E. R. Glaser, "MBE growth of AlGaN/GaN HEMTs with high power density," *Electronics Letters*, vol. 38, p. 1740, 2002.
3. N. G. Weimann, M. J. Manfra, and T. Wächtler, "Unpassivated AlGaN-GaN HEMTs with Minimal RF Dispersion Grown by Plasma-Assisted MBE on Semi-Insulating 6H-SiC Substrates," *IEEE Electron Device Letters*, vol. 24, p. 57, 2003.
4. R. Behtash, H. Tobler, M. Neuburger, A. Schurr, H. Leier, Y. Cordier, F. Semond, F. Natali, and J. Massies, "AlGaN/GaN HEMTs on Si(111) with 6.6W/mm output power density," *Electronics Letters*, vol. 39, p. 626, 2003.
5. S. Rajan, P. Waltereit, C. Poblenz, A. Chakraborty, S. J. Heikman, J, S, Speck, and U. K. Mishra, "Power Performance of AlGaN/GaN HEMTs grown on SiC by Plasma-Assisted MBE", *IEEE Electron Dewice Letters*, vol. 25(5), p. 247, 2004.
6. D. S. Green, J. S. Speck, and U. K. Mishra, "CBr_4 doping of GaN in rf plasma MBE," *J. App. Phys.*, vol 95, p. 8456, 2004.
7. C. Poblenz, P. Waltereit, S. Rajan, U. K. Mishra, and J. S. Speck, "Effect of carbon doping on buffer leakage in AlGaN/GaN high electron mobility transistors , " *J.Vac.Sc.Tech. B*, vol. 22, p. 1145, 2004.
8. B. Heying, X. H. Wu, S. Keller, Y. Li, D. Kapolnek, B. P. Keller, S. P. DenBaars, and J. S. Speck, "Control of GaN surface morphologies using plasma-assisted molecular beam epitaxy," *Appl. Phys. Lett.*, vol. 68, p. 643, 1996.
9. M. J. Manfra, N. G. Weimann, J. W. P. Hsu, L. N. Pfeiffer, and K. W. West, "Dislocation and morphology control during molecular-beam epitaxy of AlGaN/GaN heterostructures directly on sapphire substrates," *Appl. Phys. Lett.*, vol. 81, p. 1456, 2002.
10. P. Waltereit, C. Poblenz, S. Rajan, U. K. Mishra, and J. S. Speck, "Growth of GaN buffer layers on 4H-SiC(0001) by plasma assisted molecular beam epitaxy for high mobility electron transistors ," *Appl. Phys. Lett.*, submitted for publication.
11. A. Chini, D. Buttari, R. Coffi e, L. Shen, S. Heikman, A. Charaborty, S. Keller, U. K. Mishra, "Power and linearity characteristics of fi eld-plated recessed-gate AlGaN-GaN HEMTs," *IEEE Electron Device Letters*, vol. 25(5), p. 229, May 2004.

International Journal of High Speed Electronics and Systems
Vol. 14, No. 3 (2004) 738–744

World Scientific
www.worldscientific.com

STABLE HIGH POWER GaN-ON-GaN HEMT

K.K. CHU, P.C. CHAO, AND J.A. WINDYKA
Microwave Electronics Group, BAE Systems,
Nashua, NH 03060, USA

High power AlGaN/GaN HEMTs on free-standing GaN substrates with excellent stability have been demonstrated for the first time. When operated at a drain bias of 50V, devices without a field plate showed a record CW output power density of 10.0W/mm at 10GHz with an associated power-added efficiency of 45%. The efficiency reaches a maximum of 58% with an output power density of 5.5W/mm under a drain bias of 25V at 10GHz. Long-term stability of device RF operation was also examined. Under ambient conditions, devices biased at 25V and driven at 3dB gain compression remained stable at least up to 1,000 hours, degrading only by 0.35dB in output power. Such results clearly demonstrate the feasibility of GaN-on-GaN HEMT as an alternative device technology to the GaN-on-SiC HEMT in supporting reliable, high performance microwave power applications.

1 Introduction

The wide bandgap AlGaN/GaN material system inherently offers high voltage and high current capabilities, allowing excellent AlGaN/GaN HEMT power performance from S- to Q-band [1]-[4]. Successful demonstrations of hybrid amplifiers as well as MMICs have also been achieved [5]-[7]. However, in spite of the excellent device and circuit demonstrations, there have been limited published reports regarding the reliability of AlGaN/GaN HEMT technology [8]-[11]. Unlike well-established GaAs and InP-based components, GaN-based devices are almost exclusively grown on foreign substrates such as sapphire (Al2O3), SiC and Si [12]. Growth on foreign substrates is not ideal, as differences in lattice constant and coefficients of thermal expansion result in high defect densities which may compromise device reliability. AlGaN/GaN layers grown on SiC or sapphire substrates typically have threading dislocation densities on the order of 109 cm-2, increasing device leakage [13] and promoting metal diffusion along the dislocation sites [14]. The availability of a low-dislocation GaN substrate technology helps circumvent this problem.

The first demonstration of AlGaN/GaN HEMT on free-standing GaN substrates was reported by Khan et al. in 2000 [15]. Device DC results comparable to that on SiC substrates were shown, yet no microwave data were reported. Using a HVPE GaN template technique, Weimann, et al. recently demonstrated an output power density of 1.1W/mm at 2GHz with a GaN-on-GaN HEMT [16]. In this paper, we report for the first time the microwave power performance and device reliability of 0.15μm AlGaN/GaN HEMTs fabricated on low-dislocation semi-insulating free-standing GaN substrates. Low gate leakage current, state-of-the-art microwave power performance as well as stable RF operation have been demonstrated. The results illustrate the great potential of GaN substrate technology in supporting reliable AlGaN/GaN HEMTs for high power microwave applications. In this paper, the key factors resulting in the high power performance with good stability of GaN-on-GaN are also discussed.

2 Device Fabrication

The device structures in this study were grown by metal-organic chemical vapor deposition (MOCVD) on free-standing GaN substrates. The GaN substrates were produced by hydride vapor phase epitaxy (HVPE) and were Fe-doped to achieve a room temperature resistivity in excess of 106Ω-cm [17]. The epitaxial structure consisted of a 1000Å undoped GaN channel layer followed by a 270Å undoped Al0.28Ga0.72N barrier. The homoepitaxial surface was smooth, with RMS roughness of less than 2Å as measured by atomic force microscopy. Unlike SiC, Si or sapphire, the use of GaN substrates allowed high quality epitaxial layers to be grown without utilizing a thick GaN buffer or a separate growth nucleation layer, simplifying the epitaxial growth process. Based on similar wafers, it is estimated that the dislocation density of the substrate and device layer was less than 107/cm2.

Device processing was started with mesa isolation using chlorine-based reactive-ion etch (RIE), followed by Ti/Al-based ohmic contact deposition and annealing. Contact resistance was measured to be 1.5Ω-mm using standard TLM techniques, ~10X higher than the typical value of our GaN-on-SiC HEMT. The much higher ohmic contact resistance of the GaN-on-GaN HEMT may be attributed to the lower dislocation density of the epi structure, resulting in a less effective metal diffusion into the GaN epi during ohmic alloy. Isolation resistance above 1GΩ/sq was measured between separate mesas after ohmic contact formation, attesting to the excellent insulating qualities of the Fe-

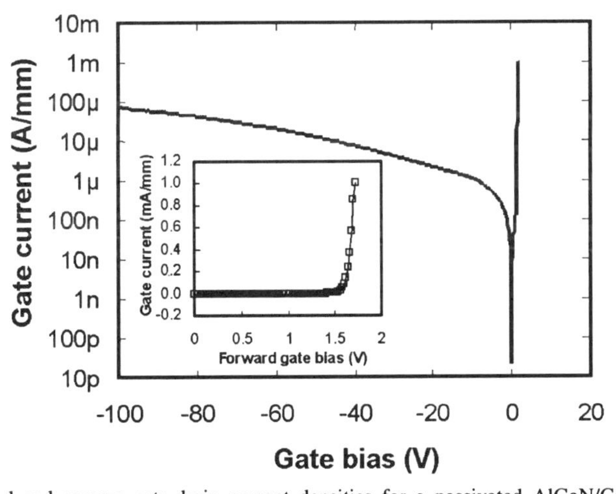

Figure 1. Forward and reverse gate-drain current densities for a passivated AlGaN/GaN HEMT on GaN substrate. Forward turn-on voltage was measured to be 1.7V (inset). Reverse gate leakage was -10μA/mm at -45V gate bias. Breakdown voltage was over 100V.

doped GaN substrate. Electron-beam lithography was then used to define 0.15μm low-resistance Ni/Au T-gates. No field plate was used in the devices. Overlay metal was deposited for device connections, followed by silicon nitride passivation using plasma-enhanced chemical vapor deposition (PECVD). The processing was typically completed with dielectric via opening and 2μm thick Au metal for source airbridging and facilitating pad bonding.

3 Electrical Performance

DC measurements were performed on the fabricated devices using an HP4145B unit. Maximum drain current was 1.1A/mm (at a gate bias of +1V) while peak transconductance was 220mS/mm. Two-terminal gate diode measurements were carried out to examine gate leakage on the devices. Figure 1 shows the forward and reverse gate diode characteristics of a fully passivated device. Forward turn-on voltage (as measured at 1mA/mm forward current) was 1.7V. Reverse gate current was -2.2μA/mm at a gate bias of -20V and -10μA/mm at -45V. Two-terminal reverse breakdown voltage (defined at 1 mA/mm reverse gate current) was higher than 100V, beyond the capability of our measurement system. Compared with AlGaN/GaN HEMTs on SiC with similar device structures, the current devices on GaN substrates have approximately two orders-of-magnitude lower gate leakage values and are at the same level as the GaN HEMT with a Cu gate [18]. The reduction in gate current is attributed to the reduced dislocation density achievable on free-standing GaN substrates.

On-wafer microwave power measurements were performed using a loadpull system. The devices were operated in Class AB mode with the input and output matches tuned for maximum output power. Figure 2 shows the 10GHz continuous-wave (CW) output power densities and associated power-added efficiencies (PAE) obtained at a drain bias of 50V for a 2 × 75μm device. As can be seen from Figure 2, a CW power density of 10.0W/mm was obtained with an associated gain of 11dB and a PAE of 45%. Figure 3 shows the device power performance under various bias conditions. At 25V drain bias, a CW output power density of 5.5W/mm was measured at 10GHz with a PAE of 58%. The

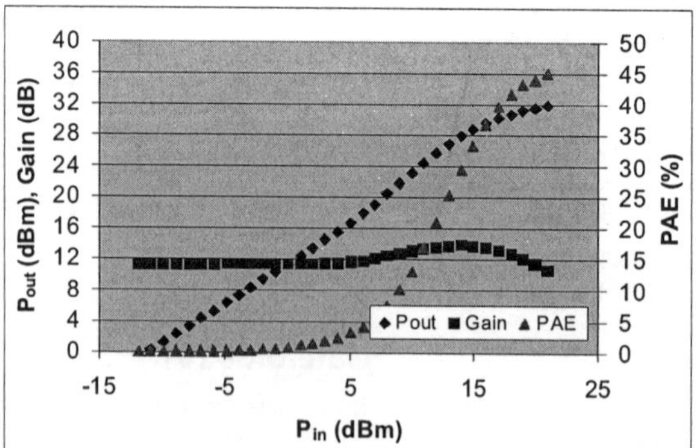

Figure 2. 10GHz CW loadpull power performance for GaN-on-GaN HEMT, showing P_{out} of 10.0W/mm (at V_{ds}=50V).

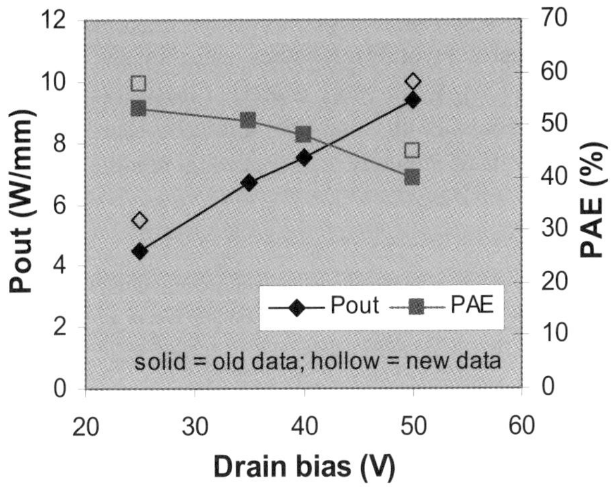

Figure 3. Bias dependence of power and PAE for GaN-on-GaN HEMT without field plate (solid: device #1, hollow: device #2).

output power increases virtually linearly with drain voltage, indicating that current dispersion was not significant in the devices. To the authors' knowledge, this is the best microwave power performance ever reported for AlGaN/GaN HEMTs based on GaN substrates.

For AlGaN/GaN HEMT structures on SiC substrates with thicker GaN buffer layers (1 - 3μm), short gate devices designed for high gain often suffer from high sub-threshold leakage without the use of a compensation-doped buffer. Consequently device operation is limited to lower drain bias (30V or below) and thus lower output power. In this work, the thin GaN buffer together with the Fe-doped semi-insulating substrate significantly improved electron confinement and allowed device operation up to 50V, thereby increasing the output power density achievable for 0.15μm gate devices. It should be noted that even with the higher thermal resistively GaN substrate, the non-optimized GaN-on-GaN HEMT has already demonstrated power performance comparable to that of the GaN-on-SiC devices with field plate.

4 Device Reliability

Devices on GaN substrates were subjected to a prolonged RF stress test at ambient to examine their long-term stability. Quiescent drain bias was set at 25V with a drain current of 200mA/mm. The same loadpull test stand was used to optimize input and output matches for maximum output power. Input power level was kept constant at the initial 3dB compression point of the devices. Measured output power and drain current for a $2 \times 75\mu$m device are shown in Figure 4 over a period of 1,000 hours. Initial device performance corresponds to a 3dB compressed power of 4.3W/mm with an associated PAE of 55%. After 1,000 hours of continuous RF stress, device output power remained stable, degrading only by 0.35dB. Device drain current was found to track closely with output power, dropping 4% during the 1,000-hour test, as shown in Figure 4. This device degradation result compares favorably to other AlGaN/GaN HEMT degradation rates reported in literature [8]-[11]. While more extensive multi-temperature testing is needed to extract actual device lifetime, this first demonstration clearly illustrates the excellent potential of GaN substrate technology in supporting reliable, high power microwave devices.

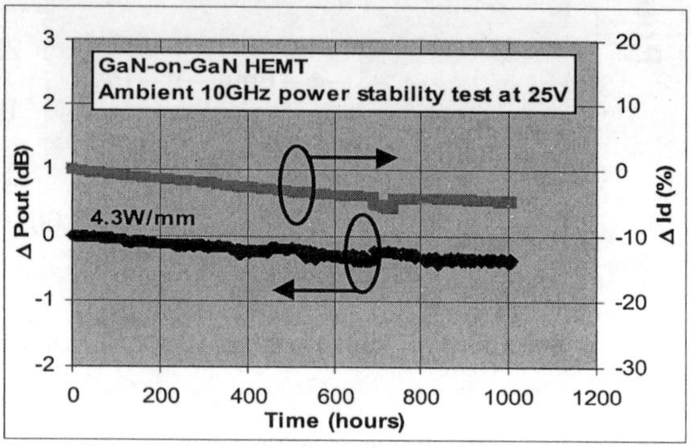

Figure 4. Ambient RF life test showing stable operation at 25V for up to 1,000 hrs. The power degradation tracks well with drain current.

Understanding failure mechanisms and achieving good reliability for GaN HEMTs are significant challenges. Figure 5 summarizes possible factors responsible for the excellent electrical stability of AlGaN/GaN HEMTs. Keys to enhancing the reliability of GaN HEMTs in general may therefore include management of material strain, reduction of electric field and leakage currents, improvement in epi quality, and reduction in trap state formation both at the surface and in bulk material. With lower dislocation density and reduced inherent strain in the epitaxial layers, AlGaN/GaN HEMTs on GaN substrates are expected to have improved device lifetime over their heteroepitaxial counterparts on SiC or sapphire substrates.

5 Conclusions

High power microwave AlGaN/GaN HEMTs have been demonstrated on free-standing GaN substrates for the first time. An excellent CW power density of 10.0W/mm with an associated PAE of 45% was achieved at 10GHz and 50V drain bias. Two-terminal gate leakage of the devices was very low, reaching -10μA/mm at a gate bias of -45V. When operated at a drain bias of 25V and 3dB output compression, devices showed a 0.35dB power degradation over 1,000 hours. Such results demonstrate the excellent potential of AlGaN/GaN HEMTs on GaN substrates in achieving high microwave power with good reliability. With the demonstrated excellent power performance and stability, GaN HEMTs on GaN substrates provide an attractive option for highly reliable high power microwave devices and also could serve as a vehicle to better understand GaN-on-SiC reliability issues.

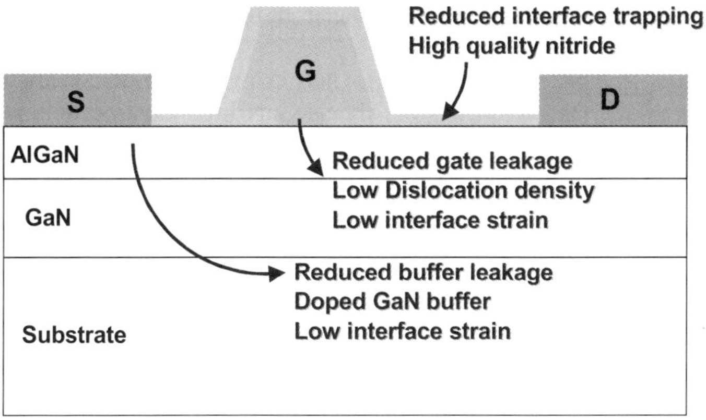

Figure 5. Possible physical factors for stability/reliability improvement of a GaN-on-GaN HEMT (in comparison to the GaN-on-SiC HEMT).

Acknowledgments

The authors would like to thank Cree(ATMI) for supplying GaN wafers and epi test data, J. F. Bass for performing electron-beam lithography and M. Pizzella and L. M. MtPleasant for RF measurements.

References

1. Y. F. Wu, A. Saxler, M. Moore, P. Smith, S. Sheppard, P. M. Chavarkar, T. Wisleder, U. K. Mishra, and P. Parikh, "30-W/mm GaN HEMTs by field plate optimization," IEEE Electron Device Lett. 25, 117 (2004).
2. V. Tilak, B. Green, V. Kaper, H. Kim, T. Prunty, J. Smart, J. Shealy, and L. Eastman, "Influence of barrier thickness on the high-power performance of AlGaN/GaN HEMTs," IEEE Electron Device Lett. 22, 504 (2001).
3. K. Kasahara, H. Miyamoto, Y. Ando, Y. Okamoto, T. Nakayama, and M. Kuzuhara, "Ka-band 2.3W power AlGaN/GaN heterojunction FET," in IEDM Tech. Digest, 677 (2002).
4. K.S. Boutros, M. Regan, P. Rowell, D. Gotthold and B. Brar, "High performance GaN HEMTs on 3-inch SI-SiC substrates," in Proc. CS ManTech Conf., 23 (2004).
5. W. L. Pribble, J. W. Palmour, S. T. Sheppard, R. P. Smith, S. T. Allen, T. J. Smith, Z. Ring, J. J. Sumakeris, A. W. Saxler, and J. W. Milligan, "Applications of SiC MESFETs and GaN HEMTs in power amplifier design," in IEEE MTT-S Dig. 3, 1819 (2002).
6. B. M. Green, V. Tilak, S. Lee, H. Kim, J. A. Smart, K. J. Webb, J. R. Shealy, and L. F. Eastman, "High-power broad-band AlGaN/GaN HEMT MMICs on SiC substrates," IEEE Trans. Microwave Theory Tech. 49, 2486 (2001).

7. R. Actis and K. Chu, "5W coplanar 4-18GHz NDPA GaN MMIC," to be published.
8. H. Kim, V. Tilak, B. M. Green, H. Cha, J. A. Smart, J. R. Shealy, L. F. Eastman, "Degradation characteristics of AlGaN-GaN high electron mobility transistors," in Proc. IEEE Int. Reliability Physics Symp. 214 (2001).
9. C. Lee, L. Witkowski, M. Muir, H. Q. Tserng, P. Saunier, H. Wang, J. Yang, and M. A. Khan, "Reliability evaluation of AlGaN/GaN HEMTs grown on SiC substrate," in Proc. IEEE Lester Eastman Conf. High Performance Devices, 436 (2002).
10. J. A. Mittereder, S. C. Binari, P. B. Klein, J. A. Roussos, D. S. Katzer, D. F. Storm, D. D. Koleske, A. E. Wickenden, and R. L. Henry, "Current collapse induced in AlGaN/GaN HEMTs by short-term DC bias stress," in Proc. IEEE Int. Reliability Physics Symp. 320 (2003).
11. T. Kikkawa, E.Mitani, K. Joshin, S. Yokokawa and Y. Tateno, "An over 100W CW output power amplifier using AlGaN/GaN HEMTs," in Proc. CS ManTech Conf. 97 (2004).
12. A.W. Hanson, R. Borges, J.D. Brown, J.W. Cook, Jr., T. Gehrke, J.W. Johnson, K. Linthicum, S. Peters, E. Piner, P. Rajagopal, J.C. Roberts, S. Singhal, R. Therrien and A. Vescan, "Development of a GaN transistor process for linear power applications," in Proc. CS ManTech Conf. 107 (2004).
13. J. W. P. Hsu, M. J. Manfra, R. J. Molnar, B. Heying, and J. S. Speck, "Direct imaging of reverse-bias leakage through pure screw dislocations in GaN films grown by molecular beam epitaxy on GaN templates," Appl. Phys. Lett. 81, 79 (2002).
14. C. Y. Hsu, W. H. Lan, and Y. C. S. Wu, "Effect of thermal annealing of Ni/Au ohmic contact on the leakage current of GaN based light emitting diodes," Appl. Phys. Lett. 83, 2447 (2003).
15. M. A Khan et al., "GaN-AlGaN heterostructure field-effect transistors over bulk GaN substrates," Appl. Phys. Lett. 76, 3807 (2000).
16. N.G. Weimann, M.J. Manfra, J.W.P Hsu, K. Baldwin, L.N. Pfeiffer and K.W. West, "AlGaN/GaN HEMTs grown by MBE on sapphire, SiC, and HVPE GaN templates," in Proc. IEEE Lester Eastman Conf. On High Performance Devices, 126 (2002).
17. R. P. Vaudo, X. Xu, A. Salant, J. Malcarne, and G. R. Brandes, "Characteristics of semi-insulating, Fe-doped GaN substrates," Phys. Stat. Sol. (a) 200, 18 (2003).
18. J.P. Ao, D. Kikuta, N. Kubota, Y. Naoi and Y. Ohno, "Copper gate AlGaN/GaN HEMT with low gate leakage current," IEEE Electron Device Lett. 24, 500 (2003).

International Journal of High Speed Electronics and Systems
Vol. 14, No. 3 (2004) 745–749

World Scientific
www.worldscientific.com

THICK GaN LAYER GROWN BY Ga VAPOR TRANSPORT TECHNIQUE

HUAQIANG WU, PHANIKUMAR KONKAPAKA,
YURI MAKAROV*, and MICHAEL G. SPENCER

School of Electrical and Computer Engineering Cornell University, Ithaca, NY
** Semiconductor Technology Research Inc. Richmond, VA, USA*

Using a novel Ga vapor transport technique, thick gallium nitride layers have been deposited on a 3-6 μm GaN film grown on sapphire substrate via hydride vapor phase epitaxy (HVPE). GaN powder was successfully used as a stable Ga source material for the growth. A growth rate of greater than 200 μm/hr was achieved by optimizing the growth temperature and the reactant gas flow rates. This growth rate is as high as that achieved by HVPE. The GaN layers were characterized by X-ray diffraction (XRD), scanning electron microscopy (SEM), and Atomic Force Microscopy (AFM).

1 Introduction

Techniques for growing several wide band gap semiconductor materials have been developed for use in new high-power and short-wavelength light-emitting diodes (LEDs) and laser diodes (LDs). GaN is one of the most promising materials among these. However, a high density of dislocations is induced by growing GaN layers on hetero-substrates because of the large lattice mismatch and thermal expansion coefficient difference. On this point, a GaN substrate with a low dislocation density is very attractive for GaN epitaxial growth. Growth rates of GaN layers achieved by conventional methods such as MOCVD or MBE are very low and are not suitable for producing bulk GaN single crystals. Other various growth techniques are now under development, such as: HVPE[1], high pressure solution growth[2], and sublimation techniques[3-5]. Free standing GaN crystals grown by sodium flux technique and ammonothermal technique also have been reported[6].

In this work, we attempt to grow single crystal GaN using a Ga vapor transport technique. Ga vapor is obtained by decomposing GaN powder instead of vaporizing metal Ga. A high growth rate has been achieved due to the unique reactor design and large surface area of GaN powder. Details of the growth system, the advantages of using GaN powder, and characterizations of the grown layers are presented and discussed.

2 Experimental Procedures

The apparatus used in this work is shown in Figure 1. The GaN powder source temperature and the seed substrate temperature can be controlled independently by two separate heating zones. The source GaN powder (99.99% metal basis) is charged in the annular space of the BN boat. This powder is commercially available. Its yellow color indicates a high oxygen concentration. The GaN powder decomposes into Ga vapor and N_2. The Ga vapor is further transported to the substrate by a N_2 carrier gas flowing over the powder. The flow rate of the N_2 carrier gas can be varied to control the amount of Ga vapor transported into the growth region. High purity (99.9999%) NH_3 is delivered to the substrate separately by another quartz tube. Since the NH_3 flow rate is very low compared to that of the N_2 carrier gas, N_2 is added to the same tube with NH_3 to produce the desired flow pattern. A sapphire substrate with a 3-6μm GaN epi-layer grown by HVPE is used as the seed. This seed substrate is cleaned using acetone, methanol and DI water. The distance between the substrate and the NH_3 inlet tube is maintained at 35mm. The pressure is kept at 600 Torr for all experiments. The average growth rate of each layer is determined by measuring cross-sectional SEM images of the layers.

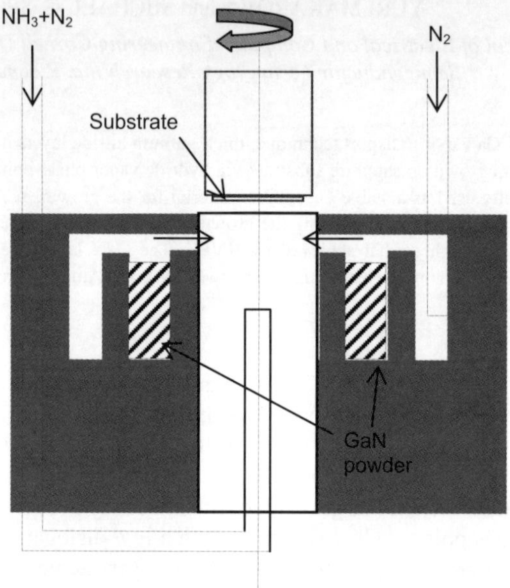

Figure 1. Schematic of the growth system. N_2 carrier gas flows over the powder and brings the Ga vapor to the seed. NH_3 is delivered by the quartz tube to the seed.

The seed substrate temperature (1050°C to 1180°C), NH_3 flow rate (10sccm to 40sccm) and N_2 carrier gas flow rate (200sccm to 600sccm) were varied to find the optimal conditions for a high growth rate. The GaN crystals grown in this study were characterized by XRD, SEM, AFM, and SIMS.

3 Results and Discussions

Ga vapor over liquid Ga has a very low equilibrium pressure, around 1Pa at 1100°C. It is well known from both published reports and our experiments that a thin GaN crust is immediately formed over the surface when NH_3 meets liquid Ga. To achieve a high growth rate, it is essential to find a stable Ga vapor source. GaN powder is chosen here to take advantage of its large surface area. In our design, the N_2 carrier gas transports the Ga vapor away fast enough to maintain a very low Ga vapor pressure over the powder. GaN powder continually decomposes since the Ga vapor partial pressure is very low in its local environment. For this reason, the N_2 carrier gas flow rate is a very important growth parameter. Ga vapor can accumulate on top of the GaN powder and form liquid Ga if the N_2 gas flow rate is too low. Also, NH_3 gas can invade the GaN source, which can make the source unstable. If the N_2 flow rate is too high, the Ga vapor and NH_3 concentrations in the gas phase will be very low, and the growth rate will be undesirable. Figure 2 shows the relationship between the powder weight loss and the N_2 carrier gas flow rate. The experimental conditions were: 1125°C for the powder temperature, 100sccm N_2 flow from the bottom quartz tube, and 600 Torr for 1 hr. The side N_2 carrier gas flow was varied from 200sccm to 600 sccm. The weight losses were obtained by measuring the powder weight before and after the experiments. Powder weight loss is a good indicator of the Ga transport rate. It is clear that the Ga transport increases proportionally with the N_2 carrier gas flow rate. The GaN powder decomposition rate is affected by the local Ga vapor partial pressure. With a higher N_2 carrier gas flow rate, there is a lower Ga partial pressure over the GaN powder, and therefore a higher GaN powder decomposition rate.

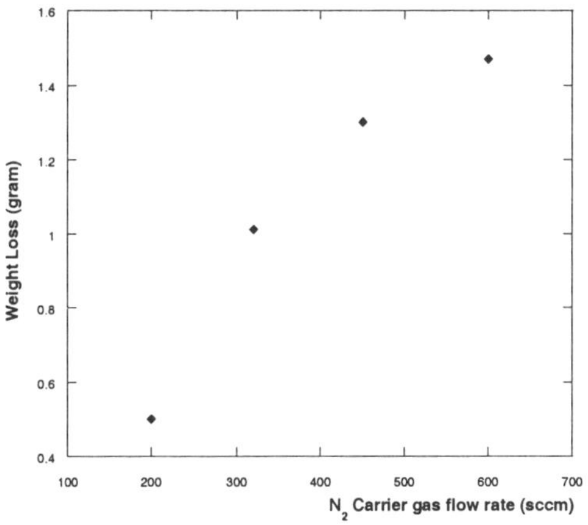

Figure 2. GaN powder weight loss versus the N_2 carrier gas flow rate.

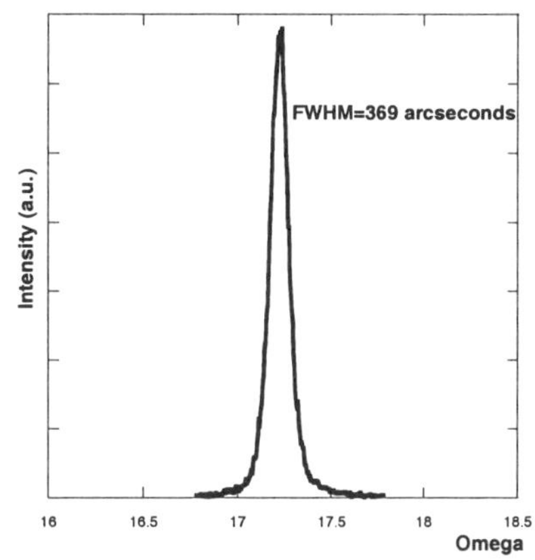

Figure 3. (0002) rocking curve of grown GaN layer,

θ-2θ XRD pattern shows only wurtzite single crystal GaN peaks. Figure 3 is the rocking curve of GaN (0002) peak. The FWHM of this peak is 369" which is smaller than 540" of GaN template. Figure 4(a) shows an SEM image of the surface morphology of a GaN layer. It is clear that spiral growth is the dominant growth mechanism in the layer. Each spiral has an average diameter of ~3μm. Figure 4(b), a 3D AFM image was taken for one spiral on the same sample. The edge is ~0.1 μm higher than the center of the spiral.

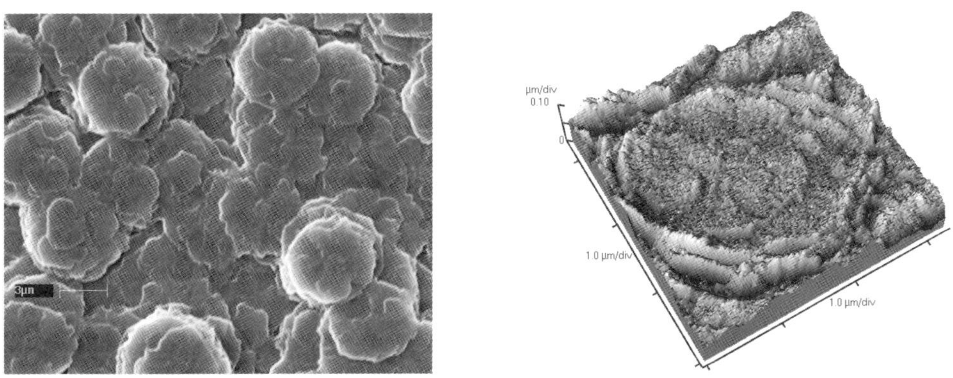

Figure 4. Surface morphology of GaN layer: (a) SEM image of the as-grown GaN layer surface; (b) AFM 3D image of one spiral.

Figure 5 shows a cross-sectional view of the GaN layer grown under optimal growth conditions: 1155°C powder temperature, 1165°C seed temperature, 450 sccm N_2 carrier gas flow rate and 10 sccm NH_3 flow rate plus 90 sccm N_2 from the bottom tube in 15 minutes. This layer is uniform and continuous. The layer is 45μm thick and the average growth rate is 180 μm/hr. With a short growth time (5minutes) under the same growth conditions, the growth rate can be even as high as 220μm/hr. This is because the commercial GaN powder has a very high oxygen concentration, and GaO and Ga_2O species have higher vapor pressures than pure liquid Ga. With this higher vapor pressure, more Ga containing gas species are transported to the substrate. The weight loss is also greater at the beginning. After the GaO and Ga_2O gas species are depleted, GaN can grow with a stable growth rate.

Figure 5 Cross-sectional SEM image of grown GaN layer.

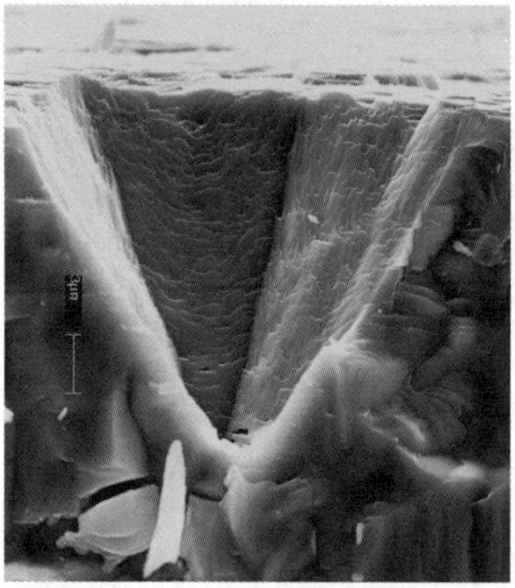

Figure 6 SEM image of V-defect cross-section.

Electrical properties of the grown layers were measured by Hall-Effect measurement technique. Electrical contacts were made using indium metal and Hall measurement was carried out using van der Pauw technique. From the measured results, the layer was found to be n-type with a carrier concentration of 6.67 x 10^{18} cm^{-3} and a mobility of 550 cm^2/Vs. This high carrier concentration was attributed to the presence of impurities in the source material.

Glow Discharge Mass Spectrometry (GDMS) was carried out to determine the impurity level of the source material: commercial GaN powder. The analysis results showed extremely high oxygen and carbon concentrations in this powder. These two elements are actually present at levels higher than the detect limit of the GDMS technique which is around 3-5%. This high oxygen concentration affects the quality of the layer. Figure 6 shows a cross-sectional SEM image of a V-defect. This V-defect is similar to those that are observed in HVPE GaN and GaN/InGaN systems.[7,8] They are concave hexagonal pyramids that are bounded by six {1-101} facets. The high oxygen impurity content in the layer (~$10^{21}/cm^3$ by SIMS analysis) is the possible origin of these V-defects. Recently, we are able to produce high purity GaN powder in our lab. This high purity lab-made powder will be used into the growth system soon. The GaN layer quality grown by the high purity lab-made powder is expected to be improved.

4 Conclusion

A novel GaN growth system has been constructed to perform bulk GaN growth by using GaN powder as a source material. A high Ga transport rate can be achieved using a N_2 carrier gas flowing over the powder and by taking advantage of GaN's large surface area. The Ga vapor transport rate can be controlled by varying the N_2 flow rate. A growth rate of 220um/hr is achieved in the first 5 minutes, and a 180um/hr growth rate in 15 minutes. A high oxygen concentration in the powder has been measured by GDMS. This high oxygen concentration causes substantial oxygen impurities in the grown layer ($\sim 10^{21}/cm^3$ by SIMS analysis). V-defects observed in the layers are probably generated by this high oxygen concentration. To achieve a high quality layer and stable growth of GaN, the oxygen concentration in the source powder needs to be reduced.

Acknowledgements

This work was supported by the Office of Naval Research under contract N00014-01-1-0716, Dr. Colin Wood, monitor. The authors would also like to express their sincere thanks to Mr. Barry Butterfield for his excellent technical assistance.

References

1. S. T. Kim, Y. J. Lee, D. C. Moon, C. H. Hong, T. K. Yoo, "Preparation and properties of free-standing HVPE grown GaN substrates" J. Crystal Growth (1998) vol. 194, 37.
2. S. Porowski, I. Grzegory, "Thermodynamical properties of III–V nitrides and crystal growth of GaN at high N2 pressure " J. Crystal Growth (1997) vol. 178, 174.
3. Yu. A. Vodakov, E. N. Mokhov, A. D. Roenkov, M. E. Boiko, P.G. Baranov, "High rate GaN epitaxial growth by sublimation sandwich method " J. Crystal Growth (1998) vol. 183, 10.
4. Y. Naoi, K. Kobatake, S. Kurai, K. Nishino, H. Sato, M. Nozaki, S. Sakai, Y. Shintani, "Characterization of bulk GaN grown by sublimation technique", J. Crystal Growth (1998) vol. 189, 163-166
5. M. Imade, S. Kobayashi, F. Kawamura, M. Yoshimura, Y. Mori, T. Sasaki, "Growth of thick GaN films with high growth rate using sublimation method under high pressure", Jpn. J. Appl. Phys. (2004) vol. 43, L486-488
6. H. Yamane, M. Shimada, T. Sekiguchi, F. J. DiSalvo, "Morphology and characterization of GaN single crystals grown in a Na flux " J. Crystal Growth (1998) vol. 186, 8.
7. X. A. Cao, J. A. Teetsov, F. Shahedipour-Sandvik and S. D. Arthur "Microstructural origin of leakage current in GaN/InGaN light-emitting diodes ", J. Crystal Growth (2004) vol. 264, 172-177.
8. X. H. Wu, C. R. Elsass, A. Abare, M. Mack, S. Keller, P. M. Petroff, S. P. DenBaars, J. S. Speck, and S. J. Rosner "Structural origin of V-defects and correlation with localized excitonic centers in InGaN/GaN multiple quantum wells", Appl. Phys. Lett. (1998) vol. 72, 692

International Journal of High Speed Electronics and Systems
Vol. 14, No. 3 (2004) 750–755

DEPENDENCE OF RF PERFORMANCE OF GaN/AlGaN HEMTS UPON AlGaN BARRIER LAYER VARIATION

ELIAS FARACLAS[1], RICHARD T. WEBSTER[2], GEORGE BRANDES[3]
AND A. F. M. ANWAR[1]

[1] *Electrical and Computer Engineering, University of Connecticut, Storrs, CT 06269, USA*

[2] *AFRL/SNHA, 80 Scott Dr, Hanscom AFB MA 01731-2909, USA*

[3] *Cree Inc., 4600 Silicon Dr, Durham, NC 27703, USA*

The dependence of microwave performance of GaN/AlGaN High Electron Mobility Transistors (HEMTs), namely the unity gain current cut-off frequency (f_T) and the maximum oscillation frequency (f_{MAX}), are reported as a function of the mole fraction of Al and the thickness of the barrier AlGaN layer. The parameters are computed using a physics-based model and compared to experimenttal results. Schrödinger and Poisson's equations are solved self-consistently to relate the applied gate bias to the channel electron concentration. The contributions of both spontaneous and piezoelectric polarizations towards f_T are explored. Finally, because of interest in using this family of devices at elevated temperatures, each simulation was repeated between 300K and 500K for comparison.

1 Introduction

The use of AlGaN/GaN HEMTs has been successfully demonstrated for applications in high power circuitry. The advantage of GaN for high power applications is due to its large band gap (3.4 eV) and high breakdown field (3×10^6 V/cm). Standard HEMTs with power density as high as 12W/mm at 10 GHz from a GaN/AlGaN/GaN HEMT on SiC substrate has been reported.[1] Power density as high as 32.2W/mm at 4 GHz from an AlGaN/AlN/GaN HEMT on SiC substrate with a field plate has been reported.[2] Advances in growth technology have enabled the realization of the superior transport properties of GaN to yield devices with an improved unity gain current cut-off frequency (f_T) and the maximum oscillation frequency (f_{MAX}). The high thermal conductivity of GaN (1.5 W/cm/K) grown on SiC (4.5 W/cm/K) also makes possible the operation of GaN HEMTs at very high temperatures. Operation of GaN HEMTs on SiC substrates at 750°C has been reported.[3]

The operation of undoped supply layer FETs is dependant upon the presence of surface charge and polarization of the constituting material.[4-7] It is assumed that the carriers induced in the channel are due to the surface charge while the polarization in the AlGaN facilitates the formation of the quantum well. The mole fraction of the AlGaN barrier layer as well as its thickness play a major role in the operation of this family of devices. As reported by Ibbetson, *et al.*, for a given trap level, there exists a minimum AlGaN barrier layer thickness for the formation of the 2DEG layer.[8] Moreover, the 2DEG concentration saturates beyond a certain AlGaN layer thickness. In this paper, the role of the AlGaN barrier layer on the microwave performance, that is f_T and f_{MAX} of AlGaN-GaN HEMT, is presented. The determination of the microwave parameters is based upon the solution of Schrödinger-Poisson equations to calculate the dependence of the 2DEG concentration upon the AlGaN barrier. The transport parameters in GaN are obtained from an ensemble Monte Carlo simulation. From these computations, the transconductance, g_m, and gate-to-source capacitance, C_{gs}, can be computed. From these, f_T can be calculated as

$$f_T = g_m/(2\pi C_{gs}). \tag{1}$$

2 Model

The small signal parameters and the characteristic frequencies are obtained by formulating the charge control and the transport parameters. Schrödinger and Poisson's equations are solved self-consistently to relate the applied gate bias to the channel electron concentration, n_{2DEG}. The formulation takes into account the contribution of both

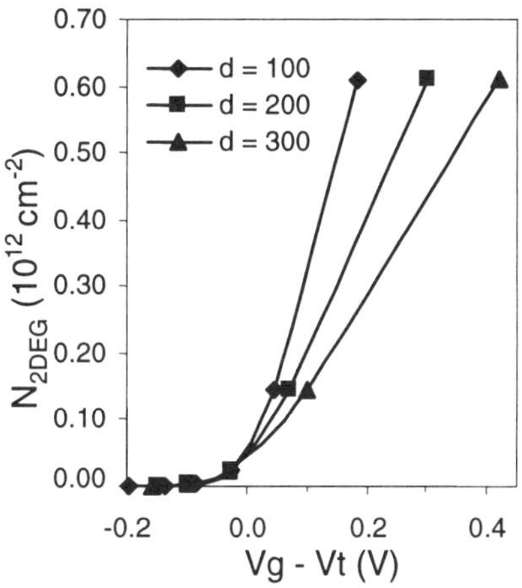

Figure 1: Variation of the 2DEG concentration as a function of normalized gate bias with barrier layer thickness as a parameter. Al-mole fraction is 0.3.

the spontaneous and piezoelectric polarizations. The quantum well parameters such as the average distance of the two dimensional electron gas (2DEG) concentration, Xav, and the position of the Fermi level, E_F, with respect to the conduction band potential in GaN at the GaN/AlGaN heterointerface are calculated and expressed as a linear function of the 2DEG concentration to facilitate the determination of the small signal parameters:

$$Xav = Xav_0 + \alpha n_{2DEG} \tag{2}$$

$$E_F = Ef_0 + \beta n_{2DEG} \tag{3}$$

where n_{2DEG} is the calculated 2DEG concentration, E_{F0} and Xav_0 are the Fermi level and average distance of the 2DEG from the heterointerface when the 2DEG concentration is zero. Figure 1 shows that with increasing AlGaN layer thickness, d_{AlGaN}, the slope of the n_{2DEG}-V_G characteristic decreases implying a decrease of the device transconductance. With increasing d_{AlGaN}, for a given gate bias, the 2DEG concentration increases resulting in a decrease in the average distance of the electron cloud from the GaN/AlGaN heterointerface. The variation of Xav as a function of 2DEG concentration with different Al-mole fractions is shown in Fig.2.

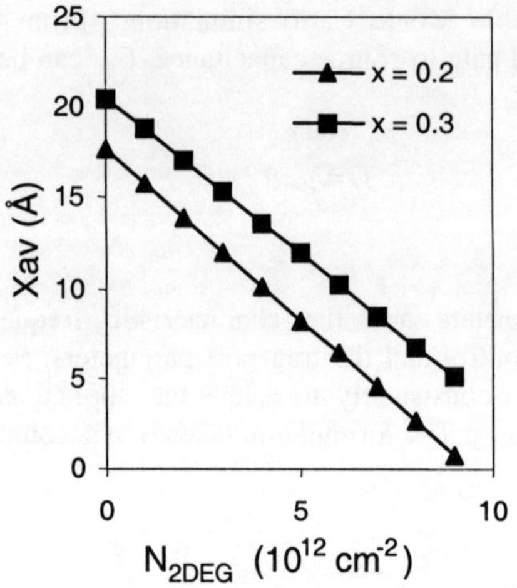

Figure 2: Average distance of the electron cloud from the heterointerface, *Xav*, as a function of 2DEG concentration for $Al_xGa_{1-x}N$ with x=0.2 and x=0.3.

Ensemble Monte Carlo simulation is carried out to determine the appropriate transport parameters such as the differential mobility and saturation velocity. In the present calculation the dependence of short channel effects on transport properties are taken into account through non-stationary transport.

3 Results and Discussion

Figure 3 shows the variation of f_T as a function of the thickness of the undoped AlGaN barrier layer with temperature and Al-mole fraction as parameters as computed from the physics-based model. The gate and drain voltages are 0V and 20V, respectively. For an Al-mole fraction of 0.2, f_T increases with barrier thickness, saturating beyond 250 Å. For thinner barriers, f_T decreases slightly with increasing temperature while for thicker

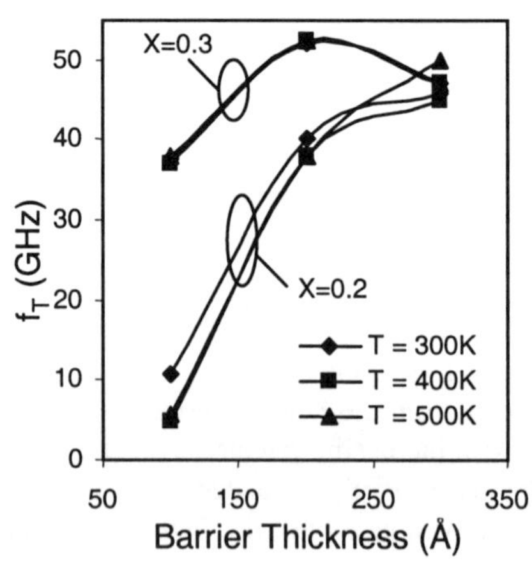

Figure 3: Current gain cutoff frequency, f_T, as a function of AlGaN-barrier layer thickness and temperature.

barriers, f_T remains constant at high temperatures. With an Al-mole fraction of 0.3, f_T reaches a maximum near 200 Å and is essentially independent of temperature.

The dependence of f_T on d_{AlGaN} is attributed to competing effects of the transconductance and gate-source capacitance. At a given gate and drain bias, g_m decreases with increasing d_{AlGaN} as discussed in connection with Fig. 1. At the same time, increasing d_{AlGaN} creates a larger separation between the gate and the channel, decreasing C_{gs}. The net result is an initial increase in f_T with d_{AlGaN} . However, C_{gs} also depends upon *Xav*, the average distance to the electron cloud, which decreases with increasing d_{AlGaN}. As the cloud moves closer to the gate, C_{gs} grows and f_T is diminished. The above results are valid at room temperature. With increasing Al-mole fraction, the 2DEG concentration increases which results in higher g_m as well as higher drive current and power. Therefore, the optimization of the composition of the barrier layer and its thickness allows one to obtain a specific speed-power product. It should be mentioned that the present calculations are isothermal and the characteristic frequencies will be lowered due to a decrease in low field mobility and saturation velocity at high drain bias.

4 Dependence upon Contributions of Polarization

The roles of spontaneous and piezoelectric polarizations are illustrated in Fig. 4, where f_T is plotted as a function of barrier thickness. The gate and drain biases are 0V and 20V, respectively. The contribution of spontaneous polarization is obtained by setting the piezoelectric component to zero in the simulation and the contribution of piezoelectric effect is obtained by setting the spontaneous component to zero. The model shows that the contribution of spontaneous polarization to microwave performance is much stronger that its piezoelectric counterpart. A similar behavior is observed when the Al-mole

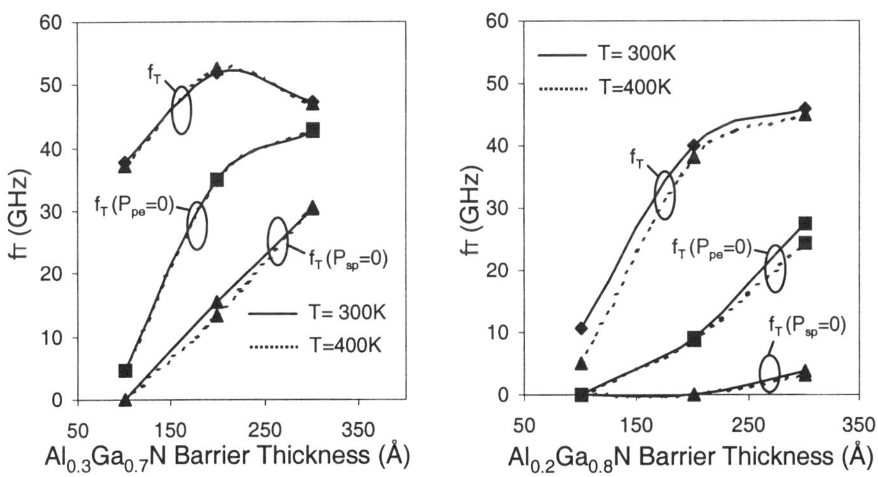

(a) (b)

Figure 4: Contributions of polarization to f_T for Al mole fractions (a) x = 0.3 and (b) x=0.2 at temperatures of 300K and 400K.

fraction is reduced from 0.3 to 0.2. The contributions in all cases are largely temperature invariant.

5 Dependence on Temperature Variation

The dependence of f_T on temperature variation is parameterized in Fig. 3. While only weak temperature dependence is seen in all cases, it is evident that an increased Al-mole fraction correlates with increased thermal stability. Barrier layers with a higher Al-mole fraction exhibit a larger conduction band discontinuity at the heterointerface. The increased depth of the resulting quantum wells result in a slightly larger separation of eigenstates yielding enhanced thermal stability.

6 Measurement

The present methodology is applied to explain the measured data of a HEMT with the following growth sequence: a 3-μm GaN layer is grown followed by a 30Å $Al_{.22}Ga_{.78}N$ spacer layer, 110 Å of doped $Al_{.22}Ga_{.78}N$ layer and 110 Å of undoped $Al_{.22}Ga_{.78}N$ capping layer. The measured threshold voltage of –2.45V is in excellent agreement with the calculated value of –2.37V. The measured threshold voltage is defined as the gate voltage required obtaining 1mA of drain current. The calculated threshold voltage is defined as the gate voltage required to induce a 2DEG concentration of 1 x $10^{12}cm^{-2}$.

The extracted unity gain current cutoff frequency is shown in Fig. 5, where f_T = 33.4 GHz. The corresponding calculated f_T is 28GHz. The extracted f_{MAX} (MAG) = 82.0 GHz, is also shown in Fig. 5. In both cases, the devices are biased at V_D=20V and V_G=-1V. Measurements were carried out using an Agilent E8361A Precision Network Analyzer.

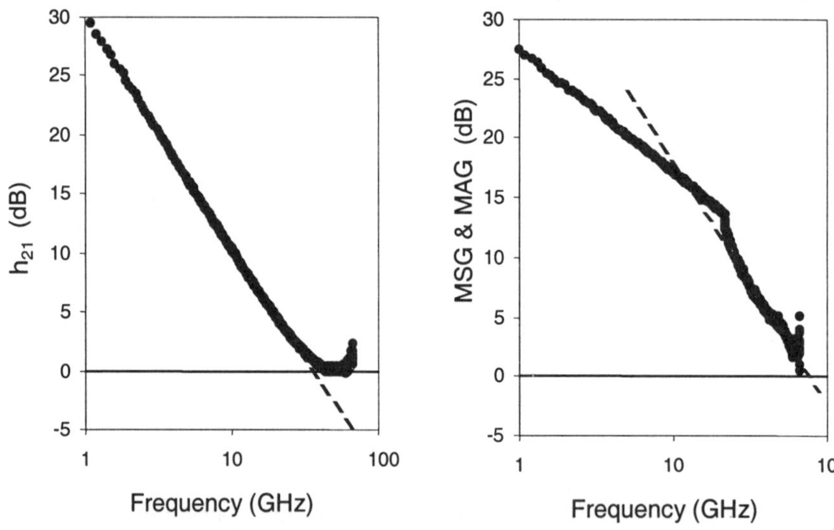

(a) (b)

Figure 5: Measured AlGaN HEMT with Drain (v): 20; Gate (v): -1.0. (a) Fit to h21 gives f_T = 33.4 GHz. (b) Fit to Maximum Available Gain gives f_{MAX} = 82 GHz.

7 Conclusions

A physics based model has shown the microwave performance of GaN/AlGaN HEMTs to be a strong function of the mole fraction of Al and thickness of the barrier AlGaN layer. The transconductance decreases more rapidly for higher Al-mole fractions while the gate capacitance decreases more slowly comparatively reducing f_T. However, devices with higher Al-mole fraction will have a higher current drive capability for a given bias and barrier thickness. Therefore, careful optimization of these growth parameters is needed to obtain a specific speed-power product.

Additionally, the contributions of both spontaneous and piezoelectric polarizations towards f_T are explored. It is shown that the contributions of spontaneous polarization have a greater impact on device speed than those of piezoelectric polarization. Finally, the RF performance of the devices is shown to be weakly temperature dependent in all cases, especially so for barrier layers with increased Al-mole fraction. This confirms that these devices are promising for robust performance in environments with greatly varied temperatures.

Acknowledgement

This work was supported in part by Air Force Office of Scientific Research under LRIR 92SN04COR.

References

1. L. Shen, R. Coffe, et al., "Polarization-Engineered GaN/AlGaN/GaN HEMTs with Record High Power without Passivation". in Device Research Conference - DRC. 2003. University of Utah, Salt Lake City, UT.
2. Y.-F. Wu, et al., "30-W/mm GaN HEMTs by Field Plate Optimization". IEEE Elec. Dev. Lett., 2004. 25(3): p. 117-119.
3. I. Daumiller,et al., "Evaluation of AlGaN/GaN HFET's up to 750oC". Device res. Conf. Dig., 1998: p. 114-115.
4. B. Jogai, "Free electron distribution in AlGaN/GaN heterojunction field-effect transistors". JAP, 2002. 91(6): p. 3721-3729.
5. B. Jogai, "Influence of surface states on the two-dimensional electron gas in AlGaN/GaN heterojunction field-effect transistors". Journal of Applied Physics, 2003. 93(3): p. 1631-1635.
6. Y. Zhang and J. Singh, "Charge control and mobility studies for an AlGaN/GaN high electron mobility transistor". JAP, 1999. 85(1): p. 587-594.
7. Y. Zhang, et al., "Charge control and mobility in AlGaN/GaN transistors: Experimental and theoretical studies". JAP, 2000. 87(11): p. 7981-7987.
8. J.P. Ibbetson, et al., "Polarization effects, surface states, and the source of electrons in AlGaN/GaN heterostructure field effect transistors". Applied Physics Letters, 2000. 77(2): p. 250-252.

International Journal of High Speed Electronics and Systems
Vol. 14, No. 3 (2004) 756–761

SELECTIVE DRY ETCHING OF GaN OVER AlGaN IN BCL_3/SF_6 MIXTURES

D. BUTTARI, A. CHINI, A. CHAKRABORTY, L. MCCARTHY, H. XING, T. PALACIOS, L. SHEN, S. KELLER, AND U. K. MISHRA

Department of Electrical and Computer Engineering,
University of California Santa Barbara, Santa Barbara, California 93106, U.S.A.

Inductively coupled plasma (ICP) etching of GaN with high selectivity over $Al_{.22}Ga_{.78}N$ in BCl_3/SF_6 mixtures has been studied. Selectivity and surface morphology were investigated over a wide range of pressures (3.75–37.5mTorr), RF powers (30–120 W), ICP powers (100–400 W), and SF_6/BCl_3 ratios (0.1–0.7). Higher pressures, lower dc biases, and higher SF_6/BCl_3 ratios increased the GaN to AlGaN selectivity. Selectivities up to 25 were measured by laser interferometry. A root mean square (rms) surface roughness of 0.67 nm was measured by atomic force microscopy (AFM) after removal of 0.5 μm from a GaN template (process selectivity: 15, as-grown rms surface roughness: 0.56 nm). A degradation in surface morphology, with the gradual formation of pits, was observed for selectivities above 10.

1 Introduction

Selective dry etch recipes are often used in semiconductor processing in order to improve process reproducibility and accuracy in etch depth. Historically the possibility of selectively dry etch n^+-GaAs on top of AlGaAs found commercial application in the processing of high electron mobility transistors (HEMTs) and other heterojunction devices [1]. We speculate that the corresponding use of an n^+-GaN cap layer on top of AlGaN/GaN HEMTs will become more and more critical when moving from X-band to Q- and higher frequency- bands, due to the stricter requirements in access resistance. A selective dry etch recipe for GaN over AlGaN is very attractive in such a perspective. The main two chemistries reported in literature in order to selectively etch GaAs over AlGaAs have been O_2/Cl_2 and fluorine/chlorine chemistries. An oxygen-based etch inhibitor chemistry has already been reported on GaN [2], [3]. In this work a BCl_3/SF_6 chemistry on gallium nitride has been extensively characterized with regard to selectivity and surface morphology over a wide range of process parameters such as pressure and RF power. The main advantage of such a chemistry compared to an oxygen based chemistry is that no micromasking (due to the volatility of SiF_4 compared to the non-volatility of SiO_2) is expected to be generated by the presence of silicon in the chamber (either originating from the sample carrier or from the epitaxial layers).

2 Experimental

Experiments were performed in a 6-inch Panasonic E640 single-wafer, load-locked ICP reactor. The load-lock is required to reduce atmospheric contamination which can drastically affect the etch characteristics of GaN and AlGaN [4]. In order to further

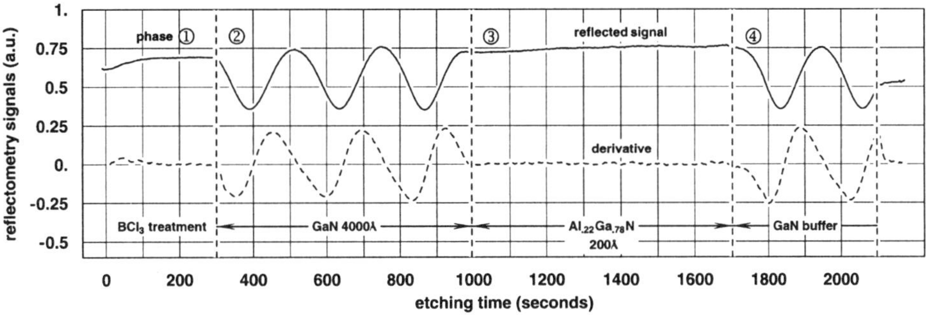

Figure 1. Example of reflectometry signal and associated derivative used to compute etch selectivity of GaN over $Al_{.22}Ga_{.78}N$. Etch conditions: BCl_3 flow: 15 sccm, SF_6 flow: 10 sccm, ICP power: 200 W, RF power: 60 W, pressure: 37.5 mTorr.

reduce the effects of residual oxygen in the chamber a low power (200W ICP, 30W RF) BCl_3 plasma pretreatment was performed in situ on the samples just before the SF_6/BCl_3 etch. Pure BCl_3 is ineffective in etching GaN, but it has a strong deoxidizing power, with beneficial effects in etch depth reproducibility [4]. Test samples were grown by metal-organic chemical vapor deposition (MOCVD) on single side polished sapphire. The epilayer structure was composed of a GaN buffer layer, followed by a 200 Å $Al_{.22}Ga_{.78}N$ etch stop layer and a 4000 Å GaN cap layer. The samples were unpatterned and photoresist free. The interferometry signal at a wavelength of 678 nm was monitored during the etch (Fig. 1). Etch rates and selectivities were calculated from the corresponding signal derivative (Fig. 1). The samples were generally 1×1 cm and placed on 6-in. silicon carrier wafers without the use of thermal grease. Selectivity data were obtained for pressure in the range 3.75–37.5 mTorr, RF power in the range 30–120W, ICP power in the range 100–400W, and SF_6/BCl_3 ratio in the range 0.1–0.7 (corresponding to SF_6 concentrations in the 10%–40% range). Under these conditions the RF power densities varied from 0.16 to 0.66 W/cm^2.

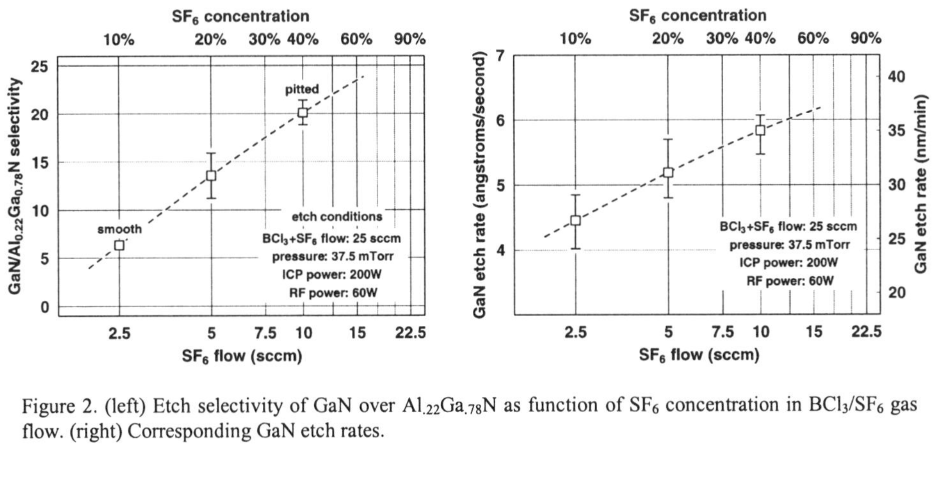

Figure 2. (left) Etch selectivity of GaN over $Al_{.22}Ga_{.78}N$ as function of SF_6 concentration in BCl_3/SF_6 gas flow. (right) Corresponding GaN etch rates.

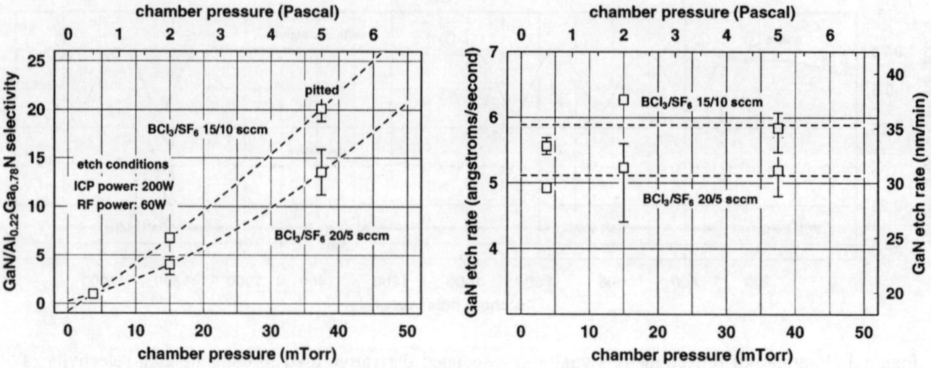

Figure 3. (left) Etch selectivity of GaN over $Al_{.22}Ga_{.78}N$ as function of chamber pressure for two different gas compositions. (right) Corresponding GaN etch rates.

GaN templates grown on sapphire were used in order to investigate surface morphology by atomic force microscopy after a 0.5 μm etch.

3 Results and discussion

As shown in Fig. 1, all etches started with a low power BCl_3 pretreatment (phase 1 in Fig. 1), followed by a BCl_3/SF_6 selective dry etch (phase 2 in Figure 1). Pure BCl_3 was found to be almost ineffective in etching GaN (phase 1 in Figure 1), probably due to the low concentration in active chlorine atoms for medium-low ICP powers [5]. Nevertheless BCl_3 has the strongest deoxidizing power among chlorine-based gases like $SiCl_4$ or CCl_4, making surface oxides efficiently reduced and improving etch depth reproducibility [4]. The intermixing of SF_6 to BCl_3 (phase 2 in Figure 1) increases the population of chlorine radicals (etching agents) and of fluorine atoms (inhibitor agents), allowing for the etching of GaN (phase 2 in Figure 1) and for the stopping on AlGaN (phase 3 in Figure 1). The enhancement in GaN etch rate due to interaction BCl_3 with SF_6 has been previously reported in literature [5]. The enhancement at low RF power can be quantified between one and two orders of magnitude. The decrease in AlGaN etch rate is due to the formation of nonvolatile AlF_3 residues on the AlGaN surface, similarly to the AlGaAs/GaAs case [6]. The non-volatility of AlF_3 reduces the etching efficiency of chlorine, allowing for the selective etch of GaN over AlGaN (and GaAs over AlGaAs [6]). Although AlF_3 is not chemically etched by chlorine, nevertheless it is efficiently removed from the surface by plasma sputtering, allowing for a sharp return to the nominal GaN etch rate after removal of the AlGaN layer (phase 4 in Figure 1).

Figure 2 (left) shows GaN to AlGaN selectivity as a function of SF_6 concentration. Etching conditions were: chamber pressure 37.5mTorr, ICP power 200W, RF power 60W, total gas flow 25 sccm. SF_6 concentrations varied between 10% and 40%. Etch selectivity increased at the increase in fluorine concentration, due to the increased

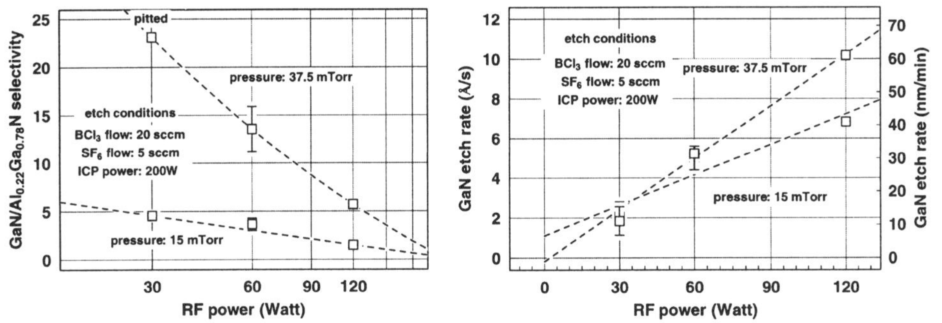

Fig. 4. (left) Etch selectivity of GaN over $Al_{.22}Ga_{.78}N$ as function of RF power for two different chamber pressures. (right) Corresponding GaN etch rates.

formation of AlF_3 [6], [7]. A modest increase in GaN etch rate was observed for the same conditions (Figure 2 right), probably due to enhanced formation of active chlorine [5].

The effects of pressure on selectivity are shown in Figure 3 (left). The pressure range from 3.75 mTorr to 37.5 mTorr has been characterized. Two different SF_6 concentrations have been analyzed: 20% (BCl_3/SF_6 20/5 sccm) and 40% (BCl_3/SF_6 15/10 sccm). Etching conditions were: ICP power 200 W and RF power 60 W. Selectivities were significantly increased at the increase in chamber pressure. Selectivities varied between 1 and 20. It is particularly interesting that GaN etch rates were negligibly affected by chamber pressure (Figure 3 right), allowing for a control in selectivity independent from etch rate.

The effects of RF power on selectivities are shown in Figure 4 (left) for two different chamber pressures (15 mTorr and 37.5 mTorr). Etch conditions were: ICP power 200W, BCl_3 flow 20 sccm, SF_6 flow 5 sccm. An increase in selectivity at the decrease in RF power was generally observed. The increase was more dramatic at high chamber pressures. No AlGaN/GaN selectivity regardless of RF power is expected for chamber pressures below 5 mTorr. The effects of RF power on GaN etch rate are shown in Fig. 4 (right). The etch rate was linearly dependent on RF power. The inability to control

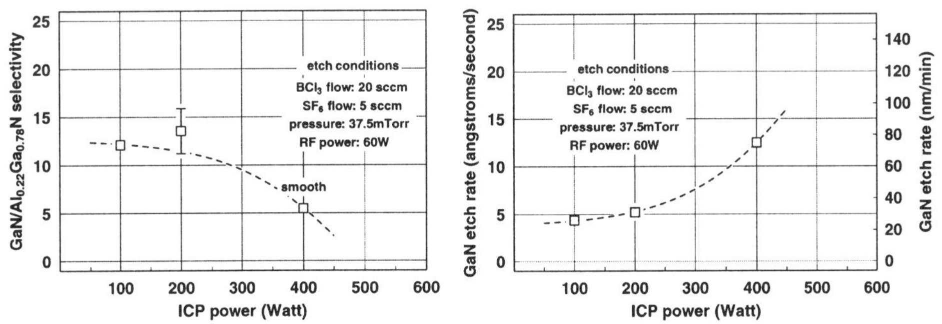

Figure 5. (left) Etch selectivity of GaN over $Al_{.22}Ga_{.78}N$ as function of ICP power. (right) Corresponding GaN etch rates.

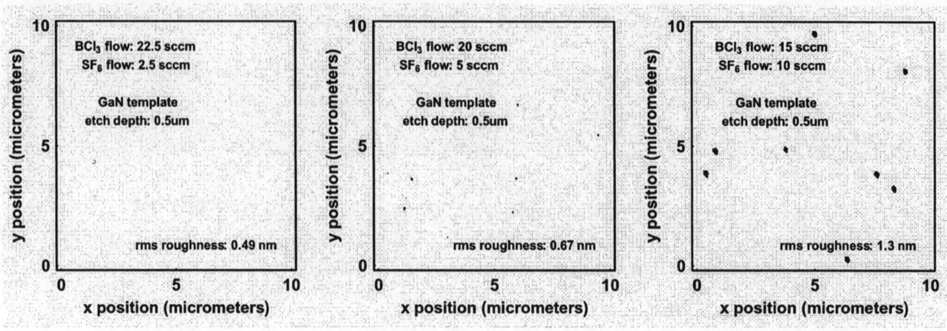

Figure 6. Surface morphology as detected by AFM after removing 0.5 μm of material on a gallium nitride template. Etch conditions: ICP power: 200 W, RF power: 60 W, chamber pressure: 37.5 mTorr. To be noted the enlargement in dislocations size at the increase in SF_6 concentration. As grown rms surface roughness was about 0.56 nm.

selectivity independently from etch rate makes the use of RF power less attractive when compared to chamber pressure. The increase in selectivity at the decrease in RF power can be explained in terms of reduced sputtering of non-volatile AlF_3 at the decrease in RF plasma powers.

Etch selectivites were roughly independent on ICP power for medium-low power levels, as shown in Figure 5 (left). At high power levels the selectivity degraded and the etch rate of GaN increased (Figure 5 right), possibly due to a higher concentration in active chlorine radicals (etching agents) compared to fluorine atoms (inhibition agents).

The reactivity of the plasma and its chemical composition were found to affect surface morphology as detected by AFM. Surface morphologies were studied on GaN templates after 0.5 μm etches. The study is significant since surface morphologies were preserved or enhanced after reaching the AlGaN etch stop layer in GaN capped AlGaN/GaN HEMTs. An increased presence of pits on GaN was detected at the increase in fluorine concentration, as shown in Figure 6. Pits density was roughly constant, but pits dimensions increased at the increase in fluorine concentration. The enlargement seemed to correlate well with the increase in GaN/AlGaN selectivity, with no pits for low selectivities and large pits for high selectivities. This correlation seemed to hold at the varying in SF_6 concentration (Figure 2 left), pressure (Figure 3 left), RF power (Figure 4 left), and ICP power (Figure 5 left). Selectivites seem to be strictly related to the chemical activity of fluorine on the sample surface, and pure SF_6 plasmas have been previously shown to produce pits on GaN [8], [9]. It is speculated that a proper choice of the etching parameters, in particular the addition of a proper third gas, will allow for high selectivities with excellent surface morphologies and reduced electrical damage.

Acknowledgments

The authors would like to thank ONR CANE MURI and DARPA programs for support. This work made use of MRL Facilities supported by National Science Foundation under award No. DMR96-32716.

References

1. J. Vatus, J. Chevrier, P. Delescluse, J. F. Rochette, *IEEE Transactions on Electron Devices*, **33 (7)**, 934 (1986).
2. Yanjun Han, Song Xue, Wenping Guo, Yi Luo, Zhibiao Hao, Changzheng Sun, *Japanese Journal of Applied Physics Part 2-Letters*, **42 (10A)**, L1139 (2003).
3. Wen-Kai Wang, Yu-Jen Li, Cheng-Kuo Lin, Yi-Jen Chan, Guan-Ting Chen, Jen-Inn Chyi, *IEEE Electron Device Letters*, **25 (2)**, 52 (2004).
4. D. Buttari, A. Chini, T. Palacios, R. Coffie, L. Shen, H. Xing, S. Heikman, L. McCarthy, A. Chakraborty, S. Keller, and U.K. Mishra, *Applied Physics Letters*, **83 (23)**, 4779 (2003).
5. Chang Seok Oh, Tai Hong Kim, Kee Young Lim and Jeon Wook Yang, *Semicond. Sci. Technol.*, **19**, 172 (2004).
6. S. Salimian and C.B. Cooper III, *J. Vac. Sci. Technol. B*, **6 (6)**, 1641 (1988).
7. S. Salimian, C.B. Cooper III, R. Norton, and J. Bacon, *Appl. Phys. Lett.*, **51 (14)**, 1083 (1987).
8. R. Cheung, R. J. Reeves, B. Rong, S. A. Brown, E. J. M. Fakkeldij, E. van der Drift, M. Kamp, *J. Vac. Sci. Technol. B*, **17 (6)**, 2759 (1999).
9. B. Rong, R. Cheung, W. Gao and M. Kamp, *Microelectronic Engineering*, **53**, 419 (2000).

International Journal of High Speed Electronics and Systems
Vol. 14, No. 3 (2004) 762–768

LOW FREQUENCY NOISE PARAMETERS IN AN AlGaN/ GaN HETEROSTRUCTURE WITH 33% AND 75% Al MOLE FRACTION

S.A.VITUSEVICH, S.V.DANYLYUK AND N.KLEIN

Institut für Schichten und Grenzflächen and CNI - Centre of Nanoelectronic Systems for Information Technology, Forschungszentrum Jülich, D-52425 Jülich, Germany

M.V.PETRYCHUK

Taras Shevchenko National University, 01033 Kiev, Ukraine

A.E.BELYAEV

Institute of Semiconductor Physics, NASU, 03028 Kiev, Ukraine

A. VERTIATCHIKH AND L.F. EASTMAN

School of Electrical Engineering, Cornell University, Ithaca, New York 14853, U.S.A

Transport and low frequency noise properties of undoped AlGaN/GaN high electron mobility transistor (HEMT) heterostructures with 33% and 75% Al mole fractions in the ohmic and non-linear regimes of applied voltages are studied. In contrast to the low Al mole fraction, the noise properties of 75 % content structures are not affected by passivation. At small voltages both kinds of structures demonstrate about the same level of 1/f excess noise. Deviations from conventional flicker noise were observed at high applied voltages. Additionally, differences in noise behaviour between the two structures were revealed. In the 75% content structures, a noise level suppression was registered in the non-linear regime, which is important for the development of low noise oscillator circuits.

1 Introduction

Recent results [1] show that the spectral purity of non-linear RF oscillators based on AlGaN/GaN high electron mobility transistors (HEMTs) is mainly due to the up-conversion of low-frequency noise from the active device. Therefore, low noise active elements possessing both high performance and high stability are needed for devices and applications. It is known that the performance of $Al_xGa_{1-x}N$/GaN HEMTs with a low value of the Al mole fraction($x<0.35$) is strongly influenced by surface conditions determined by passivation type and surface treatment methods [2,3]. The influence of the surface states can be reduced by using modulation doping [4], but this process unfortunately increases the concentration of traps resulting in an enhancement of noise level in the structures. Non-intentionally doped $Al_xGa_{1-x}N$/GaN ($x<0.35$) HEMT structures demonstrated Hooge parameters that are comparable with those of AlGaAs/GaAs, but they are found to be strongly dependent on the surface condition.

In this paper transport and noise properties of undoped AlGaN/GaN HEMT heterostructures with 33% and 75% Al mole fractions are investigated in a wide range of applied voltages. In the 75% content structures, a noise level suppression was registered in the non-linear regime, which is important for the development of low noise oscillator circuits.

2 Experimental details

The investigated structures were designed for high frequency and low noise application. The high density two dimensional electron gas was created without doping, utilizing piezo- and spontaneous polarization effects in GaN/AlGaN heterostructures. Each structure consists of the following layers grown by metal-organic vapour phase epitaxy (MOVPE) on sapphire substrate: a nucleation AlGaN (16% Al) layer, an undoped GaN buffer layer and an AlGaN (33% Al or 75% Al) undoped barrier layer. The surface of the devices was passivated with Si_3N_4 or left without any passivation. Devices with different distances from 1 to 35 μm between the ohmic contact processed by Ti/Al/Ti/Au metallization annealed for 40 s at 800 ^{0}C were investigated. A room temperature mobility μ of 1250 cm^2/Vs(33%Al) and 700 cm^2/Vs(75%Al) at a sheet carrier density n of 1x10^{13} cm^{-2} (33%Al) and 2.5x10^{13}cm^{-2} (75%Al) for both structures was measured in the 2DEG of the channel by means of the Van der Pauw method.

3 Experimental results and discussion

Self consistently simulated potential profiles for both structures under investigation are shown in Fig.1 together with corresponding electron wavefunctions for electrons on the ground quantum level in the well. It is seen that in the structure with high Al content, the high barrier results in better electron confinement and larger electron concentration (2x10^{13} cm^{-2} for 75% vs 1.1x10^{13} cm^{-2} for 33%). The electron wavefunction does not penetrate into the barrier as much as in the case of the low barrier structure and one

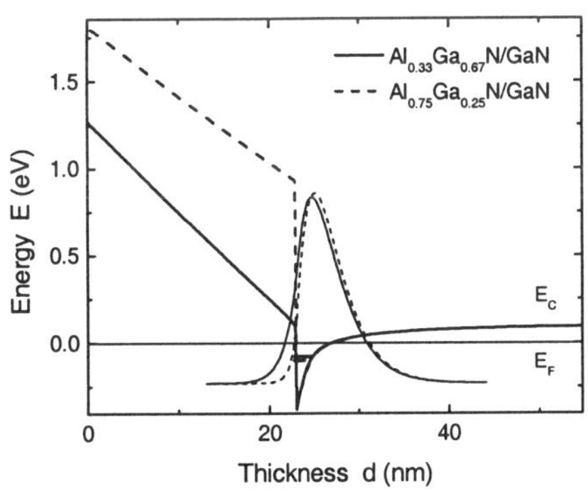

Fig.1. Self-consistently calculated potential profile and ground state electron wavefunctions for structures with low (solid lines) and high (dashed lines) Al content in the barrier.

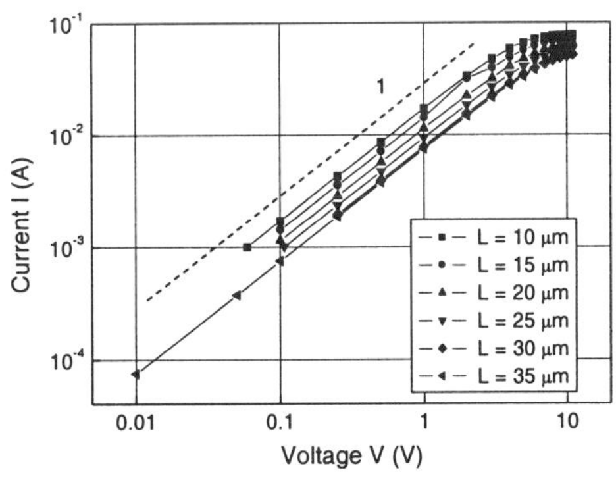

Fig.2. DC current-voltage characteristics of the $Al_{.75}Ga_{.25}N$ TLM structures with different length L between contacts.

should expect a reduction of the interaction of the channel's electrons with the barrier states.

To verify this assumption experimentally we performed a series of low-frequency noise measurements under different bias and surface conditions. DC current-voltage(I-V) characteristics were also measured to determine regions of linear and non-linear regimes of operation of the device. As is seen from the I-V characteristics, measured for structures with high Al content (Fig.2), all transmission line model (TLM) structures maintain a linear behaviour up to 1V DC bias. The DC characteristics of the structures with 33% Al content were also linear in this range of voltages. At the same, time strong deviations from linearity in I-V characteristics were observed for a further increase of applied voltage above 3V. Therefore, the separate analysis of the noise behaviour in the linear and non-linear regime is necessary.

At first we concentrate on low-frequency noise in the ohmic region of applied voltages. Measured voltage noise spectra for passivated and non-passivated devices with high Al content are shown in Fig.3. It is clear that the observed 1/f noise of the devices is only marginally better for a passivated device, while in the case of the structures with a 33% Al barrier, a passivation decrease of the noise level by one order of magnitude was observed. Therefore we can conclude that the influence of surface states on the conductivity of two-dimensional gas is greatly reduced in the devices with high Al content.

Current noise spectra being normalized to frequency for different TLM lengths are presented in Fig.4. It is remarkable that they demonstrate a non-linear dependence on the length of a structure, but the noise power follows an $S_I \sim I^2$ dependence.

If we plot the normalized noise power at 1kHz versus TLM length (Fig.5) it becomes apparent that the noise dependence on the sample length is quadratic up to L=25 μm. Usually a quadratic length dependence of the noise is observed when dominant noise sources are not distributed uniformly through the device, but localized in the part of fixed

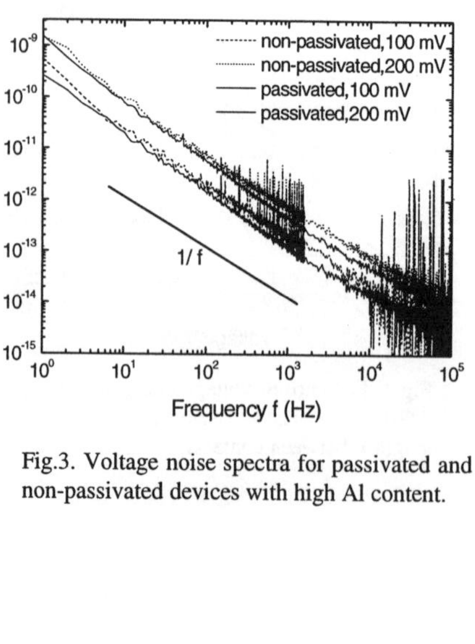

Fig.3. Voltage noise spectra for passivated and non-passivated devices with high Al content.

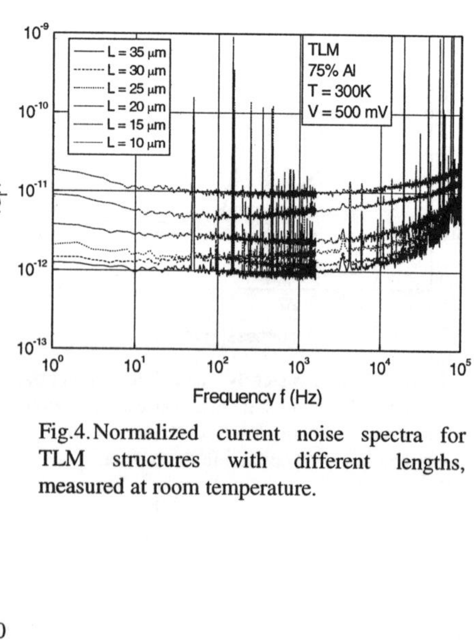

Fig.4. Normalized current noise spectra for TLM structures with different lengths, measured at room temperature.

length[5]. Most commonly, it is observed in structures where the contact barrier noise is dominant. However, for contact barrier noise, strong deviations from the quadratic dependence of the noise on current are also visible, which is not the case here. The contact resistance in the high Al content structures is the same as for low Al content ones, so we assume that in our case contact noise is generated not by the contact barrier, but in channel regions adjacent to the contacts. Their increased noisiness could be caused by a process technology not being optimized for a high Al content barrier contact. Indeed, in this case the role of the additional noise from these regions will be dominant in the short sample and should be suppressed in the long one.

The 33% Al sample's noise is not dependent on the channel length, but is more sensitive to the passivation. To compare the two structures, we calculated the dependence of the Hooge parameter [6] ($\alpha_H = \frac{S_I^0(f)}{I^2} fN$) on TLM length for a passivated low-barrier and a non-passivated high barrier device (Fig.6).

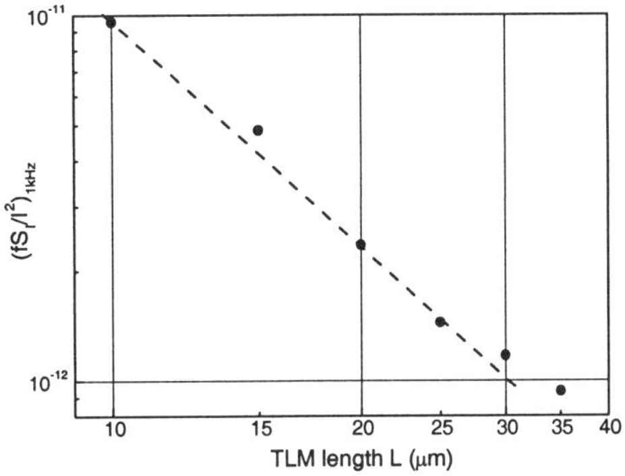

Fig.5. Dependence of normalized current noise at f=1kHz on sample length. Dashed line represents quadratic dependence.

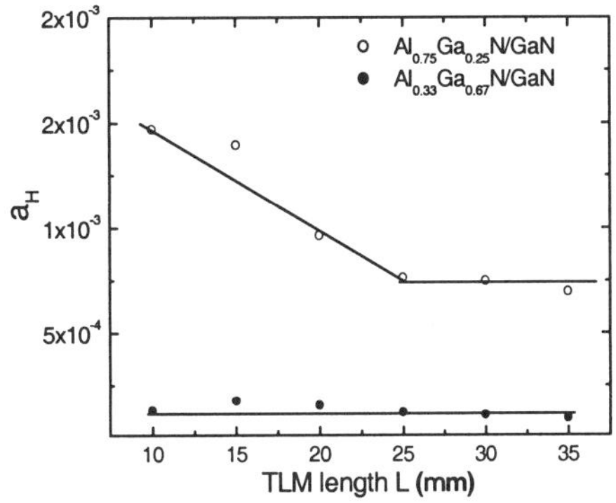

Fig.6. The Hooge coefficients for low barrier and high barrier devices as a function of device length.

The noise level of the low barrier structures does not depend on TLM length L, while for the high barrier structures the Hooge coefficient gradually decreases with TLM length for short samples (L>25 μm) and then remains constant for longer samples at a value four times higher than for conventional low barrier devices. This fact is due to the dominant role of the channel noise for longer samples while for short samples additional noisiness is introduced by the near-contact region of the device. We can assume that with an improved contact processing technology the noisiness of the high barrier devices can become independent of the length of the device and be on a par with the value for passivated low barrier structures. Combined with decreased sensitivity of the channel to surface conditions, this makes a high barrier structure a viable system for the low-field applications.

The most important range of applied voltages is the high voltage regime, where AlGaN/GaN heterostructures are utilized for high power device application. Therefore, from a practical point of view an investigation of the noise behaviour of the system in the non-linear regime is even more important. Normalized noise spectra measured under high biases for high barrier devices are presented in the Fig.7.

Strong deviations from the 1/f behaviour are observed. The noise measurement of the low barrier device also revealed deviations from 1/f behaviour. For analysis and comparison, we plot two characteristic curves for both types of devices, measured under identical conditions (Fig.8). The longest device was chosen to eliminate an influence of the contact regions and to compare a pure channel noise.

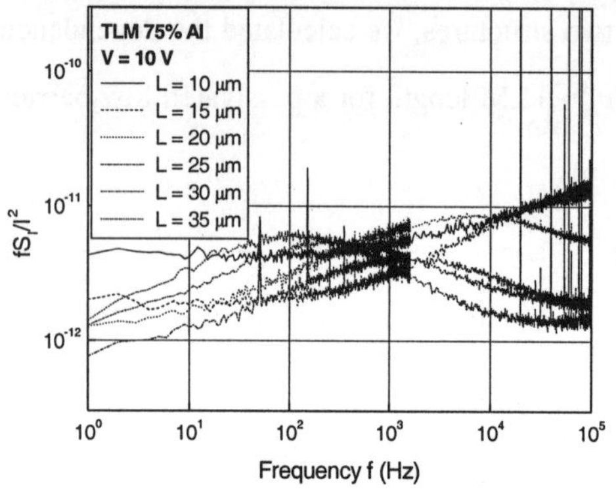

Fig.7. Low frequency noise spectra of the high barrier device under high bias conditions (U=10V) for different lengths of the device.

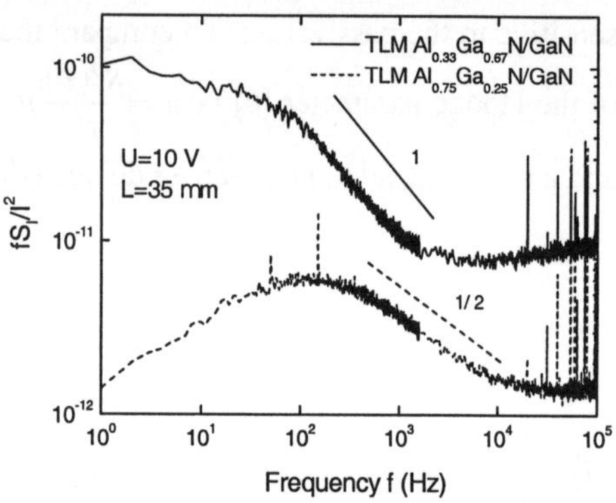

Fig.8. The low frequency noise of low Al content (solid line) and high Al content (dashed line) devices, measured for a TLM structure with length L=35 μm, at bias U=10V and T=300K.

It can be seen from Fig.8 that both devices demonstrate 1/f dependence in the high frequency part of the spectrum, but in both cases at lower frequencies additional noise appears. However, in low Al content devices the transition from low noise to high noise region is proportional to $1/f^2$ (corresponding to unity slope in normalized spectra in Fig.8), while in high Al content devices it exhibits $1/f^{3/2}$ behaviour ($f^{-1/2}$ slope in Fig.8). Additionally, in high barrier devices a suppression of the noise in the lowest frequency range is observed.

Surprisingly, the absolute value of noise for the high Al content device for the whole frequency range is at least one order of magnitude lower than for the device with 33% Al content. This was not the case for low-field measurements, where noise from the conventional device was marginally lower.

To explain this behaviour, the origin of high-field noise for both samples should be revealed. A key to this is the slope of the transition regions. It was reported earlier [7] that the $1/f^2$ dependence may be evidence of thermoactivation processes, which modulates the noise behaviour. At lower frequencies, when inverse frequency becomes comparable with charging/decharging times of deep traps in the barrier, one can indeed expect an increase of the noise level. If this trapped charge fluctuates not only between a trap and a channel, but also hopping conductivity can be realized in the barrier between traps, then a $1/f^2$ dependence of noise power could be observed. As follows from the results of simulations and low-field noise measurements, such a mechanism would be suppressed in devices with a high AlGaN barrier. Indeed, in high Al content samples instead of $1/f^2$ dependence we observe $1/f^{3/2}$ behaviour. This dependence is typical of diffusion-like noise [8] and may be caused by diffusion of a channel's charge in the GaN layer. Obviously, such noise should also be present in $Al_{.33}Ga_{.67}N$ devices but it is masked by the noise from the traps in the barrier layer. In case of high barrier devices, the latter is strongly reduced and the total noise is only influenced by diffusion-assisted processes of the device, thus resulting in lower noisiness of the structure at non-linear regimes of operation.

4 Conclusion

We have investigated the noise properties of AlGaN/GaN TLM structures with high (75%) and low (33%) Al content. The level of channel noise in the linear regime for unpassivated high barrier devices was found to be comparable with passivated low-barrier devices. The origin of high-voltage noise for low- and high barrier devices was revealed to be different. Non-equilibrium noise of the high barrier devices demonstrates $1/f^{3/2}$ dependence, which is characteristic of diffusion-assisted generation-recombination noise, while in low barrier devices noise coming from traps in the barrier is dominant. The noise level in the non-linear regime was observed to be one order of magnitude lower than for conventional devices.

Acknowledgements

This work is supported by Deutsche Forschungsgemeinschaft (Project KL 1342) and in part by the Office of Naval Research under Grant No. N00014-01-1-0828 (Project Monitor Dr. Colin Wood).

References

1. P.Rice, R.Sloan,M.Moore, A.R.Barnes, M.J.Uren, N.Malbert, N.Labat, " A 10 GHz dielectric resonator oscillator using GaN technology", 2004 MTT-S Symposium Digest,pp.1497-1500.
2. S.Arulkumaran, T.Egawa, H.Ishikawa, T.Jimbo, Y.Sano. "Surface passivation effects on AlGaN/GaN high-electron-mobility transistors with SiO_2, Si_3N_4 and silicon oxynitride", Appl.Phys.Lett., Vol.84,No.4,pp.613-615,January 2004.
3. N.A.Moser,J.K.Gillespie, G.D.Via, A.Crespo, M.J.Yannuzzi,G.H.Jessen,R.C.Fitch, B.Luo, F.Ren,B.A.Gila,A.H.Onstine, C.R.Abernathy,S.J.Pearton, "Effect of surface treatment on isolation currents in AlGaN/GaN high-electron-mobility transistors", Appl.Phys.Lett., Vol.83,No.20,pp.4178-4180,November 2003.

4. J.Bernat, M.Wolter,A.Fox,M.Marso,J.Flynn,G.Brandes,P.Kordos."Impact of layer structure on performance of unpassivated AlGaN/GaN/SiC HEMTs", Electron.Lett. Vol.40,No.1,pp.78-80,January 2004.
5. S. L. Rumyantsev, N. Pala, M.S.Shur, R.Gaska, M.E. Levinshtein, M. A. Khan, G.Simin, X. Hu and J. Yang. "Low frequency noise in GaN metal semiconductor and metal oxide semiconductor field effect transistors", J. Appl. Phys., Vol.90, No.1,pp.310-314,July 2001.
6. F.N.Hooge, T.G.M. Kleinpenning, and L.K.J.Vandamme. " Experimental studies on 1/f noise", Rep.Prog.Phys., Vol.44,No.5,pp. 479-532,1981.
7. S.A. Vitusevich, S.V. Danylyuk, N. Klein , M.V. Petrychuk, V.N. Sokolov, V.A. Kochelap, A.E. Belyaev, V. Tilak, J. Smart, A. Vertiatchikh and L. F. Eastman., "The study of excess low-frequency noise sources in nitride-based high-electron-mobility transistors". Proc. 26th International Conference on the Physics of Semiconductors (ICPS26), July 29-August 2, Edinburgh, UK, (2002),pp.171-176.
8. Z-M.Li, S.P. McAlister, D.J.Day, "Analytical Model of Low-Frequency Diffusion Noise in GaAs MESFET's" IEEE Trans. Elect.Dev. Vol.38, No.2, p.p. 232-236, February 1991.

International Journal of High Speed Electronics and Systems
Vol. 14, No. 3 (2004) 769–774

TRAP BEHAVIOR IN AlGaN/GaN HEMTs BY POST-GATE-ANNEALING

HYEONGNAM KIM, JAESUN LEE, AND WU LU†

Department of Electrical and Computer Engineering, The Ohio State University
Columbus, Ohio 43210, United States of America, †E-mail:lu@ee.eng.ohio-state.edu

Trapping effects are investigated to examine the post-gate annealing effects on AlGaN/GaN high-mobility electron transistors (HEMTs) using pulsed I-V and transient measurements. In the unannealed devices, shallow traps are identified, which have an activation of 38 meV at a drain bias of 7 V. The time constant of these traps is determined to be ~0.5 μs. Devices annealed at 400 °C for 10 minutes have a significantly smaller number of traps. However, a small number of traps with a longer time constant of 9.2 μs are created or activated during post-gate annealing. 20-minute annealing at 400 °C leads to the increase of the number of traps with emission time constants of 21.6 μs and 1.25 ms. The breakdown voltage improvement by post-gate annealing is attributed to the removal or significant reduction of the shallow level traps.

1 Introduction

GaN-based high electron mobility transistors (HEMTs) have great potential for high power and high frequency applications because of their inherent material nature as demonstrated recently.[1-3] However, critical issues including current dispersion and device reliability have limited the AlGaN/GaN HEMTs for practical applications. Current dispersion due to trapping effects associated with these devices causes decrease in the actual output drain current and voltage swing at RF operations.[4] To reduce the trapping effects, several approaches on device fabrication process and material growth have been performed including: e.g., surface passivation, a thin AlN barrier layer in AlGaN/GaN interface, application of a field plate, and Si doping on GaN/AlGaN/GaN heterostructures.[3, 5-7] Moreover, attempts to increase in breakdown voltage have been made by using a field plate and gate dielectrics.[8] Recently, we demonstrated that post-gate-annealing (PGA) process can improve AlGaN/GaN HEMTs dramatically in gate leakage current, breakdown voltage performance,[9] and even the maximum drain current under an optimized processing condition. These phenomena are mostly related to traps due to imperfection of the AlGaN/GaN heterostructures. Thus, investigations of trapping effects are of importance for any novel processing techniques for high performance AlGaN/GaN HEMTs.[10] In this paper, trapping effects are investigated to understand effects of the post-gate annealing, using short drain current transient and static/dynamic current-voltage (I-V) measurements.

2 Device Fabrication and Pulsed Measurement Methods

The AlGaN/GaN HEMT epilayer structure was grown by metal-organic chemical vapor deposition on a SiC substrate. The epilayer consists of 40 nm AlN nucleation layer, 3 μm of undoped GaN, and 20 nm undoped $Al_{0.3}Ga_{0.7}N$. Device fabrication procedures follow the standard process such as mesa isolation by dry etching, Ti/Al/Mo/Au ohmic contact metallization by electron beam evaporation and rapid thermal annealing, and Ni/Au Schottky gate with a gate-length of 0.2 μm by electron beam lithography. The devices have a gate width of 100 μm and a source to drain spacing of 3 μm. To investigate PGA effects on AlGaN/GaN HEMTs, the wafer was cleaved to three samples. Sample 1 was not annealed for reference. Samples 2 and 3 were annealed at 400 °C for 10 and 20 minutes, respectively, in a furnace with N_2 ambient. After annealing, the breakdown voltage of the annealed samples 2 and 3 increased dramatically from ~ 35 V to higher than 180 V. The maximum drain current of sample 2 in the DC I-Vs had improvement but the sample 3 had slightly degradation in comparison with the reference sample [9, 11]. The pulsed I-Vs are characterized at different QB conditions; (V_{DS0}, V_{GS0}) = (0 V, 0 V), (0 V, -5 V), and (7 V, -5 V) as shown in Fig. 1. Under these three QBs, there is essentially no drain current flowing in the channel. Electrons are depleted at QBs of (V_{DS0}, V_{GS0}) = (0 V, -5 V) and (7 V, -5 V) and some of them are trapped, while there is no electron trapping at (V_{DS0}, V_{GS0}) = (0 V, 0 V). The pulse widths (PWs) of 200 ns, 500 ns, 10 μs, and 100 μs are used with a pulse separation of 1 ms. The DC and pulsed I-V curves measured at V_{GS} = 0 V are used to investigate the trapping behavior of three samples.

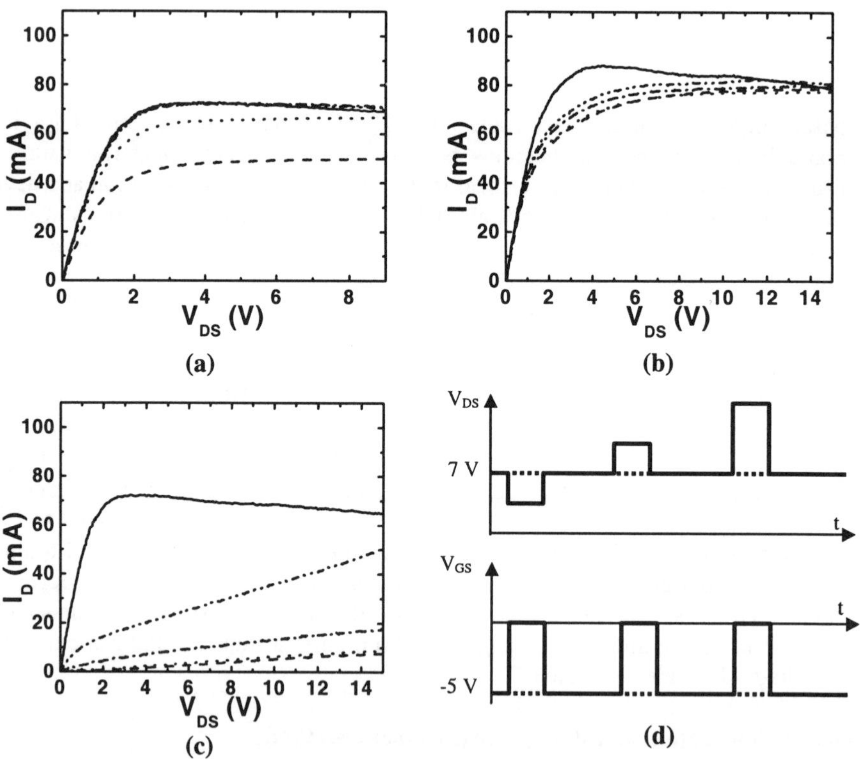

Fig. 1. DC I-V (solid) and pulsed I-V at V_{GS} = 0V and a quiescent bias point of (V_{DS0}, V_{GS0}) = (7 V, -5 V) at pulse widths of 0.2 μs (dash), 0.5 μs (dot), 10 μs (dash dot), and 100 μs (dash dot dot). [pulse separation: 1ms]: (a) unannealed HEMTs, HEMTs annealed at 400°C for 10 minutes (b) & 20 minutes (c), and (d) bias conditions

3 Results and Discussion

Figure 1 shows DC I-V and pulsed I-V with different pulse widths at V_{GS} = 0 V at a quiescent bias point of (V_{DS0}, V_{GS0}) = (7 V, -5 V). Under this quiescent bias, some electrons are captured by traps. At a longer pulse width, the drain current in pulsed I-Vs of each sample increases. If the pulse width is shorter than the emission time constant (t_E), electrons captured by traps do not have enough time to be fully emitted. But if the pulse width is long enough, all the trapped electrons are detrapped and contribute to drain current. The current dispersion between DC and the 0.2 μs pulsed I-Vs in the unannealed HEMTs (Fig.1(a)) is much larger than that in the 10-minute annealed devices, indicating a larger number of traps are removed during the post-gate annealing. The difference between DC I-V and 10/100 μs pulse I-V curves in the unannealed devices is negligible. But I_D dispersion in the annealed samples 2 and 3 is observed, which suggests that a small number of traps with a long time constant are created during the post-gate annealing, though the total number of traps is significantly smaller than that of unannealed devices. In 100 μs pulsed I-Vs, severer current dispersion in 20-minute

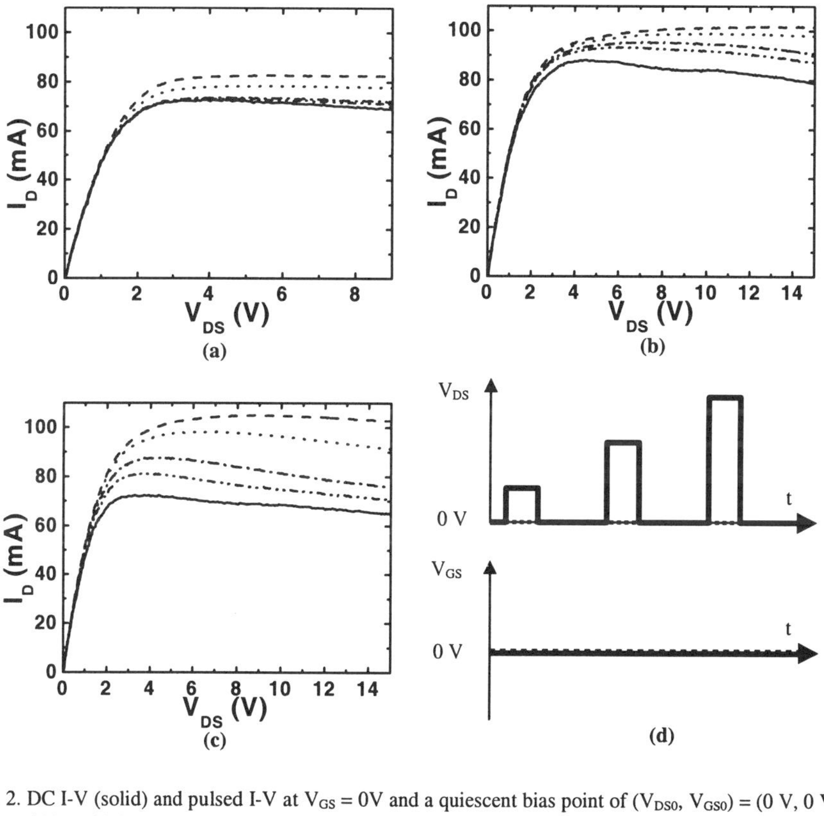

Fig. 2. DC I-V (solid) and pulsed I-V at V_{GS} = 0V and a quiescent bias point of (V_{DS0}, V_{GS0}) = (0 V, 0 V) at pulse widths of 0.2 μs (dash), 0.5 μs (dot), 10 μs (dash dot), and 100 μs (dash dot dot). [pulse separation: 1ms]: (a) unannealed HEMTs, HEMTs annealed at 400°C for 10 minutes (b) & 20 minutes (c), and (d) bias conditions

annealed HEMTs indicates that further annealing creates a much larger density of traps with a long time constant.

Fig. 2 shows DC and pulsed I-V characteristics at V_{GS} = 0 V and a quiescent bias point of (V_{DS0}, V_{GS0}) = (0 V, 0 V) with different pulse widths. As the pulse width increases, decrease in the saturation drain current is observed in the pulsed I-Vs of three samples, due to longer time available for electrons to be captured by traps. Dispersion of the saturation current between the 0.2 μs pulsed I-V and DC I-V of sample 1 and 2 is smaller than that of sample 3. This suggests that the devices annealed at 400 °C for 20 minutes have a much larger density of traps with a capture time constant (t_C) of longer than 0.2 μs. The dispersion between 10 ~ 100 μs pulsed I-Vs and DC I-Vs are observed in both annealed HEMTs. Again, this indicates that the post-gate annealing remove traps with a short capturing time constant and creates traps with a long capturing time constant.

Fig. 3 shows short drain current transient normalized by DC drain current (I_{DC}) at (V_{DS}, V_{GS}) = (7 V, 0 V). The bias was pulsed up from at (V_{DS}, V_{GS}) = (7 V, -5 V) to (7 V, 0 V). The unannealed devices show the clear transient within a few μs after pulsing-up. Based on these transient measurements, emission time constants are extracted using the following exponential decay functions:

$$I_D(t) = I_{DC}(1 - I_0 \exp[-t/\tau]) . \quad (1)$$

$$I_D(t) = I_{DC}(1 - I_1 \exp[-t/\tau_1] - I_2 \exp[-t/\tau_2]) . \quad (2)$$

where I_{DC} is the DC drain current at (V_{DS}, V_{GS}) = (7 V, 0 V), τ is time constant. The transient spectra of unannealed devices and 10-minute annealed devices fit well to Eq. (1). The transient behavior of devices annealed for 20 minutes show better fitting to Eq. (2) with double exponential decay functions. The extracted time constants of the unannealed and 10-minute annealed devices are ~ 0.5 μs and ~ 9.2 μs, respectively. The 20-minute annealed devices have emission time constants of ~ 21.6 μs and ~ 1.25 ms. These extracted time constants show clearly that the post-annealing process removes traps with a short t_E and induces traps with longer t_E.

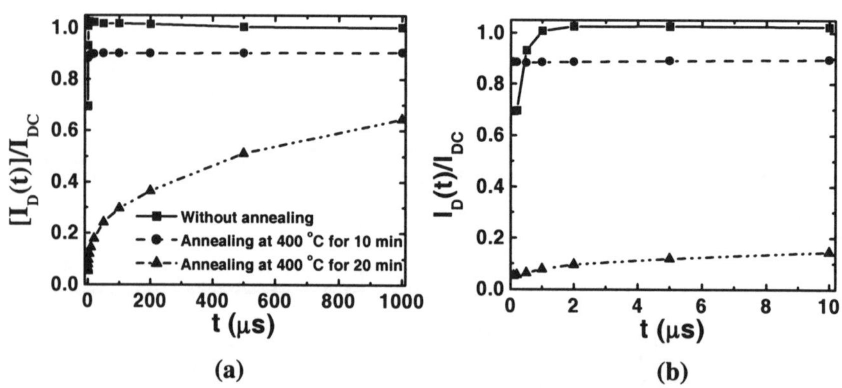

Fig. 3. (a) Normalized transient characteristics of the unannealed HEMTs (solid line), and HEMTs annealed at 400°C for 10 minutes (dashed line) & 20 minutes (dash-dot-dot line) and (b) transients in the first 10 μs. The biases are pulsed up from (V_{DS0}, V_{GS0}) = (7 V, -5 V) to (V_{DS}, V_{GS}) = (7 V, 0 V).

Fig. 4 shows normalized drain current transient of AlGaN/GaN HEMTs annealed at different conditions. The biases were pulsed down from (V_{DS0}, V_{GS0}) = (7 V, 0 V) to (V_{DS}, V_{GS}) = (7 V, -3 V). The unannealed and 10-minute annealed HEMTs don't exhibit exponential capturing process. This may be because the traps have extremely short capture time constants. However, the 20-minute annealed HEMTs show exponential decay electron capturing process. The capture time constants of ~ 0.85 μs, ~21 μs, and ~2.2 ms are extracted by fitting to a triple exponential decay function.

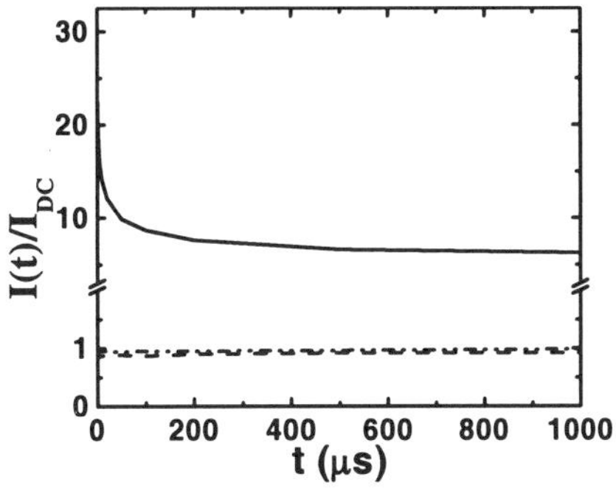

Fig. 4. (a) Normalized transient characteristics of the unannealed HEMTs (dashed line), and HEMTs annealed at 400°C for 10 minutes (dash-dot-dot line) & 20 minutes (solid line). The biases are pulsed up from (V_{DS0}, V_{GS0}) = (7 V, 0 V) to (V_{DS}, V_{GS}) = (7 V, -3 V).

Fig. 5 shows a typical short emission transient curve at room temperature and the temperature dependence of the emission time constant in the unannealed HEMTs measured from 295 K to 363 K. The temperature-dependence of t_E is shown by using the classical Arrhenius equation:

$$\tau^{-1} = CT^2 \exp(-E_A/kT)\,. \tag{3}$$

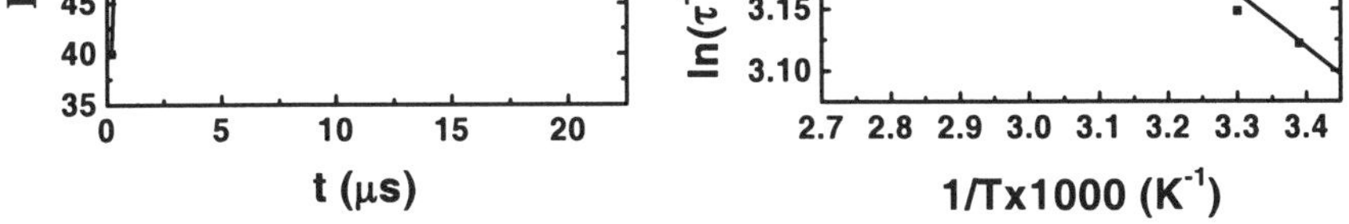

(a) (b)

Fig. 5. (a) Typical short transient characteristic of the unannealed devices at 295 K; and (b) temperature dependence of the time constant extracted from the drain current transient curves at temperatures ranging from 295 K to 363K. The gate is pulsed from V_{GS0} = -5 to V_{GS} = 0 V and the drain voltage is 7 V.

where T is the temperature in Kelvin, E_A is the activation energy, τ is the time constant, k is the Boltzmann constant, and C is a fitting factor. The extracted activation energy of those traps with short time constants is determined to be ~ 38 meV at an electric field of 0.35 MV/cm. Such shallow level traps result in the current dispersion and a low breakdown voltage of the unannealed HEMTs. So, the significant improvement in breakdown voltage of the annealed devices is attributed to the removal of these shallow traps during the post-gate annealing.

4 Conclusions

We have investigated the post-gate annealing effects on electron capturing and emission processes of AlGaN/GaN HEMTs using dynamic and transient measurements. In the unannealed HEMTs, traps have a low activation energy of ~38 meV at V_D = 7 V. The emission time constant of these traps is about 0.5 μs. After annealing, a significantly smaller number of these shallow traps are removed. However, a small number of traps with longer emission time constants (9.2 μs for devices annealed at 400 °C for 10 minutes, and 21.6 μs and 1.25 ms for devices annealed at 400 °C for 20 minutes) are created. The breakdown voltage improvement by post-gate annealing is attributed to the removal of the shallow level traps.

Acknowledgements

This work was supported by the National Science Foundation Grants DMR-0216892 and ECS-0401305, and by Accent Optical Technologies Inc. through DiVA Grant Program.

References

1. W. Lu, et al., "AlGaN/GaN HEMTs on SiC with over 100 GHz f_T and low microwave noise," *IEEE Trans. Electron Dev.* **48** (2001) 581-585.
2. V. Kumar, et al., "AlGaN/GaN hemos on SiC with f_T of over 120 GHz," *IEEE Electron Dev. Lett.* **23** (2002) 455-457.
3. Y.-F. Wu, et al., "30-W/mm GaN HEMTs by field plate optimization," *IEEE Electron Dev. Lett.* **25**, (2004) 117-119.
4. C. Nguyen, et al., "Drain current compression in GaN MODFETs Ander large-signal modulation at microwave frequencies," *Electron. Lett.* **35** (1999) 1380-1382.
5. Y. Ohno, et al., "Effects of surface passivation on breakdown of AlGaN/GaN high-electron-mobility transistors," *Appl. Phys. Lett.* **84**, (2004) 2184-2186.
6. J. S. Lee, et al., "Reduction of current collapse in AlGaN/GaN HFETs using AlN interfacial layer," *Electron. Lett.* **39**, (2003) 750-752.
7. O. Mitrofanov, et al., "Impact of Si doping on radio frequency dispersion in unpassivated GaN/AlGaN/gaN high-electron-mobility transistors grown by plasma-assisted molecular-beam epitaxy," *Appl. Phys. Lett.* **82**, (2003) 4361-4363.
8. Z. Y. Fan, et al., "Delta-doped AlGaN/GaN metal-oxide-semiconductor heterostructure field-effect transistors with high breakdown voltages," *Appl. Phys. Lett.*, vol. **81** (2002) 4649-4651.
9. J. Lee, et al., "Post Annealing Effects on Device Performance of AlGaN/GaN HFETs," *Solid State Electron.* **48** (2004) 1855-1859.
10. S. De Meyer, et al., "Mechanism of power density degradation due to trapping effects in AlGaN/GaN HEMTs," *Microwave Symposium Digest, IEEE MTT-S Digest* **1** (2003) 455-458.
11. J. Lee, et al., Submitted to Applied Physics Letters.

International Journal of High Speed Electronics and Systems
Vol. 14, No. 3 (2004) 775–784

PHOTOCAPACITANCE OF GaAs THIN-FILM STRUCTURES FABRICATED ON A SEMI-INSULATING COMPENSATED SUBSTRATE

NIKOLAI B. GOREV, INNA F. KODZHESPIROVA, EVGENY N. PRIVALOV

Department for Functional Elements of Control Systems, Institute of Technical Mechanics of the National Academy of Sciences of Ukraine, 15 Leshko-Popel' St., Dnepropetrovsk, 49005, Ukraine

NINA KHUCHUA, LEVAN KHVEDELIDZE

Research and Production Complex (RPC) "Electron Technology", Tbilisi State University, 13 Chavchavadze Ave., Tbilisi, 380079, Georgia

MICHAEL S. SHUR

Center for Broadband Data Transport, Room 9017, CII, Rensselaer Polytechnic Institute, 110 8-th Street, Troy, NY 112180-3590, USA

The results of analytical and numerical calculations of the low-frequency and the high-frequency barrier capacitance of GaAs epitaxial and ion-implanted structures fabricated on a semi-insulating compensated substrate are presented. The calculations are done for the samples in the dark and under extrinsic illumination. It is shown that the high-frequency photocapacitance of these structures exhibits a narrow peak, and the low-frequency photocapacitance has a positive peak followed by a negative valley. The underlying physical mechanisms are discussed. Methods for predicting the MESFET threshold voltage and extracting the concentration of vacant deep traps in the vicinity of the film–substrate interface and the epitaxial film thickness using capacitance–voltage measurements under extrinsic illumination are proposed.

1 Introduction

Photocapacitance measurements are widely used in semiconductor characterization. Existing photocapacitance techniques have been developed mostly for bulk semiconductors, where the Schottky barrier depletion region borders the neutral semiconductor. However, semiconductor structures with thin-film devices, such as MESFETs, feature steep dopant gradients, which give rise to built-in space charge regions. Due to the presence of such regions, the mechanism of photoresponsivity of these structures is quite different from that of bulk semiconductors [1,2], and the barrier capacitance cannot be described using the conventional depletion approximation assuming a sharp boundary for the depletion region [3]. The goal of this paper is to study the photocapacitance features caused by the space charge region at the doped layer–substrate interface and to propose ways to use these features for materials and device characterization.

2 Calculation of the Barrier Capacitance and Photocapacitance of Thin-Film Structures

The low-frequency and the high-frequency barrier capacitances $C_{LF,\,HF}$ can be expressed in terms of the electric field E_b at the metal–semiconductor interface as

$$C_{LF} = \varepsilon S_b \frac{dE_b}{dV_{rev}}, \qquad C_{HF} = \varepsilon S_b \frac{\delta E_b}{\delta V_{rev}}\bigg|\delta n_t = 0 \,. \tag{1}$$

Here, ε is the relative permittivity of the semiconductor, S_b is the barrier contact area, V_{rev} is the reverse bias, and δE_b and δn_t are the variations of E_b and of the trapped carrier density, respectively, with the variation δV_{rev} of the reverse bias.

Let us consider a GaAs epitaxial structure with a top Schottky contact. The structure consists of a low-resistivity film of thickness h grown on a semi-insulating compensated substrate of infinite thickness. The film contains only shallow donors with concentration N_d, whereas the substrate contains shallow donors and deep acceptor-type electron traps with concentrations N_s and N_t, respectively. Let a reverse bias, V_{rev}, be applied to the Schottky contact.

The distribution of the electric field E and the free carrier density n is described by the following set of equations

$$\frac{dE}{dx} = \frac{q}{\varepsilon}\begin{cases} n - N_d, & x < h \\ n + N_t \dfrac{n}{n + n_1} - N_s, & x > h \end{cases} \tag{2}$$

$$\frac{dn}{dx} = \frac{q}{kT} nE \tag{3}$$

Here, x is the coordinate across the structure, q is the electron charge, k is the Boltzmann constant, T is the absolute temperature and n_1 is the free carrier density in the conduction band when the trap energy level coincides with the Fermi level ($n_1 = N_c \exp(-W_t/kT)$ where W_t is the trap ionization energy and N_c is the effective density of states in the conduction band).

The functions E and n are continuous at $x = h$

$$E(h-0) = E(h+0), \quad n(h-0) = n(h+0), \tag{4}$$

and they satisfy the following boundary condition in the bulk of the substrate

$$\underset{x\to\infty}{E \to 0} \Leftrightarrow \underset{x\to\infty}{n \to n_{sub}} \tag{5}$$

where n_{sub} is the free carrier density in the bulk of the substrate related to N_t, N_s and n_1 by the charge neutrality condition

$$n_{sub} + N_t \frac{n_{sub}}{n_{sub} + n_1} = N_s. \tag{6}$$

In the frame of the thermionic model of Schottky barrier, the boundary condition at the metal–semiconductor interface is given by

$$V_{rev} = \frac{kT}{q} \ln \frac{N_c}{n_b} - \varphi_b / q \tag{7}$$

where φ_b is the Schottky barrier height and n_b is the free carrier density at the metal–semiconductor interface.

We assume that V_{rev} is such that the depletion regions of the Schottky barrier and of the film–substrate junction merge at some point x_m in the film. At this point, the free carrier density n reaches its peak value $n_m < N_d$.

Eliminating x in Eqs. (2) and (3) allows us to readily integrate these equations, and, accounting for Eqs. (4) and (5), we obtain the following expressions for the electric field E_b and E_j at the metal–semiconductor interface and at the film–substrate interface, respectively

$$E_b = \frac{\sqrt{2}kT}{ql_d} \left(\frac{n_b - n_m}{N_d} + \ln \frac{n_m}{n_b} \right)^{1/2}, \quad E_j = -\frac{\sqrt{2}kT}{ql_d} \left(\frac{n_j - n_m}{N_d} + \ln \frac{n_m}{n_j} \right)^{1/2}, \tag{8}$$

$$l_d \sqrt{\frac{N_d}{2}} \left(\int_{n_b}^{n_m} \frac{dn}{n\left(n - n_m + N_d \ln \frac{n_m}{n}\right)^{1/2}} + \int_{n_j}^{n_m} \frac{dn}{n\left(n - n_m + N_d \ln \frac{n_m}{n}\right)^{1/2}} \right) = h, \tag{9}$$

$$n_j = n_m \left(\frac{n_{sub}}{n_m} \right)^{\frac{N_t - N_s}{N_d + N_t - N_s}} \left(\frac{N_s}{N_t} \right)^{\frac{N_t}{N_d + N_t - N_s}} \exp\left(-\frac{n_m}{N_d + N_t - N_s} \right). \tag{10}$$

Here, $l_d = (\varepsilon kT / q^2 N_d)^{1/2}$ is the Debye length in the film and n_j is the free-carrier density at the film–substrate interface.

The integrals in Eq. (9) can be evaluated analytically if we expand the logarithm in their integrands in a power series of $n/n_m - 1$ in the vicinity of n_m and ignore n in comparison with the logarithm for $n << n_m$. Direct calculation shows that if $n = n_m/3$ is chosen as the boundary between the two approximations, the overall accuracy of approximation is reasonably good. As long as the space charge at the film–substrate interface has a dipole-like structure, the electric field E_j at this interface is mainly governed by the trapped charge in the accumulation region of the film–substrate junction, and thus it remains fixed in the high-frequency regime.

Based on the aforesaid, we can readily find both the low-frequency and the high-frequency barrier capacitance as a function of V_{rev} from Eq. (1) in the following parametric form

$$C_{LF,HF} = \frac{\sqrt{2}\varepsilon S_b}{l_d} \frac{B_{LF,HF}}{2AB_{LF,HF} + N_d/n_m - 1}, \tag{11}$$

$$V_{rev} = \frac{kT}{q}\left(\ln\frac{N_c}{n_m} + \frac{n_m}{N_d} + A^2\right) - \varphi_b/q \tag{12}$$

where

$$A = \begin{cases} A_0 + 2A_1 - \left(\ln\frac{n_m}{n_j} - \frac{n_m}{N_d}\right)^{1/2}, & n_m/n_j \geq 3 \\ A_0 + A_1 - \frac{1}{\sqrt{2(1+2p)}}\ln\frac{\left(\sqrt{1+2p-n_j/n_m} + \sqrt{(1+2p)(1-n_j/n_m)}\right)^2}{2pn_j/n_m}, & n_m/n_j < 3 \end{cases}$$

$$A_0 = \frac{h}{\sqrt{2}l_d}, \quad A_1 = \left(\ln 3 - \frac{n_m}{N_d}\right)^{1/2} - \frac{1}{\sqrt{2(1+2p)}}\ln\frac{\left(\sqrt{1+3p} + \sqrt{1+2p}\right)^2}{p},$$

$$p = 1 - \frac{n_m}{N_d}, \qquad B_{LF,HF} = -N_d\left(\frac{\partial A}{\partial n_m} + \frac{\partial A}{\partial n_j} D_{LF,HF}\right), \tag{13}$$

$$D_{LF} = \frac{dn_j}{dn_m}, \qquad D_{HF} = -\frac{\partial E_j}{\partial n_m}\Big/\frac{\partial E_j}{\partial n_j}.$$

The above expressions are valid as long as the free carrier density in the film shows a pronounced peak. Another analytic form for the low–frequency capacitance–voltage (*C-V*) characteristic can be obtained for the reverse bias range, in which the film is depleted of free carriers, i.e., the free carrier charge in the film can be ignored. In this case too, Eqs. (2) and (3) can be solved analytically to yield the following parametric form:

$$C_{LF} = \frac{\varepsilon S_b}{h - \frac{kT}{q}\frac{1}{n_j}\left(\frac{d|E_j|}{dn_j}\right)^{-1} \operatorname{sgn}(n_{sub} - n_j)}, \tag{14}$$

$$V_{rev} = \frac{kT}{q}\left[\ln\frac{N_c}{n_j} + \frac{1}{2}\left(\frac{h}{l_d}\right)^2 + \frac{qh}{kT}\left|E_j\right|\mathrm{sgn}\left(n_{sub} - n_j\right)\right] - \varphi_b/q \tag{15}$$

where

$$\left|E_j\right| = \frac{\sqrt{2}kT}{ql_d}\left(\frac{n_j - n_{sub}}{N_d} + \frac{N_t}{N_d}\ln\frac{n_j + n_1}{n_{sub} + n_1} - \frac{N_s}{N_d}\ln\frac{n_j}{n_{sub}}\right)^{1/2}. \tag{16}$$

Eqs. (14) and (15) remain valid after the disappearance of the free carrier density peak in the film, i.e. after the suppression of the film–substrate junction by the reverse bias. This makes it possible to calculate the low-frequency *C-V* characteristic over a wide voltage range by joining the dependences given by Eqs. (11)–(12) and (14)–(15).

Using the rate equation for the occupied deep traps in the presence of extrinsic illumination, it can readily be shown that extrinsic illumination can be accounted for by replacing the characteristic concentration n_1 with $n_{1I} = n_1 + S_pI/c_e$ (here, S_p is the photon capture cross-section, I is the illumination intensity and c_e is the electron capture coefficient) and replacing n_{sub} with n_{subI} related to by Eq. (6). This makes it possible to calculate the barrier capacitance C_I under extrinsic illumination and, hence, the photocapacitance $C_{ph} = C_I - C_0$ (hereinafter, the subscripts I and 0 will denote the presence and the absence of extrinsic illumination, respectively).

For comparison, the low-frequency and the high-frequency barrier capacitance and photocapacitance were also calculated by numerical integration of Eqs. (2) and (3) with the boundary conditions (5) and (7) and the continuity condition at the film-substrate interface (4). Besides, the numerical procedure employed in this calculation was also used to calculate the barrier capacitance and photocapacitance of an ion-implanted structure with due account for deep traps in the substrate. In this case, the Poisson equation had the form

$$\frac{dE}{dx} = \frac{q}{\varepsilon}\left[n + N_t\frac{n}{n + n_1} - N_d(x) - N_s\right], \tag{17}$$

the dopant concentration profile being $N_d(x) = N_p\exp\{-[(x - R_p)/2^{1/2}\sigma]^2\}$. Eqs. (17) and (3) were solved with the boundary conditions (5) and (7).

3 Results and discussion

Typical calculated low-frequency (LF) and high-frequency (HF) barrier capacitances and photocapacitances of an epitaxial structure are shown in Figs. 1(a) (analytical calculation) and 1(b) (both analytical and numerical calculation), respectively. As seen, the analytical results are in good agreement with the numerical ones. The dependence of the HF photocapacitance on the reverse bias has a narrow peak, and that of the LF photocapacitance has a positive peak followed by a negative valley, the peaks being located where the barrier capacitance drops steeply with V_{rev}. The figure clearly shows that these peaks are due to the illumination shifting the steeply decreasing region of the HF and LF *C-V* curves in the direction of increasing reverse bias. This shift may be understood as follows. The steeply decreasing region of the *C-V* curve is associated with the abrupt drop in the free carrier density in the vicinity of the film–substrate interface. Extrinsic illumination increases the free carrier density in the substrate, thus lowering the built-in voltage of the film–substrate junction and shrinking its space charge region. Hence, under extrinsic illumination, the steeply decreasing region of the *C-V* curve starts at a higher reverse bias than in the dark, leading to the photocapacitance peak. At higher reverse biases, the LF photocapacitance reverses the sign and passes through a minimum. This may be explained as follows. As the reverse bias is increased, the channel becomes nearly depleted of free carriers, and, from this point on, the predominant contribution to the LF barrier capacitance comes from emptying the deep traps at the film–substrate interface by the reverse barrier bias [4,5]. Illumination partially empties the deep traps, and thus emptying the traps by the reverse barrier bias is less pronounced, i.e., the LF photocapacitance should be negative. The trapped carrier density n_{tfs} at the film–substrate interface is related to the free carrier density n_{fs} at this interface as $n_{tfs} = N_t n_{fs}/(n_{fs} + n_1)$. The response of n_{tfs} to extrinsic illumination (i.e., to an increase Δn_1 in the characteristic concentration n_1) may be described as $\Delta n_1 \partial n_{tfs} / \partial n_1 = -\Delta n_1 N_t n_{fs}/(n_{fs} + n_1)^2$. It follows from this equation that this response is nearly equal to zero at $n_{fs} >> n_1$ (nearly all traps are occupied) and at $n_{fs} << n_1$ (nearly all traps are empty) and passes through a minimum at $n_{fs} = n_1$. Hence, the LF photocapacitance should pass through a minimum (valley) too.

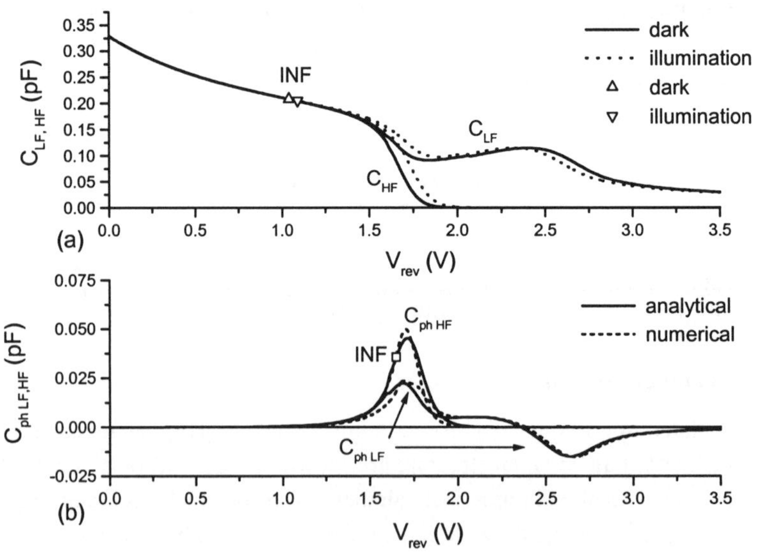

Figure 1. Calculated low- and high-frequency barrier capacitance (a) and photocapacitance (b) vs. reverse bias relationships of a GaAs epitaxial thin-film structure. $S_b = 300\ \mu m^2$; $\varphi_b = 0.8$ eV; $h = 0.2\ \mu m$; $N_d = 10^{17}$ cm^{-3}; $N_t = 10^{16}$ cm^{-3}; $N_s = 10^{15}$ cm^{-3}; $n_{10} = 9\times10^9$ cm^{-3} ($n_{sub0} = 10^9$ cm^{-3}); $T = 300$ K; $n_{1I}/n_{10} = 20$.

Typical calculated LF and HF barrier capacitances and photocapacitances of an ion-implanted structure are shown in Figs. 2(a) and 2(b), respectively (numerical calculation). As can be seen from the figure, the photocapacitance curves behave in exactly the same manner as for epitaxial structures. This suggests that the underlying mechanisms are the same as well. The only difference is that the LF photocapacitance is negative in the vicinity of zero bias. This is due to the fact that the LF photocapacitance of ion-implanted structures should show two valleys: one associated with deep trap emptying at the film–substrate interface, as in epitaxial structures, and the other associated with deep trap emptying at the metal–semiconductor interface; the latter occurs at a forward bias, and Fig. 2(b) only shows its right-hand tail.

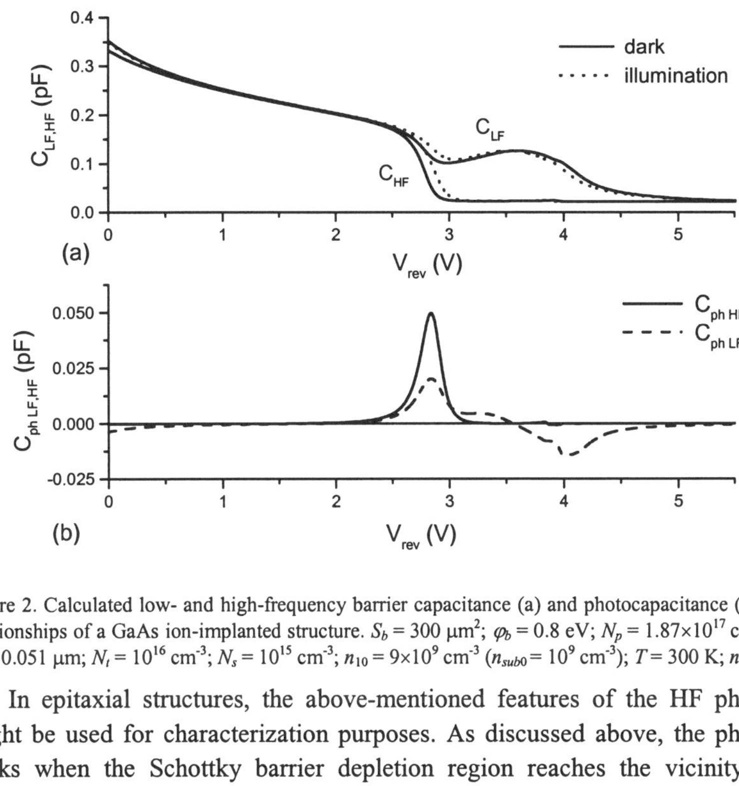

Figure 2. Calculated low- and high-frequency barrier capacitance (a) and photocapacitance (b) vs. reverse bias relationships of a GaAs ion-implanted structure. $S_b = 300$ μm^2; $\varphi_b = 0.8$ eV; $N_p = 1.87\times10^{17}$ cm^{-3}; $R_p = 0.13$ μm; $\sigma = 0.051$ μm; $N_t = 10^{16}$ cm^{-3}; $N_s = 10^{15}$ cm^{-3}; $n_{10} = 9\times10^9$ cm^{-3} ($n_{sub0} = 10^9$ cm^{-3}); $T = 300$ K; $n_{1l}/n_{10} = 20$.

In epitaxial structures, the above-mentioned features of the HF photocapacitance might be used for characterization purposes. As discussed above, the photocapacitance peaks when the Schottky barrier depletion region reaches the vicinity of the film–substrate interface, where the free carrier density in the film decreases sharply. This suggests that the vicinity of the barrier bias V_{max} at which the HF photocapacitance reaches its maximum may be related to the MESFET threshold voltage. To verify this, the MESFET threshold voltage V_T was correlated with the HF photocapacitance by experiment (all measurements whose results are presented here and below were conducted in GaAs MESFETs (gate length $l_g = 2$ μm; gate width $W_g = 320$ μm) fabricated at the Research & Production Complex "Electron Technology", Georgia, on $n^+ - n - i$- and $n^+ - n - n_b^- - i$- GaAs epitaxial structures; the structures had been grown at Kopin

Corp. by the MOCVD technique, their film doping density and as-grown film thickness being (1–2)×10^{17} cm^{-3} and 0.2 μm, respectively). The threshold voltage V_T was defined as the gate voltage at which the MESFET saturation current I_{sat} is equal to 10 μA. The measurements have shown that for n_{subI}/n_{sub0} ratios of the order of several tens and less the threshold voltage V_T can be approximated as the average of V_{max} and the voltage V_{INF} corresponding to the inflection point in the HF photocapacitance curve to the left of the maximum, at which this curve changes from being convex down to being convex up (point INF in Fig. 2(b))

$$V_T = \frac{1}{2}(V_{INF} + V_{max}). \tag{18}$$

Typical results are shown in Fig. 3. As illustrated, Eq. (18) does give a good approximation of the actual MESFET threshold voltage.

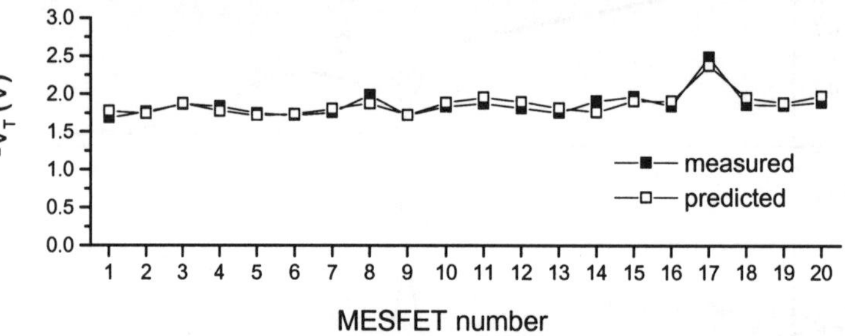

Figure 3. MESFET threshold voltage: measured and predicted from the measured HF photocapacitance.

The illumination-induced shift of the steeply decreasing region of the HF *C-V* curve responsible for the photocapacitance peak might be used for estimating the thickness of the epitaxial film and the concentration of vacant deep traps in the vicinity of the film–substrate interface $N_{vac} = N_t - N_s$, i.e., the quantity that governs the overall effect of deep traps on device performance. Using the depletion approximation, the illumination-induced increase Δh_I in the channel width can be expressed as [6]

$$\Delta h_I = d_0^+ - d_I^+ = \sqrt{\frac{2\varepsilon kTN_{vac}}{q^2 N_d (N_d + N_{vac})}}\left[\sqrt{\ln N_d/n_{sub0}} - \sqrt{\ln N_d/(n_{sub0} + \beta I)}\right]. \tag{19}$$

Here, d^+ is the width of the depletion region of the film–substrate junction, $n_{sub0} + \beta I = n_{subI}$ is the free carrier density in the substrate under illumination, and $\beta = N_s S_p / N_{vac} c_e$.

On the other hand, the steeply decreasing region of the HF *C-V* curve is associated with the abrupt drop in the free carrier density in the vicinity of the film–substrate interface, and, thus, Δh_I may be approximated as

$$\Delta h_I = x_{chanI} - x_{chan0} = \varepsilon S_b \left(\frac{1}{C_{\inf I}} - \frac{1}{C_{\inf 0}} \right) \tag{20}$$

where x_{chan} is the coordinate of the boundary between the conductive channel and the depletion region of the film–substrate junction and $C_{\inf}$ corresponds to the inflection point in the HF *C-V* curve, whereat this curve changes from being convex down (its usual behavior) to being convex up (steep decrease) (points INF in Fig.1 (a)).

Eqs. (19) and (20) make it possible to extract N_{vac} (as well as n_{sub0} and β) from HF *C-V* measurements (N_d can readily be found using the conventional *C-V* profiling) and then to estimate the film thickness h as

$$h = x_{chan0} + d_0^+ = \frac{\varepsilon S_b}{C_{\inf 0}} + \sqrt{\frac{2N_{vac}\varepsilon kT}{q^2 N_d (N_d + N_{vac})} \ln N_d / n_{sub0}} \tag{21}$$

Typical measured distributions of N_{vac} and h over MESFET wafers with and without a buffer layer are shown in Figs. 4(a) and 4(b), respectively. As seen, the distribution of N_{vac} shows wide spread, and for the wafer with a buffer layer, N_{vac} is several times lower than for the wafer without a buffer layer.

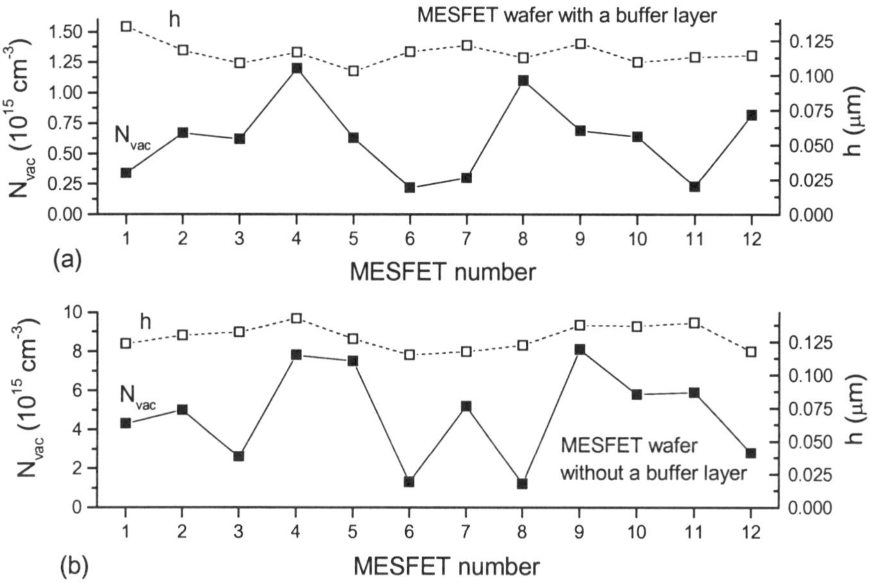

Figure 4. Measured distributions of the concentration N_{vac} of vacant deep traps in the vicinity of the film–substrate (film–buffer layer) interface and of the film thickness h over MESFET wafers with (a) and without (b) a buffer layer.

4 Conclusions

We have performed analytical and numerical calculations of the low-frequency and high-frequency barrier capacitances of GaAs thin-film structures in the dark and under extrinsic illumination with due account for the space charge region at the doped layer–substrate interface. The calculations show that the high-frequency photocapacitance–

voltage characteristic of these structures has a narrow peak, and the low-frequency photocapacitance–voltage characteristic has a positive peak followed by a negative valley. The peaks are due to the extrinsic illumination shrinking the space charge region at the doped layer–substrate interface. This shifts the steeply decreasing region of the *C-V* curve associated with the abrupt drop in the free carrier density in the vicinity of this interface to a higher reverse bias. At sufficiently high reverse biases the low-frequency barrier capacitance is governed by the emptying of the deep traps at the doped layer–substrate interface by the reverse bias. This causes the negative valley of the low-frequency photocapacitance characteristic. Hence, the above-mentioned features of the photocapacitance are governed by the built-in space charge region at the doped layer–substrate interface. For epitaxial structures, the peak of the high-frequency photocapacitance might be used for extracting the MESFET threshold voltage, and the illumination-induced shift of the steeply decreasing region of the high-frequency *C-V* curve responsible for this peak might by used for estimating the thickness of the epitaxial film and the concentration of vacant deep traps in the vicinity of the film–substrate interface. Such measurements using mercury probes can be conducted both in virgin epitaxial structures and during the MESFET fabrication process at different fabrication steps.

References

1. K. Kitahara, K. Nakai, A. Shibatomi and S. Ohkawa, *Jap. J. Appl. Phys.* **21**, 513 (1982).
2. G. J. Papaioannou, *Phys. Stat. Sol.* **96**, 99 (1986).
3. K. Lehovec and N.Zamani, *Appl. Phys. Lett.* **28**, 673 (1976).
4. S.A. Kostylev, E.F. Prokhorov, N.B. Gorev, I.F. Kodzhespirova and Yu.A. Kovalenko, *Solid-State Electron.* **43**, 169 (1999).
5. J. González-Hernández, E. Prokhorov, N.B. Gorev, I.F. Kodzhespirova and Yu.A. Kovalenko, *J. Vac. Sci. Technol. B.* **17**, 2357 (1999).
6. N.B. Gorev, I. Kodzhespirova, E.N. Privalov, L. Khvedelidze, N. Khuchua, G.G. Peradze, M.S. Shur and K. Stevens, *Solid-State Electron.* **47**, 1569 (2003).

International Journal of High Speed Electronics and Systems
Vol. 14, No. 3 (2004) 785–790

World Scientific
www.worldscientific.com

UNSTRAINED InAlN/GaN HEMT STRUCTURE

M. NEUBURGER, T. ZIMMERMANN, E. KOHN

Dept. of Electron Devices & Circuits, University of Ulm, Albert Einstein Allee 45, 89081 Ulm, Germany

A. DADGAR, F. SCHULZE, A. KRTSCHIL, M. GÜNTHER, H. WITTE, J. BLÄSING, A. KROST

Institut für Experimentelle Physik, Fakultät für Naturwissenschaften, Universitätsplatz 2, 39016 Magdeburg, Germany

I. DAUMILLER, M. KUNZE

MicroGaN GmbH, Albert Einstein Allee 45, 89081 Ulm, Germany

InAlN has been investigated as barrier layer material for GaN-HEMT structures, potentially offering higher sheet charge densities [1] and higher breakdown fields [2]. Lattice matched growth of the barrier layer can be achieved with 17 % In content, avoiding piezo polarization. In this configuration the sheet charge density is only induced by spontaneous polarization. First experimental results of unpassivated undoped samples realized on 111-Si substrate exceed a DC output current density of 1.8 A/mm for a gate length of 0.5 μm. Small signal measurements yield a f_t = 26 GHz and f_{max} = 14 GHz, still limited by the residual conductivity of the Si-substrate. A saturated output power at 2 GHz in class A bias point yielded a density of 4.1W/mm at V_{DS} = 24 V.

1 Introduction

In the AlGaN/GaN FET a highly mobile sheet charge acting as channel charge is formed by the polarization discontinuity at the interface barrier-layer / buffer, with two contributing parts, namely piezo and spontaneous polarization. For conventional AlGaN/GaN-HFETs the polarization discontinuity containing piezo and spontaneous contributions is directly correlated with the aluminum content of the barrier layer. In order to increase sheet charge density and thus the maximum DC-output current, the aluminum content of the barrier layer must be increased or modulation doping needs to be employed. However for high Al-content, the internal strain results in poor material quality.

In this contribution a new layer structure is introduced replacing the AlGaN barrier layer by an InAlN layer. This layer structure offers potentially higher sheet charge densities because of higher spontaneous polarization of InAlN compared to AlGaN [2]. Applying Vegard's law for alloys of InN, AlN and GaN as indicated in fig. 1, InAlN layers can be grown with tensile stress on GaN for In contents below 17% and with compressive stress for In contents above 17%. For a composition of 17% Indium, the

InAlN layer can be grown lattice matched on GaN and therefore has no piezo stress component. Applying this technology to InAlN/GaN HFETs, high sheet charge densities up to 2.9 x 10^{13} cm^{-2} are expected [2]. In this case the HEMT is based on spontaneous polarization only. In addition, this new HEMT-structure has been realized on Si-substrates using a buffer layer design similar to that of AlGaN/GaN-HEMTs.

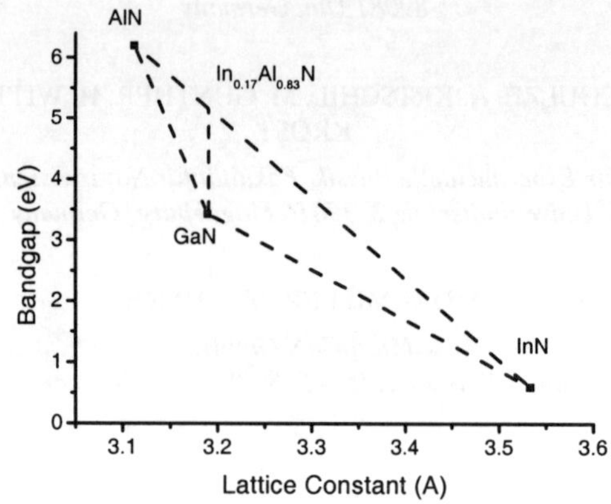

Figure 1: Bandgap over the lattice constant of InN, AlN, GaN. Lattice matched InAlN is obtained for InAlN with 17 % Indium

2 Growth and Processing

The semi-insulating GaN buffer for this structure is grown on p-type high resistive 111 Silicon (R > 3 kΩ) using a low temperature AlN nucleation layer. Also, low temperature AlN-layers are inserted for stress reduction in the iron doped 1.5 μm thick buffer. On top of the buffer a 2 nm thin AlN layer is grown, followed by a 24 nm thick InAlN barrier layer with 17 % In, verified with XRD measurements. A schematic cross section is shown in fig. 2. Details on the growth process are published elsewhere [3].

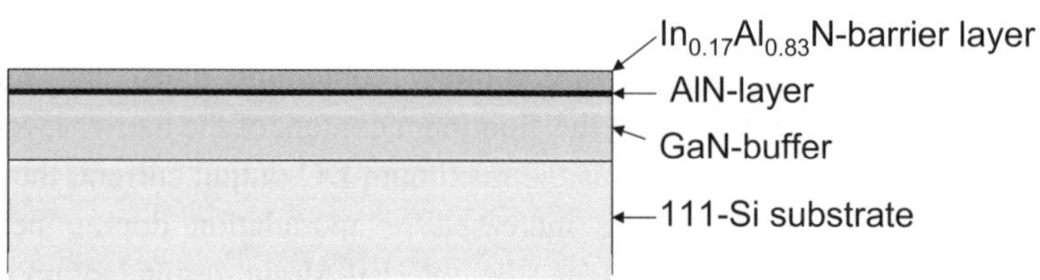

Figure 2: Cross section of an InAlN/GaN based HFET

The grown structure shows a sheet resistance of 412 Ω/ . Hall effect measurements at room temperature yield a sheet charge density of N_S = 2.8 x 10^{13} cm^{-2}, similar to what is theoretically expected, and a mobility of 309 cm^2/Vs. The low mobility can be understood by increased electron scattering at the InAlN/GaN interface as compared to AlGaN/GaN interfaces [4]. However by optimization of the thin AlN-intermediate layer in the active area, resulting in decreased scattering [5] and by epitaxial improvement of the interface quality, the mobility should be increased.

FETs are realized as follows: Mesa isolation is performed with dry etching in Argon-plasma. For the ohmic contacts standard AlGaN/GaN based contacts are employed consisting of a Ti/Al/Ni/Au metal layer structure annealed at 860 °C for 45 sec. A specific contact resistance of 7 x 10^{-5} Ωcm^2 is determined by TLM-measurements. In the FET layout, the drain-source distance is 3 μm. Ni/Au schottky gates are realized by electron beam lithography with length and width of 0.5 / 100 μm respectively and are placed asymmetric between drain and source, 1 μm apart from the source contact.

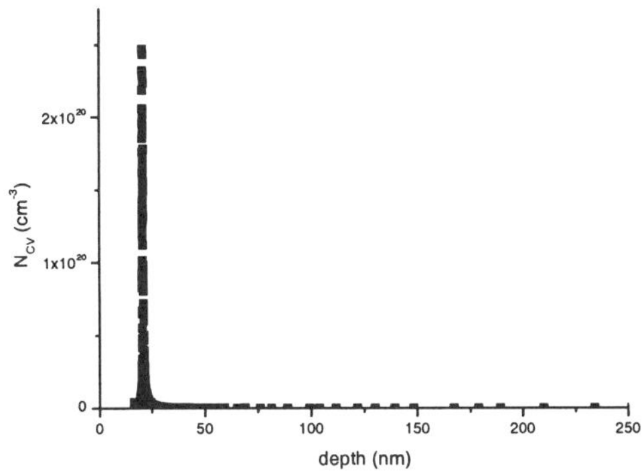

Figure 3: CV-profiling. Formation of a 2 DEG is indicated 23 nm below the surface

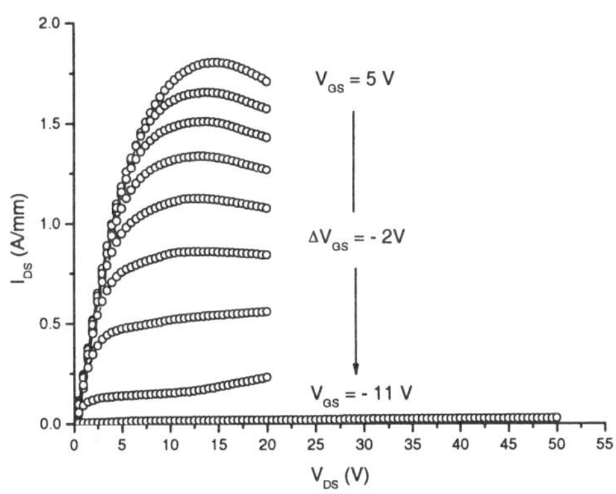

Figure 4: DC-output characteristic of an In0.17Al0.83N/GaN-HFET (LG = 0.5 μm, WG = 100 μm)

3 Experimental results

CV-profiling indicates a 2 DEG carrier density 23 nm below the surface at the InAlN/GaN interface (fig. 3), resulting in a sheet charge density of $N_{S,CV} = 2.5 \times 10^{13}$ cm^{-2} in agreement with hall measurements. From the CV analysis a barrier height approximately 1.8 eV is extracted.

DC-measurements are shown in fig. 4 ($L_G = 0.5$ μm, $W_G = 2 \times 100$ μm), yielding a maximum saturated DC-output current of I_{DS} = 1.8 A/mm, a three terminal breakdown $V_{DS,BR}$ = 52 V at pinch-off and a peak transconductance of g_m = 180 ms/mm at V_{DS} = 15 V and room temperature. Cooling the device to a substrate temperature of 150 K, the saturated DC-output current increased to I_{DS} = 2.1 A/mm. The channel is fully opened for a positive gate voltage of V_{GS} = + 5 V, indicating still a parasitic potential drop across the source-gate region.

In order to estimate the potential of this layer structure, the maximum saturated DC-output current for open channel (V_{GS} = +5 V) is measured for gate lengths ranging from 0.5 μm up to 16 μm. This is compared with the maximum saturated output current calculated according [6] (fig. 5), taking carrier drift velocity saturation into account. (V_{SAT} = 10^7 cms^{-1}). As seen in fig. 5, a good agreement between calculation and experiment is obtained, validating the model. This allows to estimate a maximum saturated output current of 2.5 A/mm for gate lengths of 100 nm.

Small signal measurements between 450 MHz and 50 GHz for a 2 x 100 μm wide device yielded a cutoff frequency f_t = 26 GHz and a maximum frequency of oscillation f_{max} = 14 GHz at the bias point V_{DS} = 15 V , V_{GS} = -6 V, shown in fig. 6. The f_{max} is still limited by the residual losses in the Si-substrate. Output power measurements at 2 GHz in class A bias point at V_{DS} = 24 V, V_{GS} = -3.8 V result in a power density of 4.1 W/mm with low power added efficiency because of a still high on-resistance.

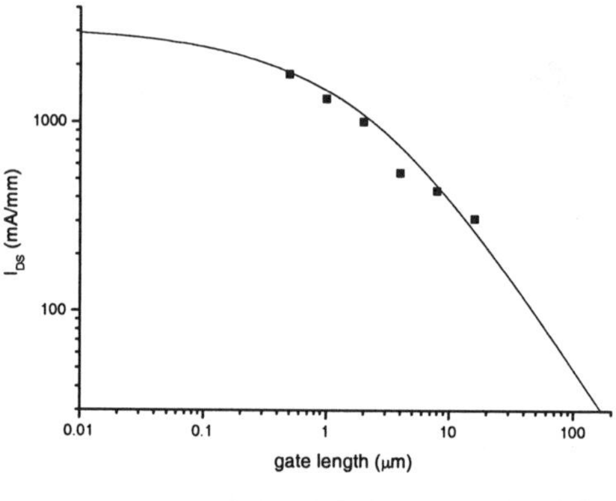

Figure 5: Calculated maximum saturated open channel current density for different gate lengths according [5]. Dots indicated are measured results for saturated open channel (V_{GS} = +5).

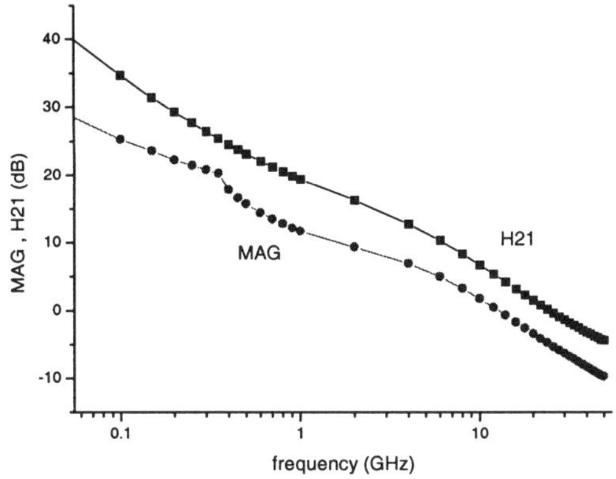

Figure 6: RF-gain plots of an InAlN/GaN-HFET. Device data: LG = 0.5 μm, WG = 2 x 100 μm, VDS = 15 V, VGS = - 6 V.

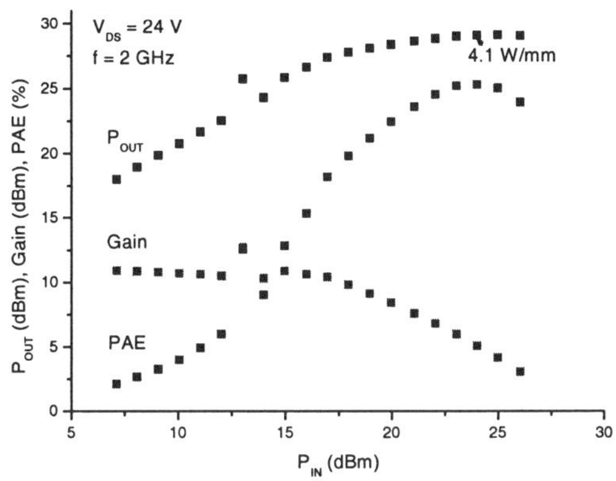

Figure 7: Output power measurement of InAlN/GaN-HFET. Device data: L_G = 0.5 μm, W_G = 2 x 50 μm, f = 2 GHz.

4 Conclusion

In this paper a new layer structure is introduced by replacing the AlGaN barrier layer of the conventional HFET by InAlN with 17 % In. Higher sheet charge densities are expected (N_S = 2.9 x 10^{13} cm^{-2}) and are verified by Hall-measurements ($N_{S,H}$ = 2.8 x 10^{13} cm^{-2}) and CV analysis ($N_{S,CV}$ = 2.5 x 10^{13} cm^{-2}). Realized unpassivated HFETs with gate lengths of 0.5 μm exceed saturated DC-output currents of 1.8 A/mm with a three terminal breakdown voltage of $V_{DS,BR}$ = 52 V in pinch-off. Although these samples still suffer from a low mobility, large signal measurements in class A bias reveal a power density of 4.1 W/mm at 2 GHz.

References

1. J. Kuzmik, Semicond. Sci. Technol., vol. 17, 540 (2002).
2. J. Kuzmik, IEEE Electron Device Letters, vol. 22, 510 (2001).
3. A. Dadgar, M. Poschenrieder, J. Blasing, O. Contreras, F. Bertram, T. Riemann, A. Reiher, M. Kunze, I. Daumiller, A. Kritschl, A. Diez, A. Kaluza, A. Modlich, M. Kamp, J. Christen, F. A. Ponce, E. Kohn, A. Krost, Journal of Crystal Growth, vol 248, pp. 556-562 (2003).
4. V. W. L. Chin, B. Zhou, T. L. Tansley, X. Li, J. Appl. Phys. vol. 77, no. 11 (1995).
5. A. Matulionis, J. Liberis, I. Matulioniene, M. Ramonas, L. F. Eastman, J. R. Shealy, V. Tilak, A. Vertiatchikh, Physical Review B-Condensed Matter, vol. 68, no. 3 (2003).
6. Y. Yaur, T. H. Ning, Cambridge University Press, Cambridge (1998).

International Journal of High Speed Electronics and Systems
Vol. 14, No. 3 (2004) 791–796

GaN MOS-HEMT USING ATOMIC LAYER DEPOSITION Al_2O_3 AS GATE DIELECTRIC AND SURFACE PASSIVATION

P.D. Ye[†], B. Yang, K.K. Ng, J. Bude
Agere Systems, 555 Union Boulevard
Allentown, PA 18109, U.S.A.

G.D. Wilk
ASM America, 3440 East University Drive
Phoenix, AZ 85034, U.S.A.

S. Halder, J.C.M. Hwang
Lehigh University, 5 E. Packer Avenue
Bethlehem, PA 18015, U.S.A.

We report on a GaN metal-oxide-semiconductor high electron mobility transistor (MOS-HEMT) using atomic layer deposition (ALD) Al_2O_3 film as a gate dielectric and for surface passivation simultaneously. Compared to the conventional AlGaN/GaN HEMT of the same design, six order of magnitude smaller gate leakage current and tripled drain current at forward gate bias demonstrate the effectiveness of ALD Al_2O_3 as a gate dielectric. The high transconductance and high effective two-dimensional electron mobility verify the high-quality of Al_2O_3/AlGaN interface with low interface trap density. The Al_2O_3 passivation effect is also studied by sheet resistance measurement and short pulse drain characterization.

1 Introduction

GaN-based high-power microwave electronic device as an emerging compound semiconductor technology attracts tremendous interest in both industry and academia. However, two key problems still remain. First, the gate leakage current for GaN metal-semiconductor field-effect-transistor (MESFET) and GaN high electron mobility transistor (HEMT) is too high due to the surface defects and finite barrier height. The high gate leakage directly impacts the drain breakdown voltage, rf performance, noise figure, and reliability of the device. Second, without proper passivation, GaN devices exhibit current collapse with a high rf-input drive on the gate, which significantly reduces rf output power and degrades the device performance. In the past years, several groups have attempted to suppress the gate leakage using the metal-insulator-semiconductor

[†] Electronic mail: peterye@agere.com

field-effect-transistor (MISFET) [1-2] or metal-oxide-semiconductor field-effect-transistor (MOSFET) [3] approaches. These devices perform quite poorly. More recently, significant progress has been made on metal-insulator-semiconductor high electron mobility transistor (MIS-HEMT) or metal-oxide-semiconductor high electron mobility transistor (MOS-HEMT) using SiO_2 [4-8], Si_3N_4 [9-10], and other exotic oxides [11-12].

In this paper, we report, for the first time, AlGaN/GaN MOS-HEMTs with atomic layer deposition (ALD) Al_2O_3 as gate dielectric and surface passivation. Similar to SiO_2 [4-8], Si_3N_4 [9-10], and Sc_2O_3 [11], Al_2O_3 can significantly reduce the gate leakage current of AlGaN/GaN HEMTs thereby improve their performance and reliability. Al_2O_3 offers additional advantages of large bandgap (9 eV), high dielectric constant (k ~ 10), high breakdown field (10^7 V/cm), thermal stability (amorphous up to at least 1000°C), and chemical stability against AlGaN (without interdiffusion and interaction of Si and Al). Further, Al_2O_3 formed by ALD has become one of the leading candidates to replace SiO_2 in future-generation Si CMOS digital ICs [13]. The ALD Al_2O_3 has demonstrated low defect density, high uniformity, and nm scalability. The ALD Al_2O_3 process is robust and highly manufacturable. We have already demonstrated excellent performance in GaAs and InGaAs MOSFETs with ALD Al_2O_3 [14-16].

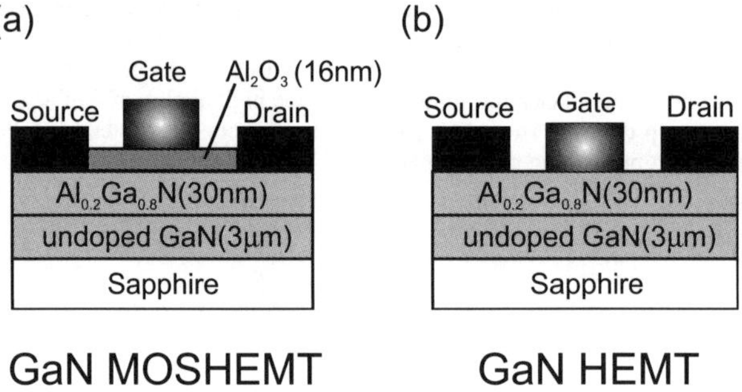

Figure 1. Schematic illustration of (a) AlGaN/GaN MOS-HEMT with ALD-grown Al_2O_3 as gate dielectric and (b) AlGaN/GaN HEMT as a baseline.

2 Device Fabrication

For the present MOS-HEMTs as shown in Fig. 1, after the AlGaN/GaN heterostructure was formed on 2-inch sapphire wafers by MOCVD, 16-nm-thick Al_2O_3 was deposited by ALD in an ASM Pulsar2000™ module, using the TMA/H_2O process. The wafers were conveniently transferred in air before and after ALD. Device isolation was achieved by nitrogen implantation. Source-drain and gate contacts were formed by Ti/Al/Ni/Au and Ni/Au, respectively. Conventional HEMTs of similar structures were processed together with the MOS-HEMTs. The gate lengths of the measured devices are 0.65, 1, 5, and 10 μm. The gate width is 100 μm. The source-to-gate and the gate-to drain spacings are ~ 2 μm. Their characteristics are compared below.

3 Device Characterization

Fig. 2(a) shows the C-V measurements at 1 MHz for 8000 μm^2 gate capacitors on the MOS-HEMT and HEMT devices. The comparable sharp rise from depletion to accumulation in CV curves for both MOS-HEMT and HEMT demonstrates the high-quality interface of Al_2O_3 to AlGaN. Fig. 2(b) shows GaN MOS-HEMT devices exhibit extremely low gate leakage currents of 10^{-6} A/cm^2, more than six orders of magnitudes lower at positive gate bias, compared to the conventional GaN HEMTs. The forward two-terminal breakdown voltage is ~ 12 V. The low gate leakage for MOS-HEMT increases the breakdown voltage and the power-added efficiency while decreasing the noise figure.

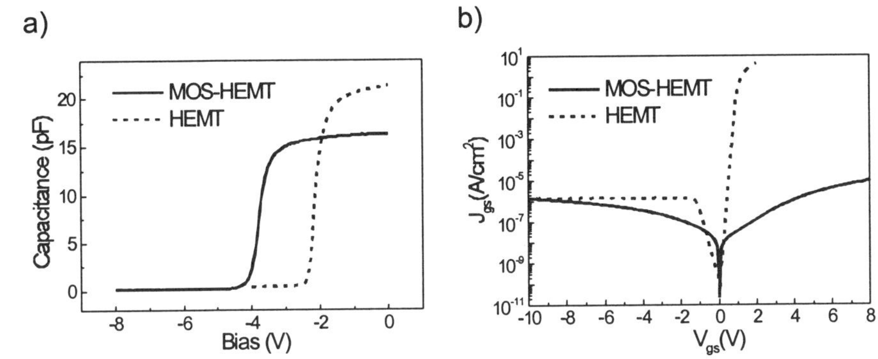

Figure 2. (a) Measured C-V characteristics of a GaN MOS-HEMT (solid line) and a HEMT (dashed line) for equal area of ~8000 μm2. (b) Gate leakage current for the MOS-HEMT (solid lines) and the baseline HEMT (dashed lines).

Fig. 3 shows well-behaved I-V characteristics of GaN MOS-HEMT operating at V_{ds}=10 V with a pinch-off voltage of -3 V. The off-state drain-source three-terminal breakdown voltage with 2 μm drain-gate spacing is around 145 V. The results indicate that ALD Al_2O_3 is an effective gate dielectric for AlGaN/GaN devices. Fig. 4(a) shows the gate-voltage dependence of the drain current for the identical geometry HEMTs and MOS-HEMTs. The drain current is ~150% higher for the MOS-HEMTs, reaching a maximum value of ~ 0.45 A/mm. Using $I_{ds}=e \cdot n_s \cdot \upsilon_{sat} \cdot W$, we estimate the maximum sheet carrier density $n_s=6\times10^{12}/cm^2$ as expected for undoped AlGaN/GaN two-dimensional channel density. Here, υ_{sat} ~6×10^6 cm/s for GaN. Fig. 4(b) shows that the intrinsic peak g_m for MOS-HEMT is ~100 mS/mm and for HEMT is ~120 mS/mm. The decrease of g_m corresponds to the added Al_2O_3 thickness. The high g_m for MOS-HEMT indicates that the Al_2O_3/AlGaN interface has low interface-trap density. The drain current and g_m can be improved by employing doped AlGaN layer or SiC substrates.

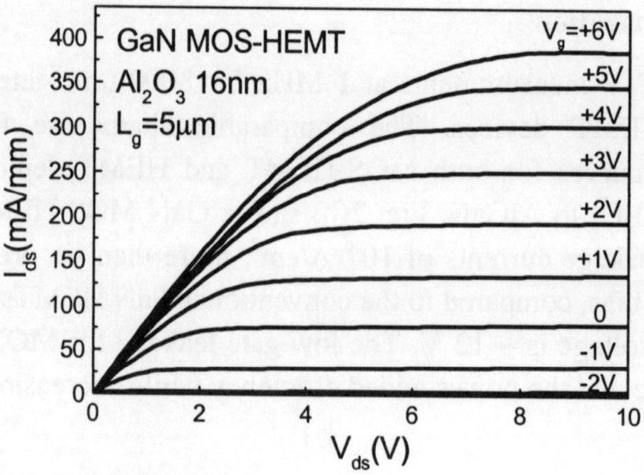

Figure 3. I-V characteristics for a MOS-HEMT.

In order to quantitatively characterize the AlGaN/GaN interface with ALD Al_2O_3 on top, we study the transverse field effect on 2D electron mobility, which is directly related to interface properties. By measuring I_{ds} vs. V_{gs} at V_{ds}=0.1 V (mobility region) of 5 and 10 μm long channel devices, we are able to extract the series resistance of the devices with a relative error of less than 5%. Fig. 5 is the summary plot of the effective 2D electron mobility vs. effective electric field. The mobility of 1200 cm^2/Vs at low transverse field is consistent with the value obtained from the Hall measurements. The effective 2D electron mobility of 640 cm^2/Vs at the high transverse field of 0.6 MV/cm is much higher than 400 cm^2/Vs, the universal surface mobility of Si MOSFETs at the same transverse field. It is also higher than the surface mobility on GaAs and InGaAs surface obtained from ALD Al_2O_3 GaAs and InGaAs MOSFETs [16]. It's mainly beneficial from the epitaxially grown AlGaN/GaN heterojunction, which has much less interface traps compared to oxide-semiconductor interfaces. Because of the high-quality ALD Al_2O_3 as gate dielectric on AlGaN (i.e., low gate leakage current and low interface trap density at Al_2O_3/AlGaN interface), it enables us to measure the effective 2D electron mobility at high electron density and high transverse field for the first time.

To study the passivation effect of ALD Al_2O_3 on AlGaN, we measure the sheet resistances of AlGaN/GaN with and without ALD Al_2O_3 on top. The lack of Al_2O_3 passivation at drain-gate and source-gate regions for the baseline HEMT could lead to increased parasitic resistance, thus degrades intrinsic g_m. The parasite resistance is determined from the transmission line model (TLM) method fabricated on the same chip. The sheet resistance measured from TLM for MOS-HEMT with ALD Al_2O_3 passivation is ~700 Ω/sq. and for HEMTs without any passivation is ~950 Ω/sq., which indicates the effectiveness of the ALD Al_2O_3 passivation on AlGaN. The preliminary study of the pulsed drain characteristics also shows interesting results. For example, after up to 80 V drain voltage stress, DC drain characteristics (solid lines in Fig. 6) show the well-known current collapse effect on drain current at V_{ds} < 15 V. It is mainly due to hot carrier injections at the oxide/semiconductor interface or even in the bulk oxide at the drain side

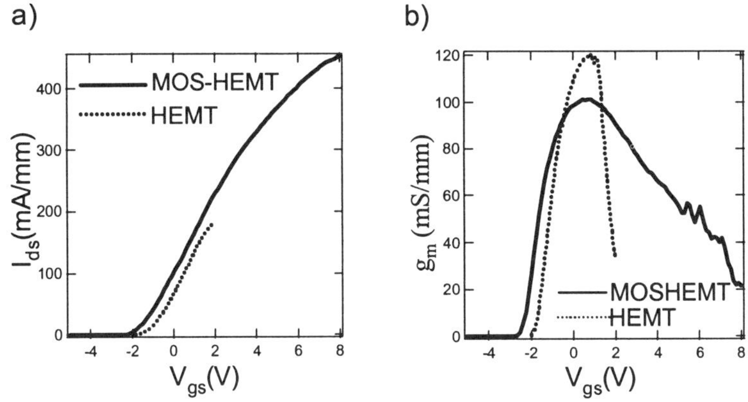

Fig. 4. (a) The drain-current Ids verses gate bias Vgs in the saturation region (Vds=10V) for the MOS-HEMT (solid line) and the baseline HEMT (dashed line) with Lg=5 μm. (b) The intrinsic transconductance gm verses gate bias Vgs at the same drain bias Vds=10V for the MOS-HEMT (solid line) and baseline HEMT (dashed line) of the same device.

under high voltage drain stress. The dashed lines in Fig. 6 show that the short pulse (1-μs) drain characteristics are fully recovered from DC characteristics at V_{gs}=4V quiescent conditions. The short pulse drain measurements at different quiescent points could be an effective method to study Al_2O_3 surface passivation. More device evaluation in terms of CW and pulsed, small- and large-signal characteristics is in progress.

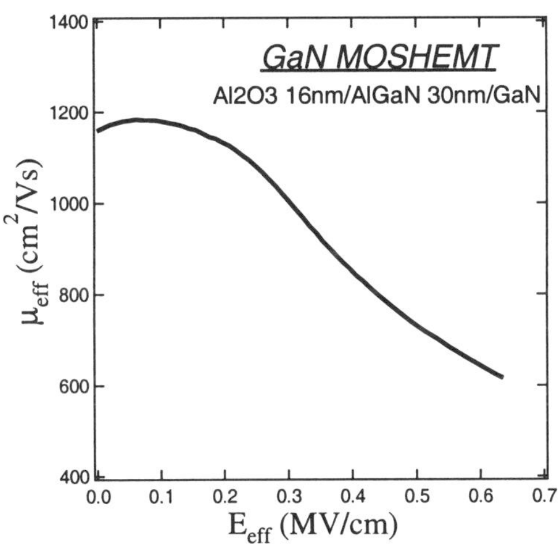

Fig. 5. Effective two-dimensional electron mobility μeff verses effective electron field Eeff. The data are obtained from 5 μm and 10 μm-gate-length devices.

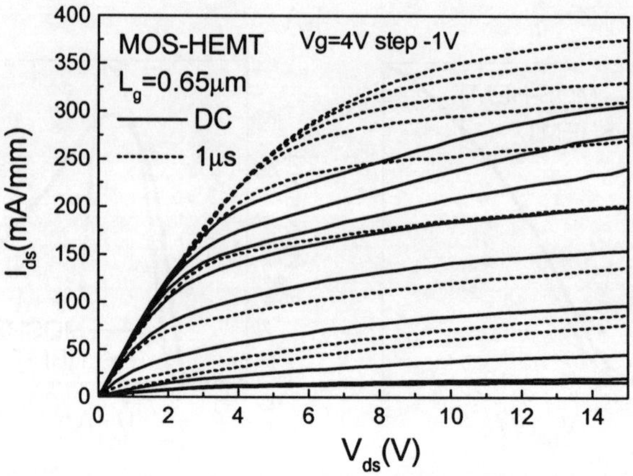

Fig. 6. Output I-V characteristics of MOS-HEMT under DC (solid line) and 1μs pulsed-gate bias (dashed line).

4 Summary

In summary, ALD Al_2O_3 process provides high-quality gate dielectric and surface passivation for AlGaN/GaN HEMTs. The resulted MOS-HEMT shows favorable characteristics when compared to MOS-HEMTs with other gate insulators such as those of [1]-[12]. This indicates excellent potential of Al_2O_3/AlGaN/GaN MOS-HEMTs for high-speed and high-power applications.

References

1. M. Asif Khan, et al., Electron. Lett., vol. 30, pp. 2175, Dec. 1994.
2. S.C. Binari, et al., International Symp. Compound Semiconductor, pp. 459, 1995.
3. F. Ren, et al., Appl. Phys. Lett., vol. 73, no. 26, pp. 3893-3895, Dec. 1998.
4. M. Asif Khan, et al., IEEE Electron Device Lett., vol. 21, no. 2, pp. 63-65, Feb. 2000.
5. M. Asif Khan, et al., Appl. Phys. Lett., vol. 77, no. 9, pp. 1339-1341, Aug. 2000.
6. G. Simon, et al., Electronics Lett., vol. 36, no. 24, pp. 2043-2044, Nov. 2000.
7. Koudymov, et al., IEEE Electron Device Lett., vol. 23, no. 8, pp. 449-451, Aug. 2002.
8. G. Simon, et al., IEEE Electron Device Lett., vol. 23, no. 8, pp. 458-460, Aug. 2002.
9. G. Simon, et al., IEEE Electron Device Lett., vol. 22, no. 2, pp. 53-55, Feb. 2001.
10. X. Hu, et al., Appl. Phys. Lett., vol. 79, no. 17, pp. 2832-2834, Oct. 2000.
11. R. Mehandru, et al., Appl. Phys. Lett., vol. 82, no. 15, pp. 2530-2532, Apr. 2003.
12. T. Hashizume et al., Appl. Phys. Lett., vol 83, no. 14, pp. 2952-2954, Oct. 2003.
13. G. D. Wilk, et al., J. Appl. Phys., vol. 89, pp. 5243-5275, 2001.
14. P. D. Ye, et al., IEEE Electron Device Lett. vol. 24, no. 4, pp. 209, Apr. 2003.
15. P. D. Ye, et al., Appl. Phys. Lett., vol. 83, no. 1, pp. 180-183, July 2003.
16. P. D. Ye, et al., Appl. Phys. Lett., vol. 84, no. 3, pp. 434-436, 2004.

International Journal of High Speed Electronics and Systems
Vol. 14, No. 3 (2004) 797–804

World Scientific
www.worldscientific.com

DEPENDENCE OF ELECTRON MOBILITY ON EPI CHANNEL DOPING IN GaN MOSFETS

J. RUAN
Rensselaer Polytechnic Institute, Troy, NY 12180 USA

K. MATOCHA
GE Global Research, Niskayuna, NY 12309 USA

W. HUANG AND T.P. CHOW
Rensselaer Polytechnic Institute, Troy, NY 12180 USA

The field-effect mobility and Hall mobility of electrons in lateral, accumulation-mode GaN MOSFETs have been comparatively studied. The devices were fabricated on three different samples, Sample A consisting of 70 nm n-type ($N_D = 5\times10^{17}$ cm^{-3}) GaN and Sample B 0.5 μm unintentionally doped ($N_D < 1\times10^{16}$ cm^{-3}) GaN, both on 2 μm undoped $Al_{0.35}Ga_{0.65}N$ grown on the n^+-SiC substrates and Sample C 3 μm unintentionally doped GaN on the sapphire substrate. The maximal field-effect mobilities are 120 cm^2/V-s, 90 cm^2/V-s and 53 cm^2/V-s respectively in the three samples. The Hall mobility is also measured for Samples A & B and reaches a maximal value of 160 cm^2/V-s.

1 INTRODUCTION

Gallium nitride (GaN) is becoming more and more popular not only for photonics application but also for power applications in recent years. Much research work has been carried out on GaN devices due to the attractive material properties: wide bandgap (3.4 eV), large critical electric field (3 MV/cm), reasonably high electron mobility (comparable to Si) and good thermal conductivity. One particular advantage of GaN material is that different kinds of semiconductor devices, like optoelectronic, microwave, sensing and power devices, can be integrated in a single GaN chip, if processing technology is advanced enough.

For GaN devices with control terminals, much of the work is focused on HEMTs due to the high electron mobility and large electron concentration achieved [1]. Some are about GaN MESFETs [2] while only a few are about GaN MISFETs. GaN MISFETs have been studied mainly in low-voltage regime for radio frequency applications. In all MISFETs, creation of a high-quality, low interface-state density is crucial for achieving good device performance. Several materials have been used as gate dielectrics on GaN, including SiO_2, Si_3N_4, Ga_2O_3, Gd_2O_3 [3-6]. Of all the materials, SiO_2 provides the largest conduction band offset which is useful in increasing the possible accumulation layer charge while preventing tunneling into the dielectric.

In this work, the field-effect mobility and Hall mobility of electrons in lateral, accumulation-mode GaN MOSFETs on different epi-layers were comparatively studied.

2 DEVICE STRUCTURE

Before the study of GaN MOSFETs, MOS capacitors were fabricated with deposited HTO oxide and characterized [7]. C-V measurement indicates that the surface state density becomes larger as deeper into the bandgap in GaN. This is one of the reasons that accumulation-mode rather than inversion-mode MOSFETs are studied since the Fermi level need only be swept through the upper half of the bandgap. Also, impurity implantation and activation are not necessary in the accumulation-mode process. The MOSFETs were fabricated on three different samples: Sample A consisting of 70 nm n-type ($N_D = 5\times10^{17}$ cm^{-3}) GaN and Sample B 0.5 μm unintentionally doped ($N_D < 1\times10^{16}$ cm^{-3}) GaN, both on 2 μm undoped $Al_{0.35}Ga_{0.65}N$ grown on the n^+-SiC substrates and Sample C 3 μm unintentionally doped GaN on sapphire substrate.

Device fabrication began with mesa isolation by plasma etching. LPCVD (HTO) deposited the gate dielectric at 900 °C. 500 nm $POCl_3$ doped polysilicon was formed over the SiO_2, acting as the gate material. Ti/Al (35/115 nm) annealed at 700 °C for 30 seconds was used for the source and drain contacts and Ti/Mo was used for interconnecting metallization. The device cross section in Sample A is shown in Fig. 1.

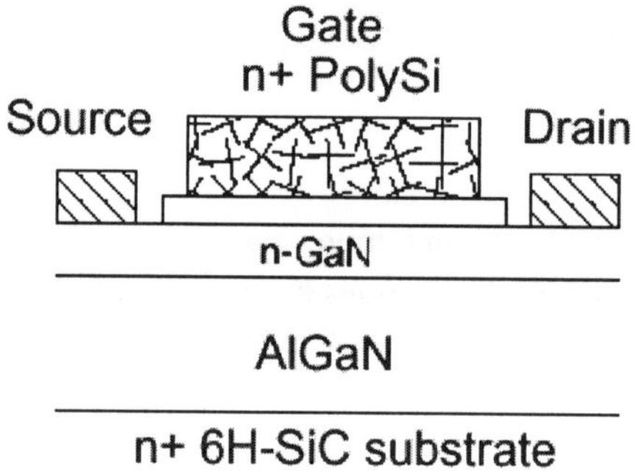

Fig. 1. GaN MOSFET Cross Section in Sample A

3 DEVICE CHARACTERIZATION AND ANALYSIS

Both the field-effect mobility and the Hall mobility are measured from the MOS-Gate Hall Bar structure shown in Fig. 2. The procedures are described as followings:

3.1 Field-Effect Mobility

1. Transfer characteristic of the MOSFET are measured in the linear region at $V_{DS} = 0.1$ with the gate bias V_{GS} swept from the negative to the positive. Source is grounded at all the time. Fig. 3 shows the measured transfer characteristic of Sample A. The drain current goes almost flat at large gate voltage, indicating large parasitic resistance at the source/drain contact.
2. While measuring $I_{DS} \sim V_{GS}$, the voltage difference between Voltage Probe 1 and Probe 2 (named V_C) is also monitored. Thus the relation between V_C and V_{GS} is obtained. This is used to eliminate the influence of huge parasitic source/drain series resistance and extract the transfer characteristic of an ideal MOSFET.

3. V_C/I_{DS} gives out the resistance between Voltage Probe 1 & Probe 2. Assume the gate length between Voltage Probe 1 & Probe 2 is L_C and the total gate length is L_G. The channel resistance of the ideal MOSFET is V_C/I_{DS} * (L_G/L_C). Since V_{DS}/I_{DS} is the total resistance of the real MOSFET, the parasitic resistance is calculated to be $V_{DS}/I_{DS} - V_C/I_{DS} * (L_G/L_C)$ as shown in Fig. 4.
4. Excluding the effect of parasitic resistances, the transfer characteristic of the ideal MOSFET can be extracted. The measured and corrected transfer characteristics (V_{DS} = 0.1V) of sample A is put together in Fig. 5.
5. The equation for an ideal MOSFET in the linear region:

$$I_d = \mu C_{ox} \frac{W}{L}(V_g - V_{th})V_{ds} \qquad (1)$$

Differentiating I_d by V_{GS} we get the tansconductance vs. V_{GS} and thereafter field effect mobility vs. V_{GS} based on the device dimension parameters.

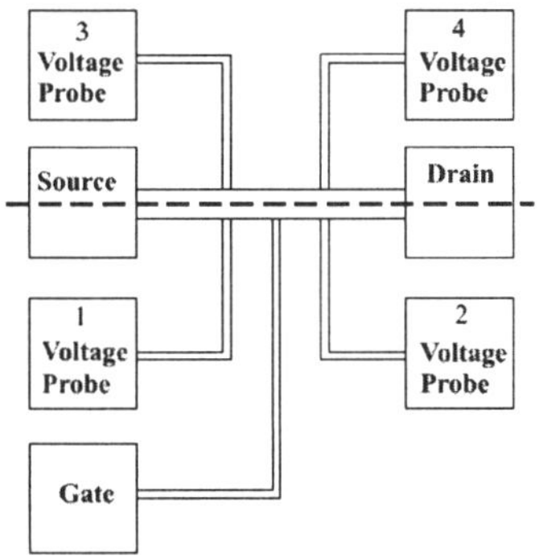

Fig. 2 MOS-Gated Hall Bar Structure

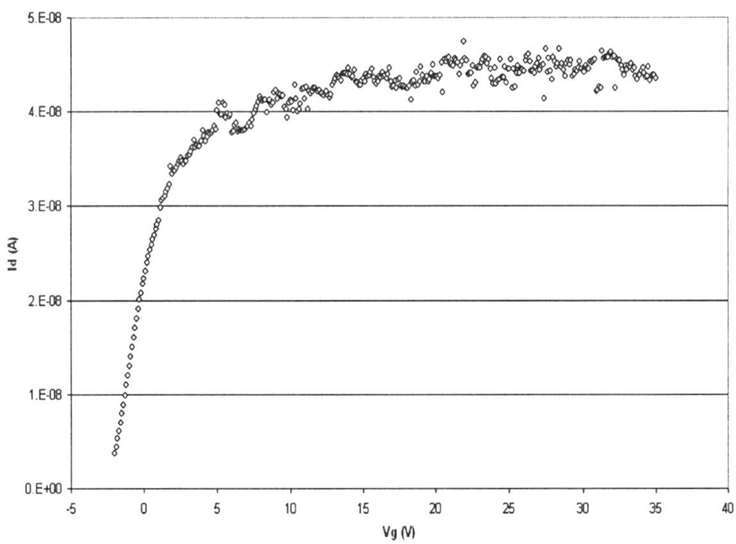

Fig. 3. Measured Transfer Characteristic (Sample A)

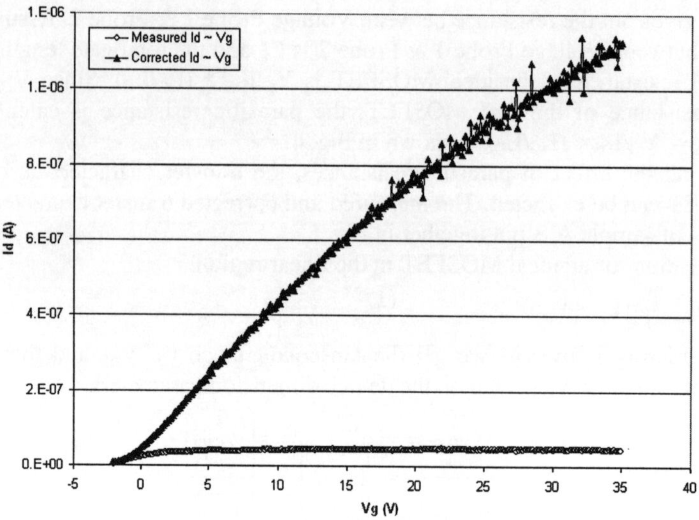

Fig. 4 Measured and Corrected Transfer Characteristics (Sample A)

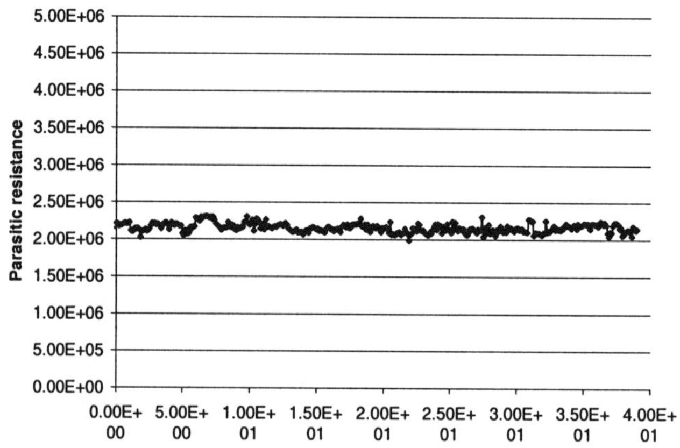

Fig. 5 Calculated parasitic resistance Vs. Gate Votage (Sample A)

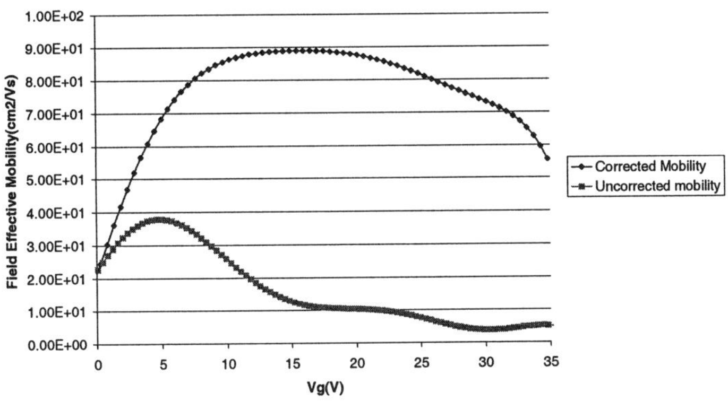

Fig.6 Field-Effect Mobility vs. Gate Voltage (Sample A)

The field-effect electron mobility reaches a maximum of 120 cm^2/V-s for Sample A and drops to 64 cm^2/V-s at V_g = 30 V (Fig. 6). Sample B has a maximum field-effect mobility of 90 cm^2/V-s which decreases to 70 cm^2/V-s at V_g = 30 V. Sample C has a maximum field-effect mobility of 53 cm^2/V-s.

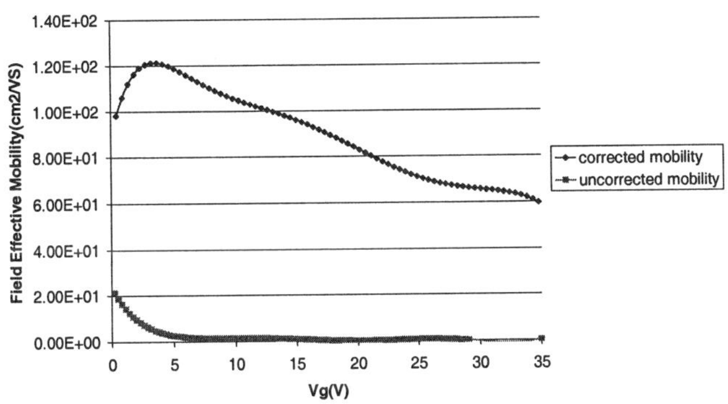

Fig. 7. Field-Effect Mobility vs. Gate Voltage (Sample B)

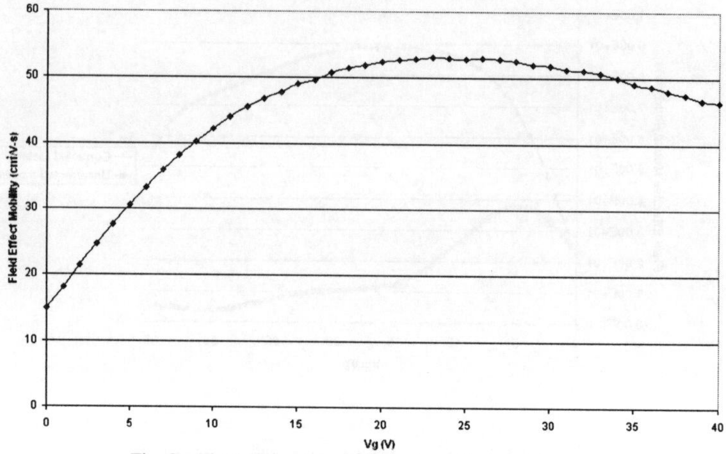

Fig. 8. Field-Effect Mobility vs. Gate Voltage (Sample C)

As shown in Fig. 4-7, there are significant difference between uncorrected field effective mobility and the corrected field effective mobility due to the existence of the large parasitic resistance in the order of 10^6 Ω including contact resistance, drain, source resistance, el. When the gate voltage increases, the channel conductivity increases, the channel resistance decrease, which become relatively small comparing to parasitic resistance. Then the field effective mobility decreases and even reach to zero as shown in Fig. 6-7. This hall bar structure can reveal the real channel resistance and gives us a more accurate value especially at high gate voltages.

3.2 Hall Mobility

The field-effect mobility extracted from the transconductance assumes that all charges induced by the gate are mobile and none are trapped at the interface, which is not the valid due to the existence of surface traps at the SiO_2/GaN interface. In order to separate mobile charges from the immobile ones, Hall measurements are utilized with the MOS-gated Hall bar structure.

Hall measurements have yielded a maximum Hall mobility of 160 cm^2/V-s with a maximum sheet carrier concentration of 3.5e12 cm^{-2} for Sample B at V_g =50 V, as shown in Fig. 9 and Fig. 10. Opposite to the case in Si, the Hall mobility increases with gate bias in GaN. Similar result is also observed in Sample A. Comparing the Field-effect and Hall mobilities, we can conclude that there are moderate carrier trapping at the SiO2/GaN interface because of the high mobile carrier concentration and < 2× difference in the mobility values.

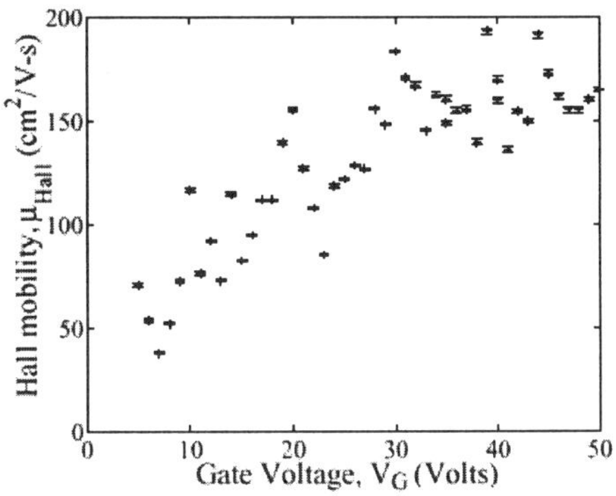

Fig. 9. Hall Mobility vs. Gate Voltage (Sample B)

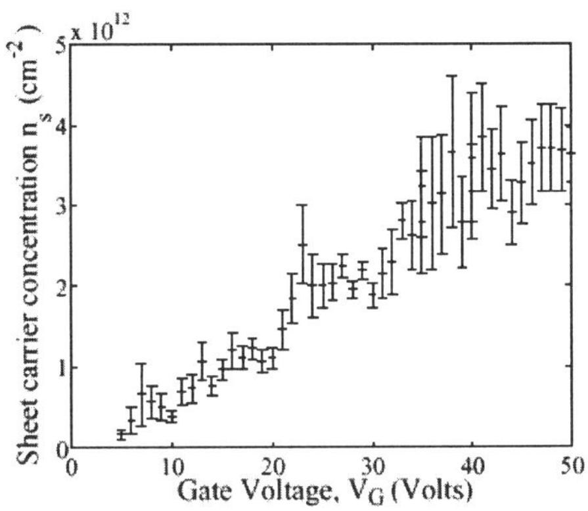

Fig. 10. Sheet Carrier Concentration vs. Gate Voltage (Sample B)

Summary

GaN MOSFETs were fabricated on three different GaN samples. Field-effect mobility and Hall mobility were extracted, based on the characterization of MOS-gate Hall bar structures. Different GaN epi-channel samples result in different mobility values.

Acknowledgment

The authors would like to thank the funding support of National Science Foundation (Award # EEC-9731677). They would also like to thank GE R&D at Schenectady and Philips Research for providing equipment for processes and measurements.

References

1. N. Q. Zhang, B. Moran, S.P. DenBaars, U.K. Mishra, X.W. Wang, and T.P. Ma, "Effects of surface traps on breakdown voltage and switching speed of GaN power switching HEMTs", *International Electron Devices Meeting Proceedings 2001*, pp. 25.5.1 – 25.5.4.

2. R. Gaska, M.S. Shur, X. Hu, J.W. Wang, A. Tarakji, G. Simin, A. Khan, J. Deng, T. Werner, S. Rumyantsev, and N. Pala, "Highly doped think-channel GaN-metal-semiconductor field-effect transistors", Applied Physics Letters, vol. 78, pp. 769-771, 2001.
3. H.C. Casey, Jr., G.G. Fountain, R.G. Alley, B.P. Keller, and S.P. DenBaars, "Low interface trap density for remote plasma deposited SiO_2 on n-type GaN", Applied Physics Letters, vol. 68, pp. 1850-1852, 1996.
4. S. Arulkumaran, T. Egawa, H. Ishikawa, T. Jimbo, and M. Umeno, "Investigations of SiO_2/n-GaN and Si_3N_4/n-GaN insulator-semiconductor interfaces with low interface state density", Applied Physics Letters, vol. 73, pp. 809-811, 1998.
5. H. Kim, S.-J. Park, and H. Hwang, 'Thermally oxidized GaN film for use as gate insulators", Journal of Vacuum Science and Technology B, vol. 19, pp. 579-581, 2001.
6. J.W. Johnson, B.P. Gila, B. Luo, K.P. Lee, C.R. Abernathy, S.J. Pearton, J.I. Chyi, T.E. Nee, C.M. Lee, C.C. Chuo, and F. Ren, "SiO2/Gd2O3/GaN metal oxide semiconductor field effect transistors", Journal of Electrochemical Society, vol. 148, pp. G303-G306, 2001.
7. Kevin Matocha, Ronold J.Gutmann, and T. Paul Chow, "Effect of annealing on GaN-Insulator interfaces Characterized by Metal-Insulator-Semiconductor Capacitors", IEEE Transaction on Electron Devices, vol. 50, No. 5, May 2003.

International Journal of High Speed Electronics and Systems
Vol. 14, No. 3 (2004) 805–809

World Scientific
www.worldscientific.com

FABRICATION OF SELF-ALIGNED T-GATE AlGaN/GaN HIGH ELECTRON MOBILITY TRANSISTORS

JAESUN LEE, DONGMIN LIU, HYEONGNAM KIM
MICHAEL L. SCHUETTE, and WU LU
Department of Electrical and Computer Engineering
The Ohio State University, Columbus, OH 43210, United States of America

JEFFREY S. FLYNN and GEORGE R. BRANDES
ATMI, Danbery, Connecticut 06810, United States of America

Self-aligned AlGaN/GaN high electron mobility transistors (HEMTs) are fabricated and the direct current and radio frequency small signal performance of self-aligned devices is characterized in comparison with non-self-aligned devices. An ultra-thin Ti/Al/Ti/Au ohmic metal scheme is used for gate to source and drain self-alignment. To suppress the gate leakage current, the ohmic contact annealing of self-aligned devices is performed in a furnace. The self-aligned devices with 0.25 μm gate-length and 100 μm gate-width exhibit good pinch-off characteristics. The maximum drain current at a gate bias of 1 V is 620 mA/mm for self-aligned HEMTs, and 400 mA/mm for non-self-aligned devices, respectively. A maximum extrinsic transconductance of 146 mS/mm is measured in self-aligned devices, while non-self-aligned HEMTs show only a peak g_m of 92 mS/mm. The self-aligned devices exhibit an extrinsic f_T of 39 GHz and an f_{MAX} of 130 GHz, whereas non-self-aligned HEMTs show an f_T of 15 GHz and an f_{MAX} of 35 GHz.

1 Introduction

AlGaN/GaN HEMTs have excellent potential for high frequency, high power, and high temperature applications because of their unique material properties such as high electron peak velocity, high breakdown voltage, high electron density, and high thermal and chemical stability. Significant progress in device performance has been demonstrated in the last few years.[1-8] However, further improvements of AlGaN/GaN HEMTs rely on the improvement of material quality and further reduction of parasitic resistance. To minimize the source access resistance, Ching-Hui Chen et al. re-grew an n+ GaN layer self-aligned to the gate to reduce the source and drain ohmic contacts using a re-grown process.[9] Another effective way is to reduce the distances between the gate to source and drain as closely as possible, which has been proved to be an effective methodology to improve device speed performance by Si, III-V devices. Self-aligned AlGaN/GaN HEMTs are very attractive because of the minimized source access resistance due to the reduction of separation from gate to source and drain. In the self-aligned process, the thin ohmic level follows the gate metallization. It is crucial that the ohmic annealing temperature should not exceed some critical point above which the Schottky gate characteristics are significantly degraded. However, the thick metal scheme and high processing temperature of ohmic contacts on III-nitrides hinder the realization of self-aligned devices. So far, little has been reported on self-aligned AlGaN/GaN HEMTs. Recently, it was demonstrated that the breakdown voltage of AlGaN/GaN HEMTs with Ni/Au gate could be significantly improved after a thermal annealing process even at 750

°C in a furnace,[10, 11] which makes the fabrication of self-aligned devices feasible. In this letter, we report self-aligned AlGaN/GaN HEMTs with thin ohmic contacts annealed at 750 °C and compare direct current (DC) and radio frequency (RF) small signal characteristics of self-aligned with those of non-self-aligned AlGaN/GaN HEMTs fabricated on the same material.

2 Fabrication

Fig. 1 shows the cross-sectional view of self-aligned T-gate AlGaN/GaN HEMT. The epilayer of AlGaN/GaN HEMT structure was grown by metal-organic chemical vapor deposition on (0001) sapphire substrate. The epilayer consists of 40 nm AlN nucleation layer, 3 μm of undoped GaN, and 20 nm undoped $Al_{0.3}Ga_{0.7}N$. The sheet resistance of epilayer is 432 Ω/sq. For device fabrication, device isolation was obtained with shallow mesa dry etching in a chlorine-based plasma. The drain-source first-level ohmic contacts with 3.5-μm spaces were deposited with Ti/Al/Pd/Au and annealed at 850 °C for 30 s in a rapid thermal annealing system. This level was included to reduce the contact resistance of the ohmic contacts. Electron-beam lithography and Ni/Au metallization were performed for gate contacts with a gate-length of 0.25 μm. The Ti/Al/Ti/Au thin ohmic layer with a total thickness of 80 nm was deposited and annealed. The thin ohmic layer is self-aligned to the T-gate using gate overhangs as a shadow mask. The annealing processing condition is chosen to make sure the formation of ohmic contact and the suppression of gate leakage current. Ni/Au layer was deposited for overlay. Fig. 2 is a SEM photograph of a typical 0.25 μm self-aligned AlGaN/GaN HEMT. It shows very smooth morphology of the thin ohmic layer even after annealing.

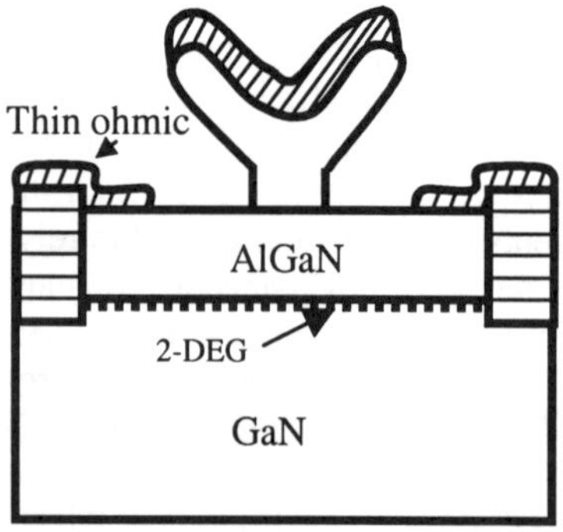

Fig. 1. Cross sectional view of self- aligned T-gate AlGaN/GaN HEMT.

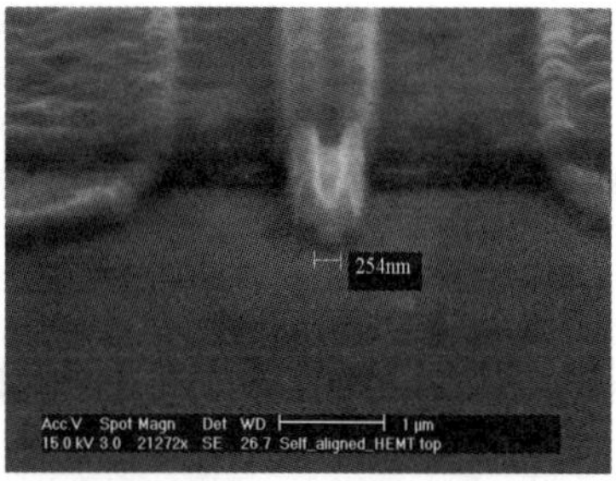

Fig. 2. SEM photograph of self-aligned T-gate AlGaN/GaN HEMT

3 Results and Discussion

In fabrication of self-aligned HEMTs, the formation of thin ohmic layer is crucial. Transmission line measurement (TLM) with Ti/Al/Ti/Au metal scheme was used to optimize annealing condition. Fig. 3 shows the current-voltage (I-V) characteristics between two contact pads with a spacing of 4 μm after annealing. It exhibits a linear relationship between current and bias voltage, showing good ohmic characteristics. The pad-to-pad resistances calculated from I-V data for different pad spaces are given in Fig 4. The ohmic contact resistivity and contact resistance of the thin ohmic layer are determined to be 3.8×10^{-6} Ωcm^2 and 0.8 Ωmm, respectively.

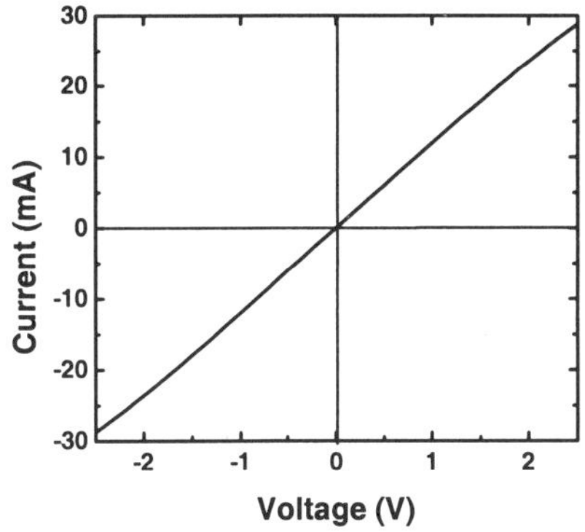

Fig. 3. Typical I-V characteristic between two ohmic contacts on AlGaN/GaN heterostructures with ultrathin Ti/Al/Ti/Au metallization.

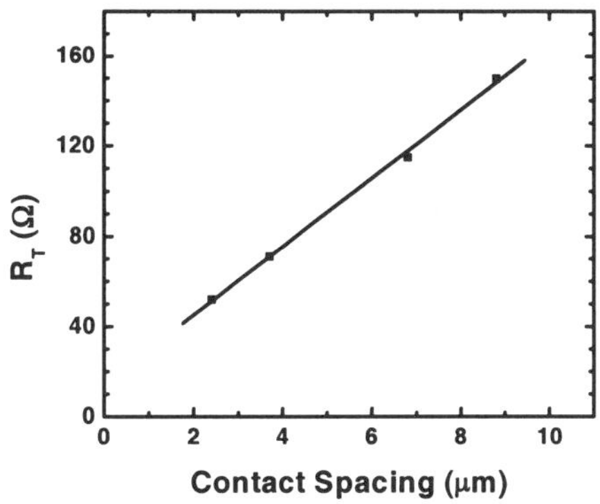

Fig. 4. Resistances of TLM pattern as a function of contact spacings.

Fig. 5 shows the typical drain current characteristics of self-aligned (solid lines) and non-self aligned AlGaN/GaN HEMTs with 0.25-μm gate-length and 100-μm gate-width. Despite the small separation between gate to source and drain, self-aligned devices exhibit excellent pinch-off behaviors. The gate biases range from –7 V to 1 V with a step of 1 V. The maximum drain current at a gate bias of 1 V and pinch-off voltage are 620 mA/mm and –7 V for self-aligned HEMTs, and 400 mA/mm and – 6 V for non-self-

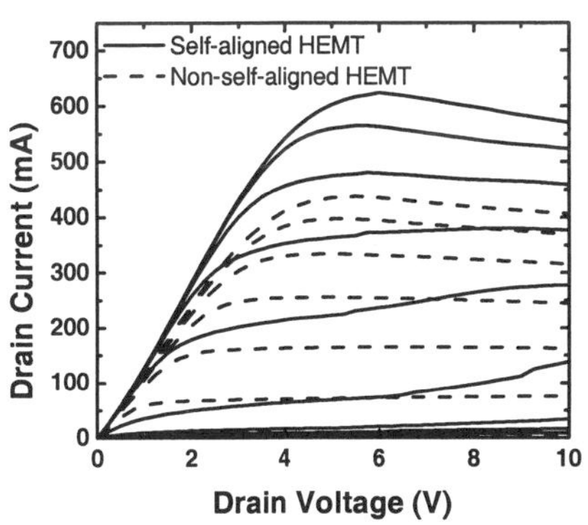

Fig. 5. I_D-V_D characteristics of self-aligned (solid lines) and non-self-aligned (dashed lines) AlGaN/GaN HEMTs. The gate voltage is in the range of -7 V to 1 V in a step of 1 V.

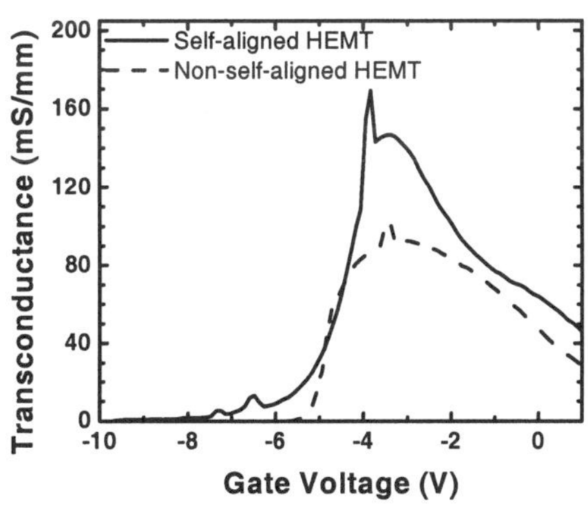

Fig. 6. Transconductance of self-aligned (solid line) and non-self-aligned (dashed line) AlGaN/GaN HEMTs. The drain bias is 8 V.

aligned devices, respectively. The knee voltages of both type devices are 4 V. At gate biases of 1, 0, and–1 V, the current-voltage characteristics of both type devices at drain bias voltage higher than 8 V exhibit negative differential resistance characteristic due to the poor thermal conductivity of sapphire substrates. Fig. 6 shows the typical transfer characteristics of self-aligned (solid lines) and non-self-aligned devices (dashed lines). As a result of reduced spacing between gate to source and drain, self-aligned devices exhibited a maximum extrinsic transconductance (g_m) of 146 mS/mm at a gate voltage of – 3.4 V and at a drain bias voltage of 6 V, while non-self-aligned HEMTs exhibited only a peak g_m of 92 mS/mm at a gate of – 3.1 V. The g_m improvement is attributed to the smaller source access resistance of self-aligned devices, $R_{acc} = R_{sh} \times L_{gs} = 1\ \Omega$, where R_{sh} is the sheet resistance and L_{gs} is the gate-to-source distance. For comparison, the non-self-aligned devices have a R_{acc} of 6.3 Ω. From the gate-bias intercept of the extrapolation of drain current curve, the threshold voltages of self-aligned and non-self-aligned devices are determined to be – 5.5 V and – 4.8 V, respectively. Fig. 7 shows gate-drain Schottky diode current characteristics of self-aligned devices. During measurements, the drain was shorted to the source. The gate-to-drain leakage current was determined to be 2×10^{-8} A at a gate bias of – 40 V, even though the gate to drain and source space is as small as only about 0.2 μm. This remarkably small gate leakage current is attributed to the reduction of trapping effects due to thermal annealing after formation of gate metallization.[12]

For microwave characteristics, on-wafer measurements of small signal S-parameters were performed from 1 to 50 GHz to determine f_T and f_{MAX}. Fig. 8 shows the plots of current gain $|h_{21}|$ and maximum stable power gain (MSG) and maximum available gain (MAG) versus frequency for self-aligned (solid lines) and non-self-aligned (dashed lines) devices. f_T and f_{MAX} were determined by the extrapolation of $|h_{21}|$ and MSG/MAG with –20 dB/decade slope. The self-aligned devices exhibit an extrinsic f_T of 39 GHz and an f_{MAX} of 130 GHz at a gate bias of – 3.5 V and a drain bias of 5 V, whereas non-self-aligned HEMTs show an f_T of 15 GHz and an f_{MAX} of 35 GHz at a gate bias of – 3 V and a drain bias of 6 V.

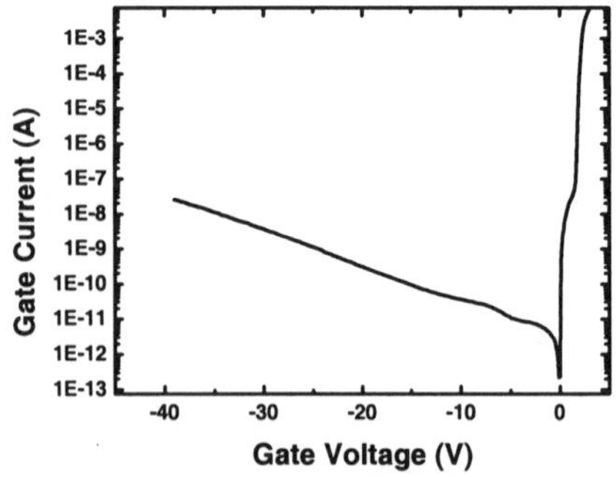

Fig. 7. Gate current characteristics of self-aligned AlGaN/GaN HEMTs as a function of gate bias.

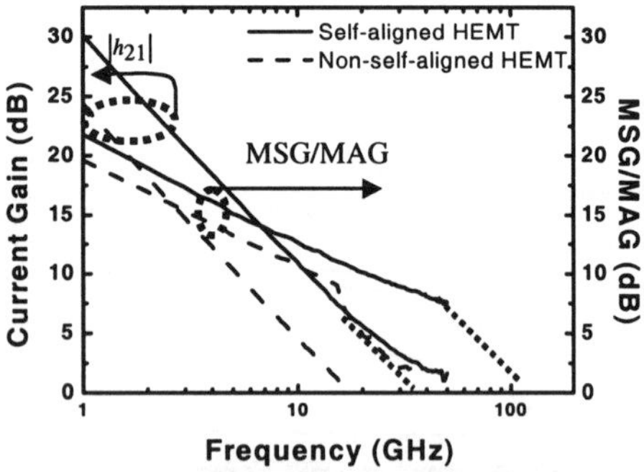

Fig. 8. Current gain and maximum stable/available power gain of 0.25 μm self-aligned and non-self-aligned AlGaN/GaN HEMTs.

4 Conclusions

We have fabricated self-aligned T-gate AlGaN/GaN HEMTs with 0.25 μm gate-length on a sapphire substrate. The self-aligned devices exhibit good pinch-off characteristics and very low gate leakage current. The maximum drain current at $V_g = 1$ V, g_m, f_T, and f_{MAX} of self-aligned devices are improved from 400 mA/mm to 620mA/mm, from 92 mS/mm to 146 mS/mm, from 15 GHz to 39 GHz, and 35 GHz to 130 GHz, respectively. All of these improvements are attributed to the smaller parasitics of self-aligned devices.

Acknowledgements

The authors would like to thank Dr. P. R. Berger and Dr. J. Bae for technical assistance. This work was partially supported by the National Science Foundation Grants DMR-0216892 and DMR-0313468.

References

1. W. Lu, V. Kumar, R. Schwindt, E. Piner, and I. Adesida, "DC, RF, and Microwave Noise Performances of AlGaN/GaN HEMTs on Sapphire Substrates," *IEEE Trans. Micro. Theory Tech.* **50** (2002) 2499-2504.
2. Y.-F. Wu, A. Saxler, M. Moore, R. P. Smith, S. Sheppard, P. M. Chavarkar, T. Wisleder, U. K. Mishra, and P. Parikh, "30-W/mm GaN HEMTs by Field Plate Optimization," *IEEE Electron Device lett,* **25** (2004) 117-119.
3. V. Kaper, V. Tilak, H. Kim, R. Thompson, T. Prunty, J. Smart, L.F. Eastman, J. R. Shealy, "High power monolithic AlGaN/GaN HEMT oscillator," *Gallium Arsenide Integrated Circuit (GaAs IC) Symposium, 2002. 24th Annual Technical Digest,* 20-23 Oct. 2002 251 – 254.
4. C. Lee, P. Saunier, Jinwei Yang, M. A. Khan, "AlGaN-GaN HEMTs on SiC with CW power performance of >4 W/mm and 23% PAE at 35 GHz," *IEEE Electron Device Lett.*, **24** (2003) 616 – 618.
5. Z. Y. Fan, J. Li, J. Y. Lin, and H. X. Jiang, "Delta-doped AlGaN/GaN metal-oxide-semiconductor heterostructure field-effect transistors with high breakdown voltages," *Appl. Phys. Lett.*, vol. **81** (2002) 4649-4651.
6. W. Lu, J. Yang, M. A. Khan, I. Adesida, "AlGaN/GaN HEMTs on SiC with over 100 GHz f_T and low microwave noise," *IEEE Trans. Electron Dev.* **48** (2001) 581-585.
7. V. Kumar, W. Lu, R. Schwindt, A. Kuliev, G. Simin, J. Yang, M. A. Khan, and I. Adesida, "AlGaN/GaN hemos on SiC with f_T of over 120 GHz," *IEEE Electron Dev. Lett.* **23** (2002) 455-457.
8. Y.-F. Wu, A. Saxler, M. Moore, R. P. Smith, S. Sheppard, P. M. Chavarkar, T. Wisleder, U. K. Mishra, P. Parikh, "30-W/mm GaN HEMTs by field plate optimization," *IEEE Electron Dev. Lett.* **25**, (2004) 117-119.
9. C. H. Chen, S. Keller, G. Parish, R. Vetury, P. Kozodoy, E. L. Hu, S. P. Denbaars, and U. K. Mishra, "High-Transconductance self-aligned AlGaN/GaN modulation-doped field-effect transistors with regrown ohmic contacts," *Appl. Phys. Lett.* **73** (1998) 3147-3149.
10. J. Lee, D. Liu, H. Kim, and W. Lu, "Post Annealing Effects on Device Performance of AlGaN/GaN HFETs," *Solid State Electron.* **48** (2004) 1855-1859.
11. J. Lee, D. Liu, H. Kim, and W. Lu, "Post-processing Annealing Effects on Direct Current and Microwave Performance of AlGaN/GaN High Electron Mobility Transistors," submitted to *Appl. Phys. Lett.*
12. H. Kim, J. Lee, and W. Lu, "Post-annealing effects on trapping behaviors in AlGaN/GaN HEMTs," submitted to *Phys. Stas. Sol.*

International Journal of High Speed Electronics and Systems
Vol. 14, No. 3 (2004) 810–815

World Scientific
www.worldscientific.com

A NEW FIELD-PLATED GaN HEMT STRUCTURE WITH IMPROVED POWER AND NOISE PERFORMANCE

HONGTAO XU, CHRISTOPHER SANABRIA, ALESSANDRO CHINI, YUN WEI, STEN HEIKMAN, STACIA KELLER, UMESH K. MISHRA AND ROBERT A. YORK

Department of Electrical and Computer Engineering

University of California, Santa Barbara, California 93106, USA

Field-plated structures can dramatically improve power capacity of GaN HEMT devices. In this paper, two different field-plated GaN HEMT structures will be demonstrated and compared to each other. The results show that a new GaN HEMT structure improves both power and noise performance without additional processing or costs.

1 Introduction

AlGaN/GaN HEMTs have attracted considerable interest as power devices in microwave applications, promising greater than a tenfold increase in power-density as compared with GaAs devices. Recently, the power capacity of AlGaN/GaN HEMT was dramatically improved by applying field-plated structures. It not only increases device breakdown voltage, but also reduces the high-field trapping effect. Using field-plates, AlGaN/GaN HEMTs with a CW power density of 32.2 W/mm at 4 GHz was reported [1]. AlGaN/GaN HEMTs have not only been identified as the technology for high power applications, but have also shown excellent noise characteristics for LNA circuits. The high survivability of GaN HEMT allows the reduction or elimination of front-end protection, and the fabrication of both low noise and high power circuits on the same wafer [2]. Noise characterization of field-plated GaN HEMTs has not been well studied yet.

In this paper, we will demonstrate a new field-plated GaN HEMT structure. Power and high frequency noise were characterized for both the new and conventional field-plated devices. With this new structure, both power and noise performance were found to be improved simultaneously.

2 Device Structures and Fabrication

In previously reported field-plated devices [1, 3], the field-plate was deposited on the SiN_x passivation layer and connected to the gate only through the common path of the gate feeder and pad in the extrinsic device region (Fig. 1(b)). Because of the RF delay on the gate fingers, the phase difference between the gate and the field-plate could weaken the enhancement of breakdown and the reduction of high-field trapping, especially when the gate finger width is comparable to the signal wavelength. Here we present a new structure that shorts the field-plate to the gate at the ends of the gate fingers (Fig. 1(c)). SiN_x, which covered the gate at the ends of the fingers, was etched away while patterning the passivation layer for bond metal. The field-plate metal and the exposed gate metal are then electrically connected at the ends by standard metal evaporation. The active area was still protected completely by the SiN_x passivation layer to reduce the high frequency dispersion.

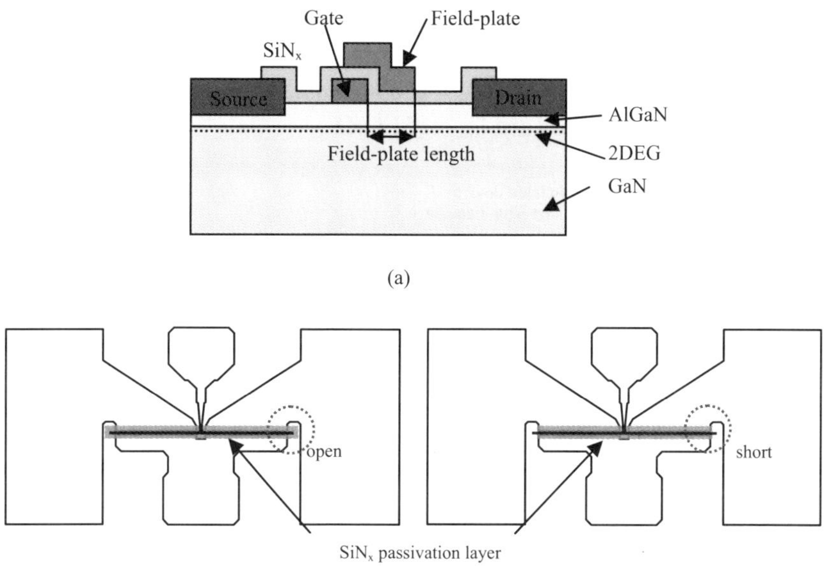

(a)

(b) (c)

Figure 1. (a) Cross-section of the GaN HEMT with a field-plate. (b) Conventional field-plate device structure. (c) New field-plated device structure with field-plates shorted to their gates at the ends of gate fingers.

In order to understand the impact of this new structure, conventional field-plated devices and new field-plated devices were designed and fabricated on the same substrate with an identical scale. The AlGaN/GaN HEMT devices were grown by metal organic chemical vapor deposition (MOCVD) on SiC substrates. The epitaxial structure consisted of a semi-insulating Fe-doped GaN base layer, followed by a 290 Å thick $Al_{0.27}Ga_{0.73}N$ barrier layer. The room temperature sheet electron concentration and Hall mobility were ~9.96×10^{12} cm^{-2} and ~1450 cm^2/Vs, respectively. The structure schematic of the field-plated AlGaN/GaN HEMT is shown in Fig. 1(a). They were fabricated using our standard process described elsewhere [3]. The GaN HEMT device has a gate length and width of L_g = 0.7 μm and W_g = 2x75 μm. Conventional and new structures with the same field-plate extensions (0.7 μm) to the drain will be compared and discussed in the paper.

3 DC and S-parameter Characterization

The fabricated devices were subjected to pulsed I-V and small-signal measurement. In Fig. 2, the output I-V characteristics obtained with pulsed width of 200 ns are shown. Passivated devices were capable of reaching a current density of I_{ds} = 1.25 A/mm at V_{gs} = 1 V. The different field-plated devices had very similar pulsed I-V curves, thus the dispersion effects were well reduced by the SiN_x passivation layer. S-parameters were measured at a bias of V_{ds} = 20 V and I_{ds} = 200 mA/mm. As shown in Fig. 3, two kinds of devices had similar f_t of 18 GHz and f_{max} of 51 GHz.

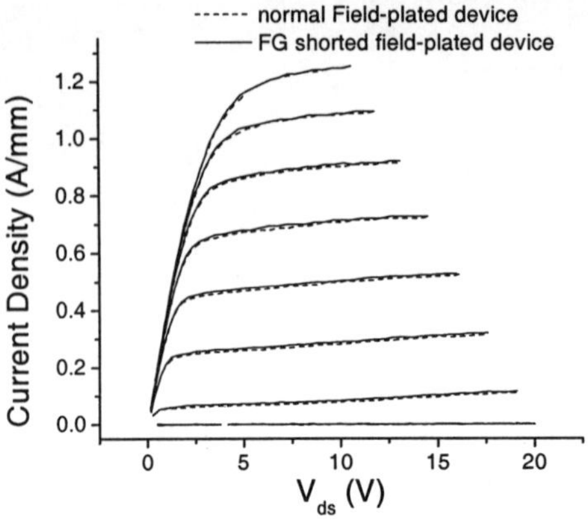

Figure 2. 200ns pulsed I-V curves of field-plated GaN HEMTs

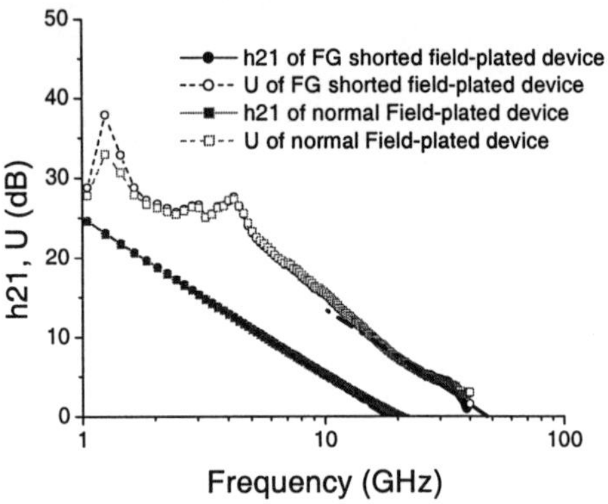

Figure 3. Small-signal characterization of field-plated GaN HEMTs

Small-signal parameters of both devices were extracted from s-parameter measurement data, based on HEMT model in Fig. 4. The charging resistance R_i is included in the equivalent circuit primarily to improve the match to S_{11}. The sum of R_g and R_i presented the gate parasitic resistance and channel resistance between the gate and the source. The simulated parameters were optimized by curve fitting. Simulation errors of s-parameters for both types of devices are less than 1 %. The simulated small-signal parameters are listed in Table 1. From the table, we see the output resistances of R_d and R_{ds} are not affected by the shorted-ends field-plated structure. But the total input resistances of R_g and R_i changed from 8.1 Ω to 7 Ω. In the standard field-plated device, the resistance of the gate fingers is in parallel with the resistance of the field-plate. When the field-plate metal is connected at the end of the gate fingers, the effective gate parasitic resistance is reduced. Also the ac current flows in the gate from both sides of each finger. It helps to improve the uniformity of the ac current on the gate.

4 Power Characterization

Single-tone power measurements were done at 4 GHz. During the measurement, conventional field-plated devices were failed to sustain high RF power with drain bias over 50 V, while devices with new modified field-plated structure can work up to 90 V. The uniformity of current on the gate could account for the improvement of the breakdown voltage. With additional reduction of the gate resistance, both saturated power and PAE were improved slightly.

	$R_i(\Omega)$	$R_{ds}(\Omega)$	$R_{gd}(\Omega)$	C_{gd}(fF)	C_{ds}(fF)	C_{gs}(fF)	g_m(S)
Normal FP device	7.178	1297.600	22.882	45.497	14.842	270.716	0.040
FG shorted FP device	6.068	1327.278	22.357	47.431	15.220	265.019	0.040
	$R_s(\Omega)$	$R_g(\Omega)$	$R_d(\Omega)$	L_s(pH)	L_g(pH)	L_d(pH)	tau(ps)
Normal FP device	6.531	0.924	5.503	12.656	49.223	13.463	2.606
FG shorted FP device	6.519	0.907	5.443	11.283	52.032	16.314	2.611

Table 1. Simulated small-signal parameters of field-plate HEMTs

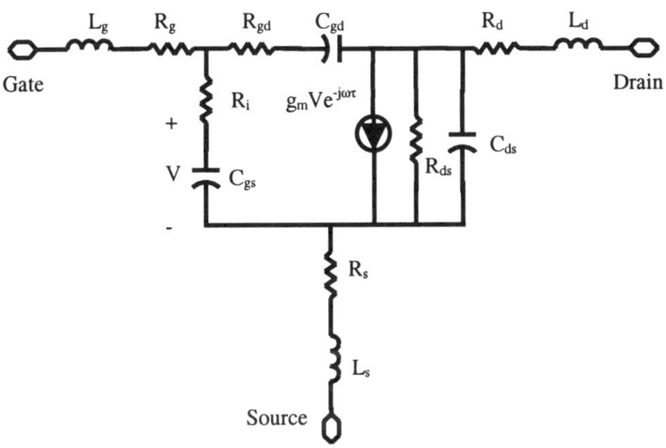

Figure 4. HEMT small-signal model including parasitic elements

Biased at V_{ds} = 30 V and I_{ds} = 38 mA/mm, the saturated output power was 7.2 W/mm while peak PAE was 75.6 % (Fig. 5). The output power density at peak PAE was 6.6 W/mm. With a higher drain bias of V_{ds} = 80 V and I_{ds} = 38 mA/mm, the saturated output power was 23.3 W/mm, while the peak PAE was 53.8 % (Fig. 6). Both output power density and PAE were improved from the previous reported results [3] with similar drain voltages. Fig. 7 shows measured saturated output powers and PAEs at different drain biases. The saturated output power kept increasing from 7.2 W/mm at 30 V drain voltage to 23.3 W/mm at 80 V drain voltage. But at 90 V drain voltage, the saturated output power dropped back slightly to 22.7 W/mm. PAE decreased almost linearly from 75 % to 46 % as the drain voltage increased from 30 V to 90 V.

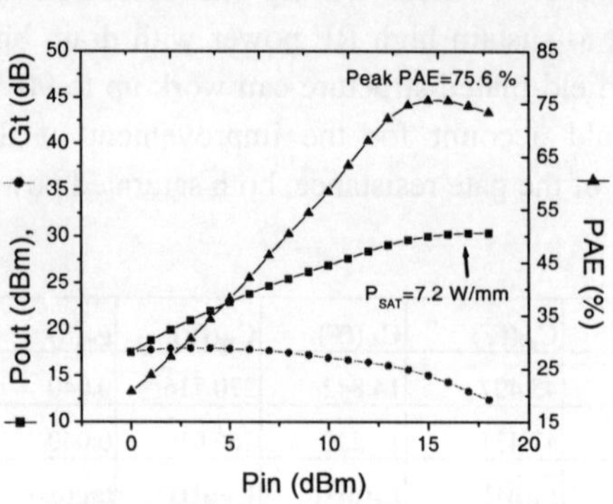

Figure 5. Single-tone power measurement of new field-plated device with 0.7 μm field-plate length at 4 GHz. Bias point was V_{ds}=30 V and I_{ds}= 38 mA/mm.

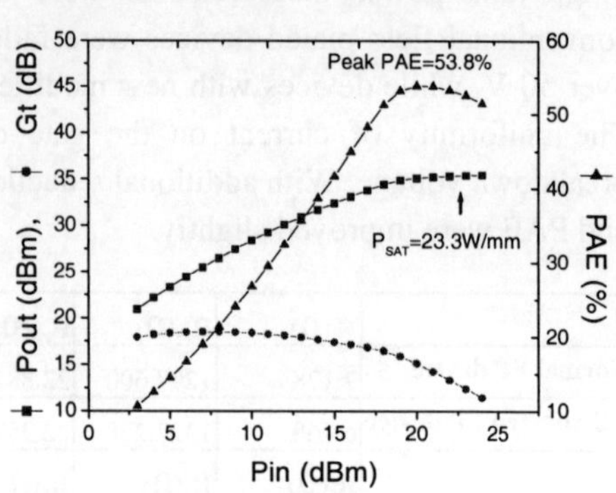

Figure 6. Single-tone power measurement of new field-plated device with 0.7 μm field-plate length at 4 GHz. Bias point was V_{ds}=80 V and I_{ds}= 38 mA/mm.

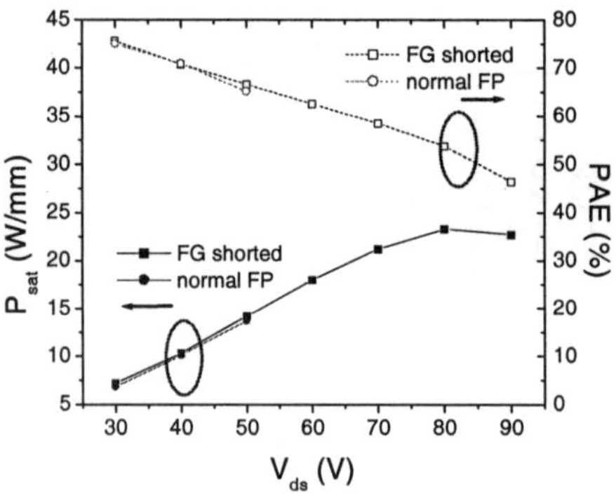

Figure 7. 4GHz saturated output power and PAE measurements of normal field-plated devices and devices with field-plates shorted to their gates at ends.

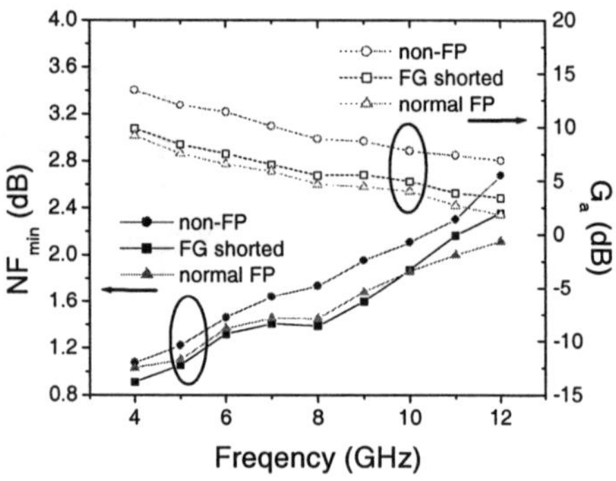

Figure 8. Noise performance of non-field-plated devices, normal field-plated devices and devices with field-plates shorted to their gates at ends.

5 Noise Characterization

As stated earlier, GaN HEMTs are also a candidate for low-noise applications at microwave frequencies. We characterized the noise performance with a Maury load-pull system. To further understand the impact of field-plates on the noise performance, a non-field-plate device with the same structure and on the same sample was tested with these two field-plated devices. High frequency noise measurements were done from 4 to 12 GHz. The optimum biasing point was found for each device by sweeping drain voltages and drain currents. In Fig. 8, both field-plated devices were found to have lower minimum noise figures, possibly because of lower gate resistances for field-plated devices. The associated gains of field-plated devices are lower because of larger gate to drain capacitances. Devices with field-plates shorted to their gates exhibit slightly better noise figure and gain than conventional field-plated devices, since the gate resistance is further reduced.

6 Summary

We have demonstrated a new field-plated GaN HEMT structure with the field-plate shorted to the gate at the finger ends. Using this new structure, the breakdown voltage of field-plated devices increased from 60 V to 100 V. At 4 GHz, field-plated GaN HEMT devices exhibited saturated output power of 23.3 W/mm with a peak PAE of 53.8 % with the drain voltage of 90 V. Both PAE and saturated output power are slightly higher because of the reduction of the gate resistance. Also, with this new field-plated structure, the minimum noise figure and associated gain are improved simultaneously. These results show that this new GaN HEMT structure improves both power and noise performance without additional processing and only a slight change to one photolithography mask..

Acknowledgments

This work was supported by ONR MURI (contract #N00014-01-1-0764) program monitored by Dr. H. Dietrich.

Reference

1. Y.-F. Wu, A. Saxler, M. Moore, R. P. Smith, S. Sheppard, P. M. Chavarkar, T. Wisleder, U. K. Mishra, and P. Parikh, "30-W/mm GaN HEMTs by Field Plate Optimization," IEEE Electron Device Lett., vol. 25, no. 3, pp. 117 -119, March. 2004.
2. H. Xu, C. Sanabria, A. Chini, S. Keller, U. K. Mishra and R. A. York, "A C-band high-dynamic range GaN HEMT low-noise amplifier," Microwave and Wireless Components Lett., vol. 14, no. 6, pp. 262-264, June. 2004.
3. A. Chini, D. Buttari, R. Coffie, L. Shen, S. Heikman, A. Chakraborty, S. Keller, and U. K. Mishra, "Power and Linearity Characteristics of Field-Plated Recessed-Gate AlGaN–GaN HEMTs," IEEE Electron Device Lett., vol. 25, no. 5, pp. 229 - 231, May. 2004.

International Journal of High Speed Electronics and Systems
Vol. 14, No. 3 (2004) 816–818

World Scientific
www.worldscientific.com

NOISE CHARACTERISTICS OF FIELD-PLATED GAN HEMTS

Y.-F. WU, M. MOORE, T. WISLEDER, P.M. CHAVARKAR, AND P. PARIKH

Cree Santa Barbara Technology Center, 340 Storke Road, Goleta, CA 93117, USA
Email: yifeng_wu@cree.com; Tel. 805 968 9460; Fax 805 968 9811

A. SAXLER

Cree Inc., 4600 Silicon Drive, Durham, NC 27703, USA

Investigation into field-plated GaN HEMTs has revealed that, although addition of the field plate reduces cutoff frequencies, noise figure actually improves. An analysis is presented to explain this phenomenon, which is related to the trade-off between gate-drain capacitance and gate conductance.

1 Introduction

Recently, electric-field modification by field plates has resulted in dramatic improvement in the current-voltage handling capabilities of GaN-based high-mobility-transistors (HEMTs). Operation at drain biases up to 120 V and power densities exceeding 30 W/mm have been reported[I]. This is a factor of 2-3x improvement over regular GaN HEMT structures. However, the tradeoff is reduction in cutoff frequencies due to the increased gate-drain capacitance, especially at lower biases where a low-noise device typically operates. Although the enhanced robustness by the field plate is also attractive for receiver applications to eliminate the front-end limiter circuit, the reduced cutoff frequencies are an important concern, since traditionally a lower cutoff frequency directly translates to a higher device noise figure. A systematic study is needed to investigate this issue.

2 Device Fabrication and Characteristics

The structure and fabrication of the GaN HEMTs in this study were reported before[I]. Field plates, which were electrically connected to the gate, were placed on top of the SiN-passivated gate with extension length L_f ranging from 0-1.1 μm. The gate dimensions were 0.55 x 246 μm^2, the on-resistance was 3 Ω-mm, the pinch-off voltage was –4V and the current density was 1 A/mm. The peak current-gain cutoff frequency and the associated power-gain cutoff frequency were 29.4 and 64.4 for L_f =0, and decreased gradually to 22.9 and 44.6 GHz for L_f =1.1 μm.

3 Noise Performance and Discussion

Noise figures at X-band were found to be minimum at 8-12 V drain bias for all devices. Figure 1 shows the results for a non-field-plate device and a device with field plate length of 0.9 μm. Contrary to conventional behavior that device noise figure increases with reduction in cutoff frequencies, the field-plated device clearly exhibits improved noise characteristics. A more detailed comparison for devices with increasing L_f values is showed in Figure 2. The noise figure systematically reduces with increase in L_f up to 0.9 μm.

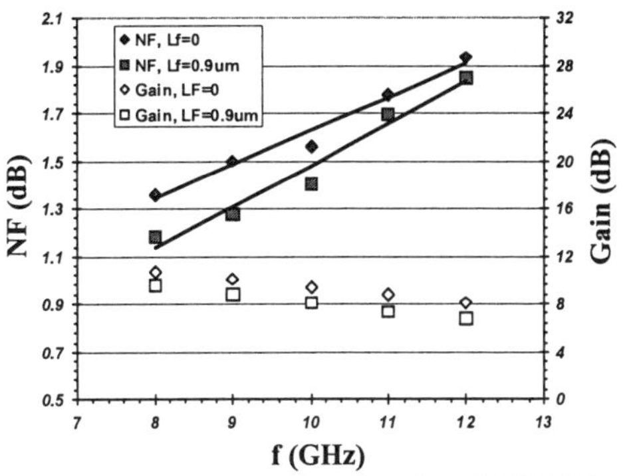

Figure 1, Noise figure vs. frequency for a GaN HEMT without a field plate (L_f=0) and and one with a field-plate length L_f=0.9μm. The field-plated device shows lower noise figure throughout the frequency range.

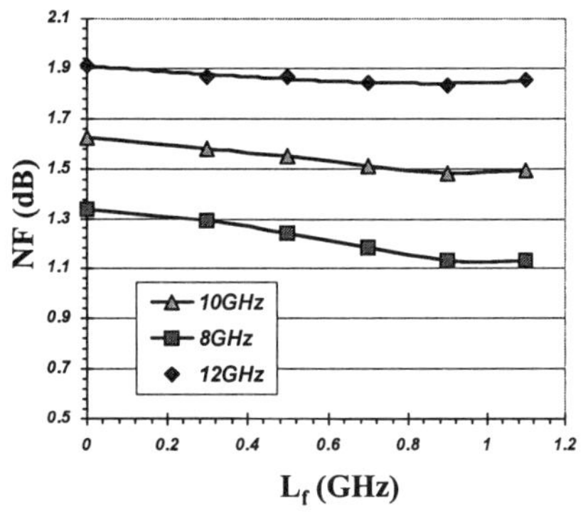

Figure 2, Noise figure at 8, 10 and 12 GHz as a function of field-plate length L_f. Systematic reduction of noise figure is seen with increase in L_f up to 0.9 μm.

To understand this phenomenon, circuit model elements were extracted from the scattering parameters by a vector network analyzer as shown in Table 1. The major parameters to determine intrinsic device gain include transconductance g_m, gate capacitance C_{gs} and output resistance R_{ds}, which are practically not affected by the field plate at such low bias voltage. Except for the output capacitance C_{ds}, which could be tuned out by output matching, two components sensitive to the field plate are the gate-drain capacitance C_{gd} and the total input resistance R_g+R_i, where R_g is the conductive resistance of the metal gate line and R_i is the channel resistance in series with C_{gs} (see Figure 3). As expected, increasing L_f increases C_{gd}, which reduces external cutoff frequencies and decreases gain. However, addition of the field plate results in an enhancement in gate conductance, which reduces device input resistance and decreases

Table 1, Device circuit elements extracted from S-parameters (device width: 246 μm; Bias: 10V x 180mA/mm; Lf: field-plate length; Rg+Ri: total input resistance.)

Lf (μm)	Gm (mS)	Cgs (fF)	Cgd (fF)	Cds (fF)	Rds (ohm)	Rg+Ri (ohm)	Ft (GHz)	Fmax (GHz)
0	*80.0*	*341.8*	*49.0*	*65.6*	*633.8*	*8.44*	*29.4*	*64.4*
0.5	*79.5*	*346.5*	*73.4*	*50.2*	*669.8*	*7.07*	*25.6*	*55.1*
0.7	*79.4*	*342.1*	*87.3*	*46.9*	*666.4*	*6.47*	*24.7*	*52.3*
0.9	*78.7*	*340.5*	*99.9*	*41.4*	*621.8*	*6.20*	*23.7*	*48.8*
1.1	*79.8*	*335.8*	*117.6*	*35.5*	*624.8*	*5.85*	*22.9*	*44.6*

input Johnson noise.

The noise sources of a device could be separated into an input part and an output part. The input noise directly contributes to the overall noise while the output noise could be suppressed by the gain of the device. As L_f increases, the input noise reduces but the decreased gain (due to an increase in C_{gd}) should increase the contribution of the output noise. Apparently, at L_f >0.9 μm, the reduction in input noise cannot not overcome the increased contribution from the output noise. Hence the overall device noise figure starts to increase. This is also consistent with the observation that the improvement in noise figure is more pronounced at the lower frequency regime where gain is high and the contribution from the output noise is less.

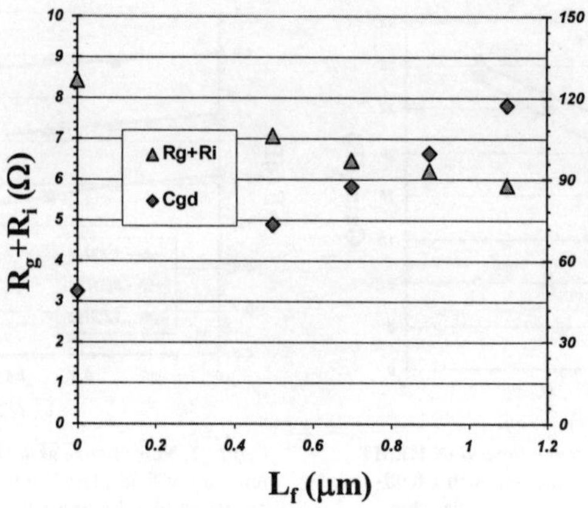

Figure 3, Input resistance R_g+R_i and gate-drain capacitance C_{gd} vs. field-plate length L_f.

4 Conclusion

In summary, our analysis has shown that the field plate does not compromise the intrinsic gain characteristics of the GaN HEMT. Although it increases the gate-drain capacitance, it also increases gate-conductance, hence reduces noise figure. This, coupled with the inherent high-power handling capabilities of the field-plated devices, makes them promising for robust low-noise front-end receiver applications.

[I] Y.-F. Wu, A. Saxler, M. Moore, R.P. Smith, S. Sheppard, P.M. Chavarkar, T. Wisleder, U.K. Mishra, and P. Parikh, "30-W/mm GaN HEMTs by Field Plate Optimization", *IEEE Electron Device Letters*, Vol. 25, No. 3, pp. 117-119, March 2004.

International Journal of High Speed Electronics and Systems
Vol. 14, No. 3 (2004) 819–824

TEMPERATURE DEPENDENT I-V CHARACTERISTICS OF AlGaN/GaN HBTS AND GaN BJTS

HUILI G. XING*, UMESH K. MISHRA
Electrical and Computer Engineering Department
University of California, Santa Barbara, CA 93106, USA

DC I-V characteristics of AlGaN/GaN heterojunction bipolar transistors (HBTs) and GaN homojunction bipolar transistors (BJTs) are analyzed in the temperature range of 200-450 K. At low current levels, the adverse effects of poor ohmic contacts coupled with paths of high leakage make it difficult to extract intrinsic device operation ["Explanation of anomalous current gain observed in GaN based bipolar transistors", Xing et al. IEEE Elect. Dev. Lett. 24(1) 2003:p.4-6]. At intermediate current levels, owing to enhanced ionization of Mg in the base, the HBTs show an increase in current gain resulting from mitigated current crowding, and the BJTs show a decrease in current gain resulting from reduction of emitter injection coefficient. The offset voltage dependence on temperature is also explained.

1 Introduction

Both experimental and theoretical work has been published on the operation of GaN-based bipolar transistors in the past decade. Some groups have reported an increased current gain β of AlGaN/GaN heterojunction bipolar transistors (HBTs) at higher temperatures due to enhanced ionization of deep Mg accepters in the base [1], while others have reported an increase in β at lower temperatures [2]. In this paper, we show that this discrepancy can be largely attributed to the different operating current levels of the devices. DC characteristics of both GaN BJTs and AlGaN/GaN graded HBTs are presented in the temperature range of 200 – 450 K, in order to have a better understanding of the device operation mechanisms.

2 Experiments

The transistors were grown by metalorganic chemical vapor deposition (MOCVD) on c-plane sapphire substrates, and fabricated using the technique of selective emitter regrowth. The details of growth and device fabrication can be found elsewhere [3, 4].

The device structures are shown in Fig. 1: both the GaN BJTs and AlGaN/GaN HBTs employed a 100 nm thick Mg-doped GaN base and the HBTs incorporated a graded AlGaN emitter structure. The completed devices have an emitter size of 20 μm x 50 μm.

* Email: hxing@nd.edu, now with Electrical Engineering Department, University of Notre Dame, IN 46556

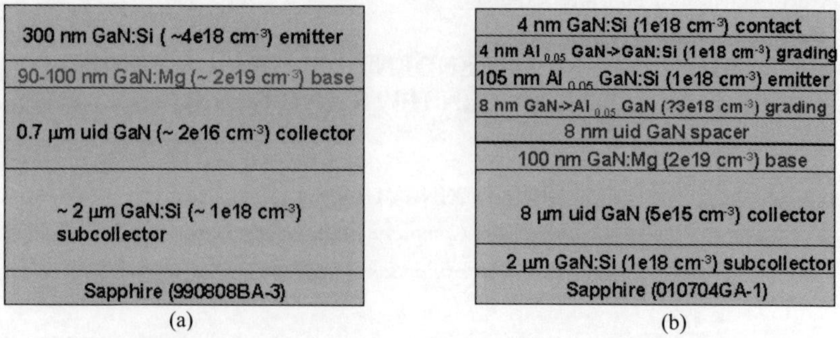

Fig.1 Device layer structures of (a) GaN BJT and (b) AlGaN/GaN HBT

3 Results and discussions

Both common emitter and Gummel Plot I-Vs were measured at varying temperatures. For GaN-based bipolar transistors, it is generally difficult to form ohmic contacts to the p-type base [5]. Sometimes the Schottky-like contacts exhibit a burn-in phenomenon in that their I-V characteristics improve during device testing. In order to minimize such effects, the devices were tested for several temperature cycles between 200 K and 450K and their characteristics were found to be reproducible. At 300 K, the BJTs show a current gain β of ~ 5 and the HBTs show β of ~ 20 at 10 mA, which corresponds to an average output current density ~ 1 $kAcm^{-2}$.

Figures 2 shows the common emitter family curves of the BJTs and HBTs at several temperatures: 200 K, 300 K and 400 K. First, it is obvious that the maximum output current I_C drops with the increasing temperature for both type of devices at a given set of base currents I_B, indicating a current gain drop. However, it is also noticed that it drops in very different fashions for the BJTs and HBTs. For the BJTs, I_C, thus the current gain, drops for almost the entire base current levels (I_B = 0.5 – 5 mA). While for HBTs, I_C remains almost constant up to 10 mA with I_B = 0.1 - 0.5 mA for all the temperatures, indicating a constant current gain; for a higher I_C, a pronounced current gain drop is observed. This difference can be explained by the classical behavior of the homojunction and heterojunction bipolar transistors along with the nature of Mg in GaN as a deep acceptor. The emitter injection coefficient increases with increasing ratio of the electron current component I_{nE} and the hole current component I_{pE} of the emitter current: $I_{nE}/I_{pE} \propto n_E/p_B \exp(\Delta E_g/kT)$. Here, n_E is the electron concentration in the emitter, p_B the hole concentration in the base, and ΔE_g the bandgap barrier seen by the holes back injecting from the base. Using an activation energy of 160 meV and a Mg concentration of 2 x 10^{19} cm^{-3}, it is calculated that the hole concentration increases from ~ 10^{16} cm^{-3} at 200 K to ~ 10^{18} cm^{-3} at 450 K. For BJTs, this will cause a large change (two orders of magnitude) in I_{nE}/I_{pE} with respect to temperature since $\Delta E_g = 0$. Therefore, a large drop in current gain is expected at elevated temperatures. For HBTs with a graded $Al_{0.05}Ga_{0.95}N$ emitter, $\Delta E_g = 140 meV$, i.e. $\Delta E_g/kT = 3.6 \sim 8.1$ for $T = 200 \sim 450K$. The exponential term

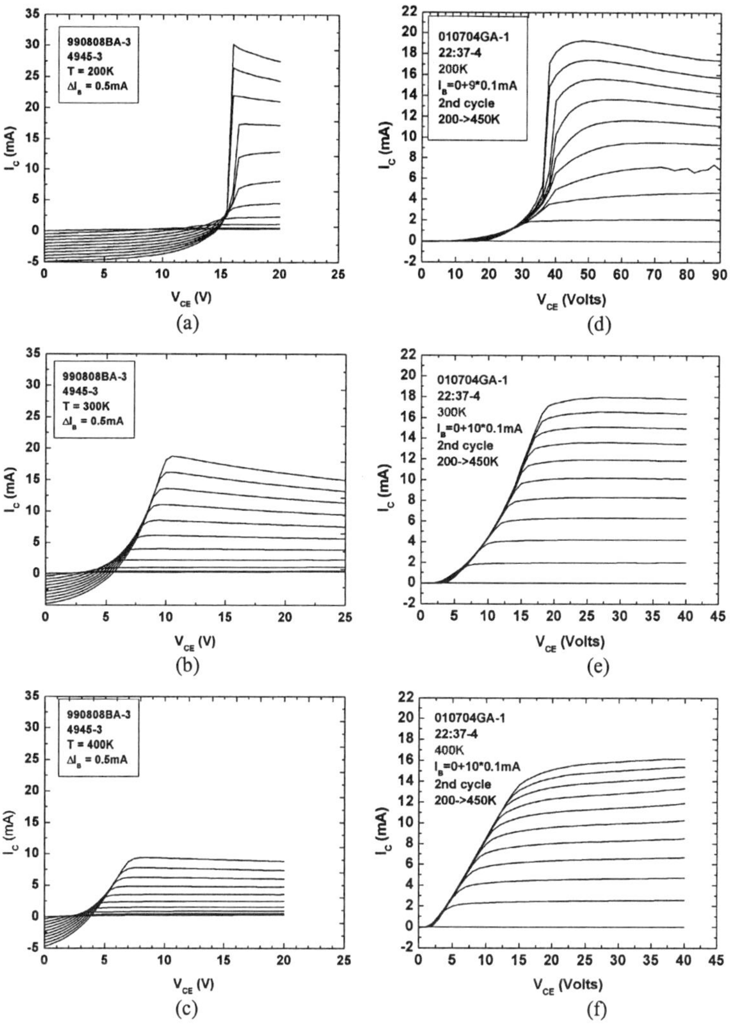

Fig.2 Common emitter characteristics in the temperature range of 200-400 K of: (left) GaN BJTs for $\Delta I_B = 0.5mA$, (a) 200 K, (b) 300 K, and (c) 400 K; (right) AlGaN/GaN HBTs for $\Delta I_B = 0.1mA$: (d) 200 K, (e) 300 K, and (f) 400 K.

clearly dominates the ratio I_{nE}/I_{pE} , thus, the current gain is expected to remain constant in this temperature range.

Another pronounced feature in Fig. 2 is the increase of the offset voltage V_{offset} with decreasing temperature. For a bipolar transistor with many external elements including non-ideal ohmic contacts and highly resistive base (i.e. high base access resistance), V_{offset} consists of three major components [6]:

$$V_{offset} \approx \Delta V_{B,cont} + (1-\alpha_{offset}) I_B R_{B,access} + I_B R_E ,$$

where $\Delta V_{B,cont}$: lateral voltage drop under the base contacts on the emitter side owing to the base current crowding around this edge of the base contacts,

α_{offset} : common base current gain of the intrinsic transistors at V_{offset},

$R_{B,access}$: base access resistance owing to the finite distance between the base contacts and the intrinsic transistor,

$I_B R_E$: voltage drop across the parasitic emitter resistance including the non-ohmic behavior of the emitter contacts.

At low temperatures, $R_{B,access}$ increases because of the decrease of hole concentration, R_E also increases since those devices employed non-alloyed emitter contacts. It is also worth noting that V_{offset} depends on the current gain and I_B. Therefore, at similar I_C, the BJT show a higher V_{offset} than the HBT. Owing to this extraordinarily high V_{offset} and I_B for the BJT, its base-collector diode is under severe forward bias when $V_{CE} < V_{offset}$ and $I_B > 1mA$. This causes the measured collector current to be negative, shown in Fig. 2.

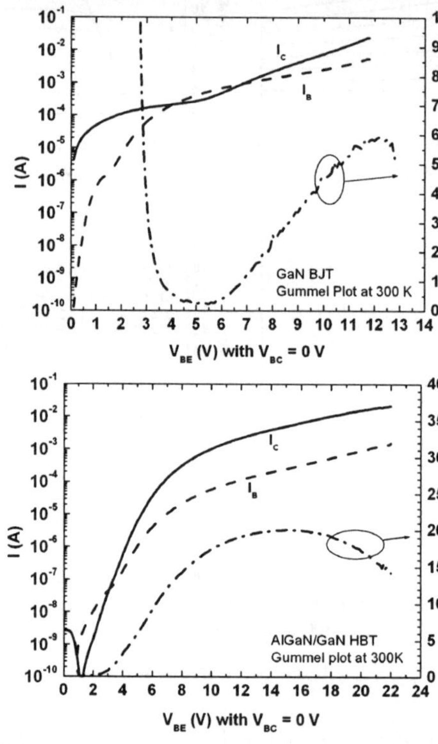

(a) (b)

Fig. 3 Gummel plot at 300 K of (a) GaN BJT and (b) AlGaN/GaN HBT.

Figure 3 shows the Gummel plot of both GaN BJT and AlGaN/GaN HBT at 300 K. The base-emitter diode of the BJT turns on about 3 - 4 V, and the base-emitter diode of the HBT turns on about 5 - 6 V, which is higher than the GaN bandgap owing to the poor ohmic contacts. Similarly, due to the high parasitic resistance, a base-emitter voltage of around 10 V is required to obtain a current level of mA range. To illustrate the current gain dependence on temperature at different current levels, Fig. 4 plots β determined from Gummel plots versus I_C for both devices at the temperature range of 200 - 450 K. For the BJT, in the low current region (< 10 A/cm^2), the apparent β increases dramatically with decreasing I_C or decreasing temperature. For example, at $I_C = 10^{-4}$ A, the apparent β increases from 1 at 450 K to 100 at 200 K (out of scale). However, the measurements at low currents do not describe the intrinsic device operation since the apparent high β results from the high leakage currents coupled with poor base contacts, and the detailed explanation can be found in Reference 7. In the intermediate current region, β shows an increase with increasing I_C, a classic behavior owing to the recombination current in the emitter-base junction. As explained above, the decrease of β with increasing temperature is mainly attributed to the increase of hole concentration in the base. In the high current region, the trend of β vs. temperature is further complicated by competing mechanisms including self-heating and emitter crowding induced Kirk effect [6].

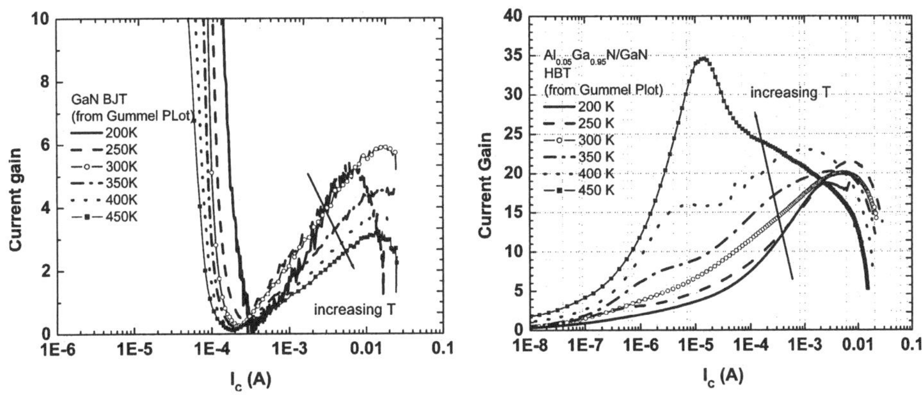

(a) (b)

Fig.4 Current gain vs. collector current in the temperature range of 200 – 450 K for (a) GaN BJTs and (b) $Al_{0.05}Ga_{0.95}N$ HBTs, with an emitter size of 20 x 50 μm^2.

Similarly for the HBT, only the intermediate current region describes the intrinsic device operation. In the I_C range of 10^{-4}-10^{-3} A, β increases with increasing temperature. This is consistent with the simulation result [1], which was attributed to the spreading of 2 dimensional base current at higher temperature owing to the higher degree of Mg activation. At higher currents (~ 10^{-3} A) where the base-emitter recombination current is negligible, the device gain shows little change with temperature, as expected for HBTs.

4 Conclusions

Temperature dependent I-Vs of AlGaN/GaN HBTs and GaN BJTs are analyzed. At intermediate current levels, owing to enhanced ionization of Mg in the base, HBTs show an increase in β resulting from mitigated current crowding and BJTs show a decrease resulting from deduction of emitter injection coefficient. The sources of offset voltage are also presented.

This work was supported by the DoD Multidisciplinary University Research Initiative (MURI) program administrated by the office of Naval Research under Dr. Harry Dietrich.

References

1. Monier C., S. J. Pearton, P.C. Chang, A.G. Baca and F. Ren, Appl. Phys. Lett., 76(21): P. 3115-7, 2000
2. Huang J.J., M. Hattendorf, M. Feng, D.J.H. Lambert, B.S. Shelton, M.M. Wong, U. Chowdhury, T.G. Zhu, H.K. Kwon and R. Dupuis, IEEE Elect. Dev. Lett., 22(4): P. 157-9, 2001
3. Xing, H., L. McCarthy, S. Keller, S. P. DenBaars and U. K. Mishra, proceeding of Int. Symp. on Compound Semiconductor, Monterey, CA, USA, October 2000
4. Xing, H., P. Chavarkar, S. Keller, S. P. DenBaars and U. K. Mishra , IEEE Elect. Dev. Lett. 24(1): P.4-6, 2003
5. Xing, H., D. S. Green, L. McCarthy, I. P. Smorchkova, P. Chavarkar, T. Mates, S. Keller, S. P. DenBaars, J. Speck and U. K. Mishra, proceeding of BIPOLAR/BiCMOS Circuits and Technology Meeting, Minneapolis, MN, USA, September 2001.
6. Xing, H. and U. K. Mishra, Int. Conf. on Nitride Semiconductors, Nara, Japan, 2003.
7. Xing, H., D. Jena, M. J. Rodwell and U. K. Mishra, IEEE Elect. Dev. Lett. 24 (1) 2003: p.4-6

International Journal of High Speed Electronics and Systems
Vol. 14, No. 3 (2004) 825–830

World Scientific
www.worldscientific.com

ANALYSIS OF HIGH DC CURRENT GAIN STRUCTURES FOR GaN/InGaN/GaN HBTs

JAMES C. LI, DAVID M. KEOGH, SOUROBH RAYCHAUDHURI,
ADAM CONWAY, DONGJIANG QIAO, AND PETER M. ASBECK

Electrical and Computer Engineering Department, University of California - San Deigo, 9500 Gilman Drive, La Jolla, CA 92093, USA

AlGaN/GaN HBTs with DC current gains in excess of 10 have been demonstrated; however, Makimoto *et al* have recently obtained a DC current gain value exceeding 2000 for a GaN/InGaN/GaN triple mesa DHBT.[1,2,3] The experimental demonstration of a GaN-based HBT with such extraordinary DC current gain has motivated the search for new device structures conducive to high DC current gain. Simulations in this work indicate that high DC current gain is difficult to achieve with a uniform, defect-free base layer. We also simulate the electrical characteristics of a new class of DHBT where the base layer is non-uniform. By non-uniformly thinning or perforating the base, the majority of the base would remain thick to minimize the negative impact to base resistance. However, small thinned regions achieve extremely high DC current gain, which can be used to significantly increase the overall DC current gain. This device structure could naturally occur when "V" defects form or Indium is non-uniformly distributed during the base layer growth. The sensitivity of DC current gain to the base thickness range, duty cycle, defect geometry, and defect type is also investigated.

I Introduction

Due to the favorable materials characteristics of GaN and its related materials, GaN-based HBTs are being developed worldwide for high voltage and switching applications. AlGaN/GaN HBTs with DC current gains (β) in excess of 10 have been demonstrated; however, Makimoto *et al* have recently obtained a β value exceeding 2000 for a GaN/InGaN/GaN triple mesa DHBT.[1,2,3] The experimental demonstration of a GaN-based HBT with such extraordinary β has motivated the search for new device structures conducive to high β. Here, a two-dimensional drift-diffusion model of a GaN/InGaN/GaN DHBT was used in the DESSIS device simulator to investigate device issues specific to GaN and its related materials. Simulations using traditional device structures and design parameters were found to yield β values barely over 100. However, it was found that various structures with nonuniform base regions could have significantly higher β.

2 Device Structure and Simulation

A two-dimensional structure of the triple mesa GaN/InGaN/GaN DHBT was considered. To reduce the computational burden, the HBT was assumed symmetric about a plane bisecting the emitter. The emitter mesa half width is 5μm with a 4.5μm wide contact; the base-collector mesa half width is 8μm with a 2μm wide contact; and the sub-collector mesa half width is 13μm with a 3μm wide contact. The vertical structure is composed of a 1000nm thick n+ GaN sub-collector ($4x10^{18}cm^{-3}$); a 500nm n-type GaN collector ($1x10^{17}cm^{-3}$); a 30nm n-type InGaN compositionally graded region ($2x10^{18}cm^{-3}$); a 100nm p+ InGaN base ($2x10^{19}cm^{-3}$); and a 50nm n+ GaN emitter ($4x10^{19}cm^{-3}$). Since this investigation is limited to DC behavior, no attempt is made to model the substrate, metal interconnections, or encapsulating dielectric materials.

Compared to a traditional HBT, the InGaN graded region and GaN emitter are highly doped. This doping profile ensures good metallurgical definition of the base-collector and base-emitter junctions despite Mg out-diffusion during layer growth. In addition, higher donor concentration in the InGaN graded region is required to compensate for the acceptor-like polarization charge resulting from compositional grading. Although the magnitude of the polarization charge is dependent on the Indium concentration, the donor concentration in the GaN emitter and InGaN graded region is kept constant throughout all simulated cases. The emitter, base, and collector specific contact resistances ($R_{C,SPEC}$) are assumed to be $1x10^{-6}\Omega cm^2$ since previous works have shown excellent $R_{C,SPEC}$ to p-type InGaN and n-type GaN.[4,5]

Due to the high value and variability of the Magnesium activation energy (E_A) in GaN and InGaN films, E_A must be properly defined in all layers, and incomplete ionization of the acceptors must be considered in the base.[6] Appropriate values of E_A were inserted manually into the structure definition. Parameters associated with the band structure and the polarization charge densities were taken from the work of Zhang *et al.*[7] The remaining material parameters were taken from the *Ioffe Physio-Technical Institute* electronic archive.[8] Carrier transport was limited to the solutions of the drift-diffusion equations. The impacts of bulk defects were modeled through the degradation of material parameters such as mobility and minority carrier lifetime.

3 Device Design

First, we consider the InGaN films to be used in the base. Since the acceptor E_A is approximately 160meV in GaN, the hole population in the base is approximately $7x10^{17}cm^{-3}$ at room temperature for the $2x10^{19}cm^{-3}$ Mg doping. Coupled with the low hole mobility (<$10cm^2/Vs$), the base sheet resistance can exceed 100kΩ/❑ for a 100nm thick GaN base. The use of InGaN, whose acceptor E_A can be several kT lower, could significantly increase the base hole population and reduce the base resistance (R_B). However, the increased hole population and the polarization charge now present tends to reduce β; therefore, a balance must be reached between R_B and β.

Figure 1 shows the effects on β and hole population of a uniformly doped InGaN base. In all cases, the acceptor concentration (N_A) is $2x10^{19}cm^{-3}$ and the nonradiative minority carrier lifetime (τ_n) is chosen to be 0.5ns. As the Indium concentration is increased from 0, 3, 7, and 11%, E_A drops from 160, 146, 122, and 94 meV, respectively. The reduction in E_A causes the peak hole concentration to rise from $6.75x10^{17}$, $8.85x10^{17}$,

$1.38x10^{18}$, and $2.32x10^{18}cm^{-3}$, respectively. There is an associated change in overall recombination lifetime τ_{net}, since the radiative recombination lifetime τ_{rad} varies according to $\tau_{rad}^{-1}=C \bullet p$, where $C=1.02x10^{-8}cm^3/s$. The overall lifetime τ_{net} drops from 110ps to 40ps, and as a result, the peak β drops from 33 to 13.

Figure 2 shows the effect of N_A on β, where N_A ranges from $5x10^{18}cm^{-3}$ to $8x10^{19}cm^{-3}$. The base is composed of $In_{0.07}Ga_{0.93}N$, and both E_A and τ_n are fixed at 122meV and 0.5ns, respectively. Although the lowest doped case improves β by a factor of two over the nominal $2x10^{19}cm^{-3}$ case, the penalty to R_B is likely to negate any benefit of improved β. The final uniform InGaN base comparison is summarized in Figure 3, which shows the impact of τ_n on β. Values of τ_n greater than 5ns may seem physically unreasonable, but τ_{rad} will dominate unless the base is very lightly doped.

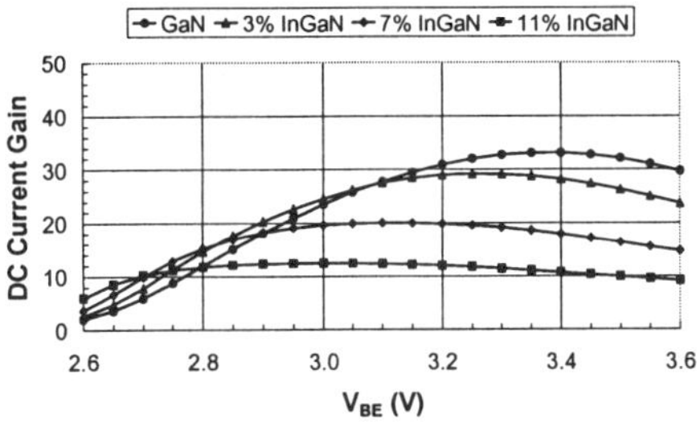

Fig. 1. β versus V_{BE} for various In concentrations in the base where $N_A=2x10^{19}cm^{-3}$ and $\tau_n=0.5ns$.

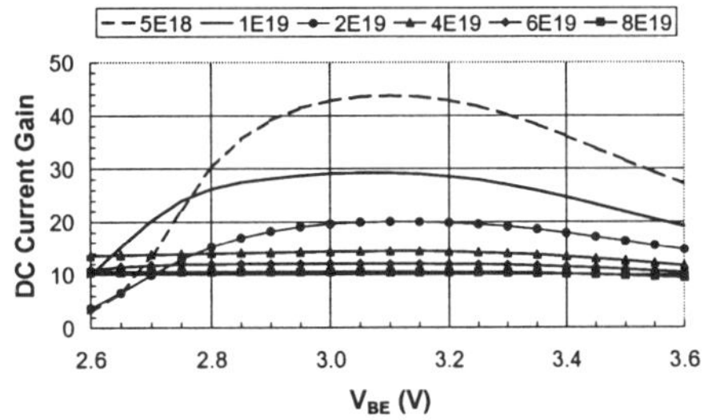

Fig. 2. β versus V_{BE} for various N_A where the base is composed of $In_{0.07}Ga_{0.93}N$ and $\tau_n=0.5ns$.

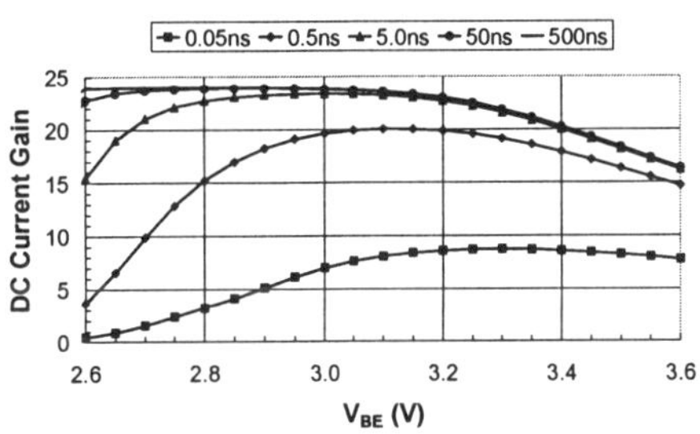

Fig. 3. β versus V_{BE} for various τ_n where the base is composed of $In_{0.07}Ga_{0.93}N$ and $N_A=2x10^{19}cm^{-3}$.

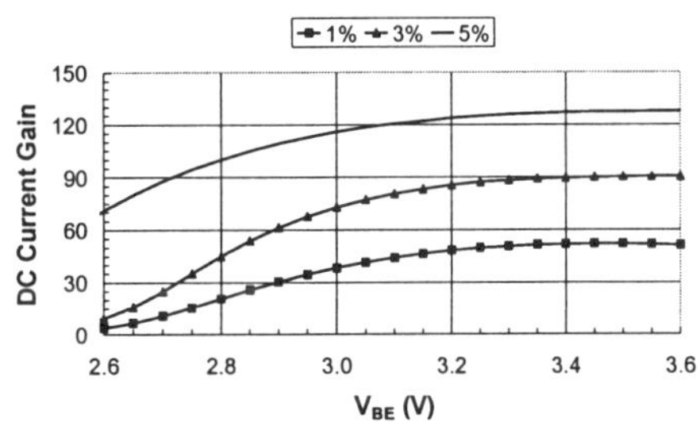

Fig. 4. β versus V_{BE} for various InGaN base grades where $N_A=2x10^{19}cm^{-3}$ and $\tau_n=0.5ns$.

With the introduction of InGaN thin films, it is possible to compositionally grade the base and provide an electric field to drive electrons across the base. Figure 4 shows the β increase in response to progressively increasing InGaN base grades. The Indium percentage at the base-emitter junction remains zero while it increases towards the base-collector junction. As expected, significant improvement in β is achieved through grading the base. The simulated cases in Figure 5 go one step further and introduce an abrupt InGaN/GaN hetero-interface at the base-emitter junction. Due to the abrupt interface, a large donor-like polarization sheet charge will be present at the base-emitter junction. Much of the β improvement gained from the base grade can be lost with the introduction of this hetero-interface. A large valence band discontinuity is not needed if an n+ emitter is required to suppress Mg out-diffusion, but a small amount of Indium should be introduced in the base to compensate for bandgap narrowing in the n+ emitter.

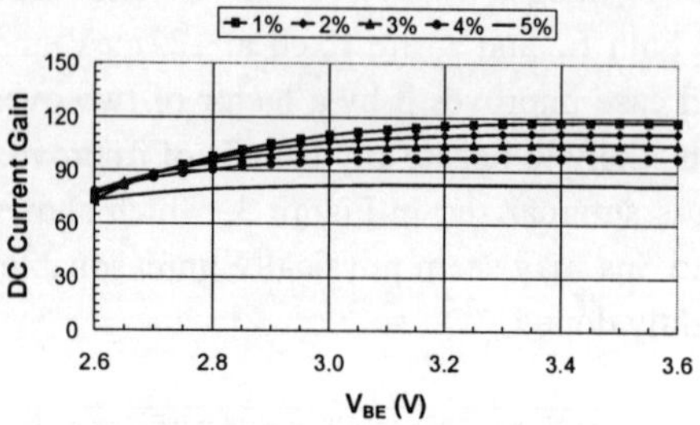

Fig. 5. β versus V_{BE} for various InGaN/GaN hetero-interfaces at the base-emitter junction.

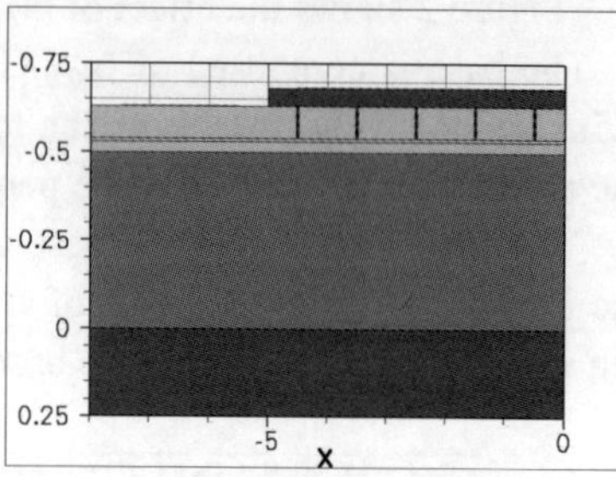

Fig. 6. Cross-section of the 2D perforated base HBT with 10nm wide and 90nm deep perforations.

4 Non-Uniform Base

The previous simulation results indicate that high β is difficult to achieve with a uniform, defect-free base layer. We have also investigated the electrical characteristics of a new class of DHBT, where the base thickness is non-uniform, designated the Perforated Base HBT (PB-HBT). Figure 6 shows a device structure that closely resembles a triple mesa HBT, but the base thickness varies digitally between 10nm and 100nm at a 96% duty cycle. By non-uniformly thinning or perforating the base, the majority of the base would remain 100nm thick to minimize R_B. However, small regions are thinned to achieve extremely high local β, which can be used to significantly increase the overall β.

This device structure might naturally occur when "V" defects form during the base layer growth.[10,11] The subsequent emitter layer growth fills in the "V" defects creating the high β regions. If a re-grown extrinsic base is utilized, the subsequent high percentage InGaN can also fill the underlying defects and prevent a large quantity of base-collector shorts in the extrinsic base region. For computational efficiency, the "V" defects are modeled as orthogonal posts or perforations. It is expected that the sloped sidewalls of the "V" defect contribute to enhancing β, but the simpler orthogonal structure is adequate to prove this hypothesis. Related structures could form when Indium is non-uniformly distributed during the base growth.

When the device structure in Figure 6 is simulated, a significant enhancement in β is observed. Figure 7 shows the effects of the perforation depth and τ_n on β for a 10nm wide and 90nm deep perforation (leaving a 10nm Mg doped base). The majority of the base remains a $In_{0.07}Ga_{0.93}N$ layer with $N_A=2x10^{19}cm^{-3}$; however, the thinned base regions account for only one percent of the total emitter area and have improved β three fold at V_{BE}=2.8V. The greatest β enhancement is achieved at low values of V_{BE}. Figure 8 shows the variation of β due to the perforation width when the depth is fixed at 90nm. Although a perforation width of 100nm may be unreasonable for a single "V" defect, the structure represents the situation where the number of defects is large. More than an order of magnitude increase in β is observed. Figure 9 shows the representative forward Gummel curves of DHBTs with a uniform, 40nm perforated, and 100nm perforated base at a fixed V_{CE}=5.0V and an emitter half width of 5μm.

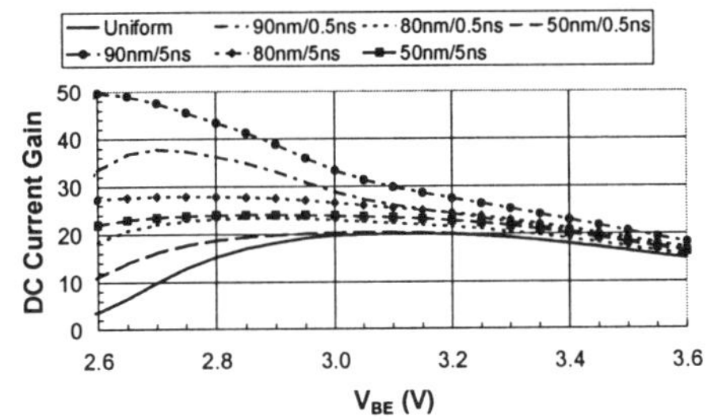

Fig. 7. β versus V_{BE} for a 10nm wide perforation at 50, 80, and 90nm depths with τ_n of 0.5 and 5.0ns.

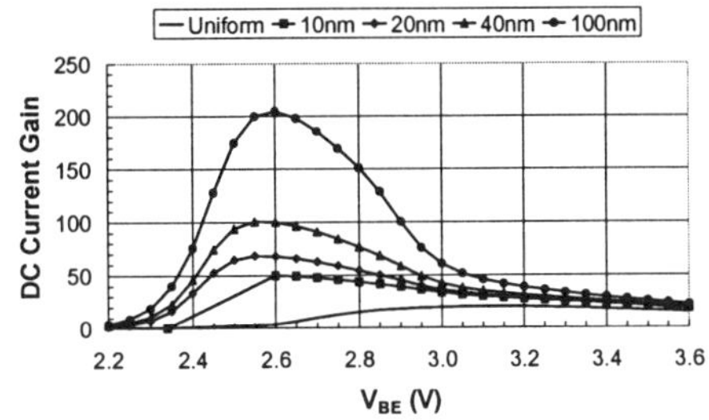

Fig. 8. β versus V_{BE} for a 90nm deep perforation at 10nm, 20nm, 40nm, and 100nm widths.

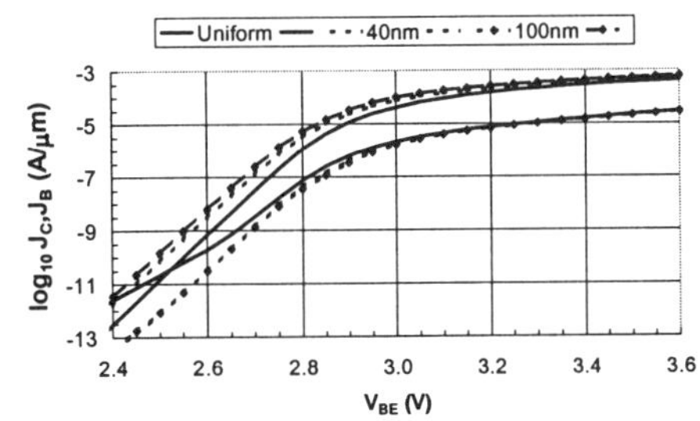

Fig. 9. Forward gummel curves of DHBTs with a uniform, 40nm perforated, and 100nm perforated base at a fixed V_{CE}=5.0V and $W_E/2$=5μm.

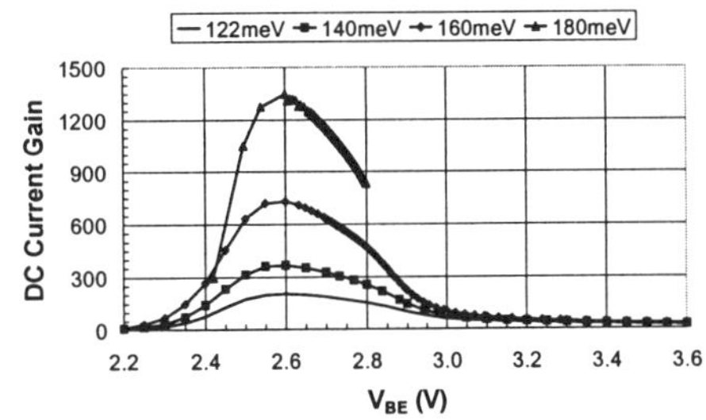

Fig. 10. β improvement with increasing E_A near the base-collector junction for a 100nm perforation at a 90nm depth.

Another level of β enhancement for the InGaN/GaN/InGaN DHBT can be made by varying the vertical properties of the base. β can be improved by reducing the hole population in the base, through its effect on τ_{net}. For a uniform base, lowering N_A would be unacceptable due to the increase in R_B. However, for the PB-HBT, the hole concentration need only be lowered under the perforations. Simulations show that decreasing N_A can lead to higher β. However, in order to prevent base width modulation and punch-through, N_A should be not be lowered excessively. An additional technique is to drive the acceptor E_A deeper. From the work of Kozodoy *et al*, it has been shown that the growth conditions and N_A can alter the E_A value significantly.[9] It is plausible that InGaN could experience a similar E_A shift based on N_A and growth conditions.

Figures 10 shows the β improvement for a 100nm wide perforation, where the acceptor E_A is increased in a thin 10nm region immediately adjacent to the base-collector junction. The perforation depth is 90nm; N_A is fixed at $2x10^{19}cm^{-3}$; and τ_n is fixed at

5.0ns. For the 40nm wide perforation, β jumps from 100 to over 600 when E_A is increased from 122meV to 180meV, and for the 100nm wide perforation, β increases from 200 to over 300 for an E_A shift from 122meV to 180meV.

An additional structure that can be considered for high β corresponds to a "piped-base" device, in which there is an n-type region of small width connecting the emitter and the collector. Such a situation could arise, for example, from a V-defect extending though the entire Mg-doped base, or from a dislocation through which Si donors have diffused. Such a device operates in a manner similar to a JFET, where the base voltage modulates the conductivity of the channel.

5 Summary

The effects of traditional device parameters can improve β for an InGaN/GaN/InGaN DHBT, but the extent of the improvement is limited. By introducing a non-uniform base and creating a PB-HBT, β can be improved by a factor of 100. However, the high-speed characteristics of the device may not improve in corresponding fashion. The junction capacitances associated with the regions of the base where collector current is not high will adversely affect frequency response.

Acknowledgements

The authors would like to thank DARPA for partial support for this work.

References

1. D.J.H. Lambert *et al*, *J. of Crys. Growth* **221**, 730 (2000).
2. L.S. McCarthy *et al*, *IEEE Trans. Elec. Dev.* **48**, no. 3, 543 (2001).
3. T. Makimoto *et al*, *Appd. Phys. Lttrs.* **83**, no. 5, 1035 (2003).
4. K. Kumakura *et al*, *Appd. Phys. Lttrs.* **79**, no. 16, 2588 (2001).
5. D. Qiao *et al*, *Appl. Phys. Lttrs.* **80**, no. 6, 992 (2002).
6. K. Kumakura *et al*, *Jpn. J. Appl. Phys.* **39**, L337 (2000).
7. H. Zhang *et al*, *Appl. Phys. Lttrs.* **84**, no. 23, 4644 (2004).
8. http://www.ioffe.rssi.ru/SVA/NSM/Semicond/
9. P. Kozodoy *et al*, *J. Appl. Phys.* **87**, no. 4, 1832 (2000).
10. F. A. Ponce *et al*, *Phys. Stat. Sol. B* **240**, no. 2, 273 (2003).
11. A.N. Bright *et al*, *J. of Elec. Microscropy* **50**, no. 6, 489 (2001).

International Journal of High Speed Electronics and Systems
Vol. 14, No. 3 (2004) 831–836

World Scientific
www.worldscientific.com

ANALYSIS OF GaN HBT STRUCTURES FOR HIGH POWER, HIGH EFFICIENCY MICROWAVE AMPLIFIERS

D.M. KEOGH, J.C. LI, A.M. CONWAY, D. QIAO,
S. RAYCHAUDHURI, and P.M. ASBECK
University of California at San Diego, La Jolla, CA 92037-0407, USA

R.D. DUPUIS
Georgia Institute of Technology, Atlanta, GA 30332, USA

M. FENG
University of Illinois at Urbana–Champaign, Urbana, Illinois 61801, USA

GaN has become increasingly important for microwave applications up to K_a band as a result of its wide band-gap, which provides a high critical breakdown field and good thermal stability, yielding excellent potential for high power and high voltage operation. It is of major interest to understand the device structures that will lead to high efficiency, high power microwave amplifiers. In this paper, we investigate by simulation the microwave performance of InGaN/GaN Heterojunction Bipolar Transistors (HBTs), with proper device geometry to account for the effects of current crowding. We provide an analysis of both emitter-up and collector-up InGaN/GaN HBT structures, based on a distributed HBT model implemented in ADS. We simulate their performance in a fully matched Class B high power amplifier. The results show that an important issue for high power operation is the mismatch between optimum load conditions for peak output power and peak gain. To increase the output impedance, it is important to decrease base-collector capacitance, which can be achieved with a collector-up structure. Furthermore, current crowding in the emitter caused by the sheet resistance in the base becomes more important at higher power operation, and is therefore a key design consideration.

1. Introduction

Over the past several years, HFETs have attracted much of the research focus in the area of GaN based electronics, but HBTs offer important potential advantages in terms of high transconductance, controlled linearity, freedom from surface and trapping effects, and controllable tradeoff of f_T and breakdown voltage. HBT technology therefore represents an important complement to that of HFETs. Recently, significant progress in the area of HBTs has been demonstrated recently with the use of GaN/InGaN/GaN HBT structures [1], yielding common-emitter current gains in excess of 2000, and low common-emitter offset voltages of 1V through the use of p-InGaN re-grown on the extrinsic base. The large breakdown voltage capability of GaN is of particular interest for high power, high voltage microwave power amplifiers.

In order to take full advantage of this high power and high voltage capability, proper device design is critical. For example, a major issue facing GaN based HBT devices is the large base sheet resistance which results from the large acceptor depth of Mg. The distributed nature of the base resistance leads to significant current crowding and requires careful attention. In this work, simulation results of InGaN/GaN HBT microwave power amplifiers based on such a distributed model are presented. The model is implemented within the Advanced Design System (ADS) circuit simulator, and attempts to accurately capture the details of base resistance, contact resistance, junction capacitance, and the distribution of current underneath the emitter contact. Simulations were performed in a Class B amplifier operating at 1 GHz, with full input and output matching, for both Emitter-Up and Collector-Up [2] device structures.

A representative device structure for both the emitter-Up and collector-Up HBT is presented in Fig. 1, an InGaN/GaN DHBT with 80 nm graded base (x_{IN}=.08), with self-aligned emitter (collector) and base contacts, and 0.25 μm width emitter (collector).

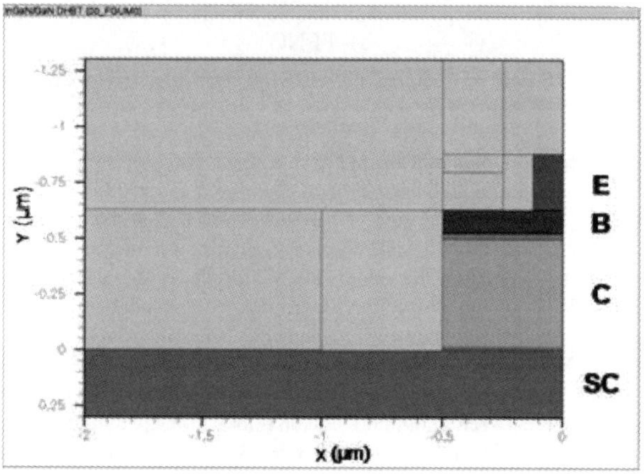

Fig. 1. Schematic of HBT structure used in microwave power amplifier simulations.

Based on this device structure, a comprehensive parameter estimation is performed, and the results inputted into a distributed HBT model within ADS, presented in Fig. 2. The model accounts for the distributed resistances and capacitances of the device by dividing the intrinsic device into 10 segments of equal widths, and the extrinsic device into 5 segments. This model is then used in DC, S-parameter, and Harmonic Balance simulations to estimate the microwave performance at 1 GHz.

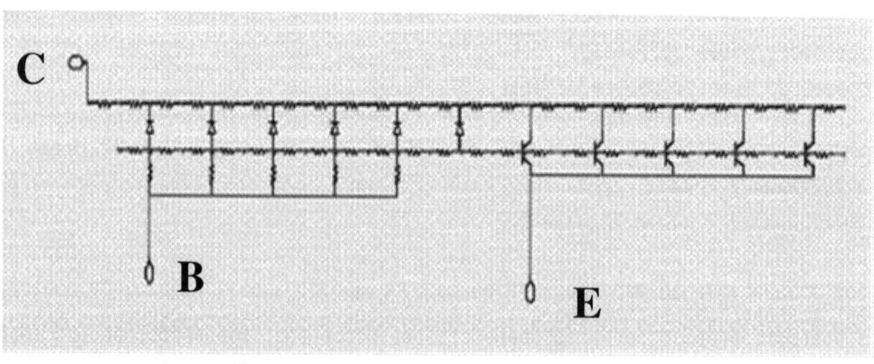

Fig. 2. Distributed model of InGaN/GaN HBT

2. Results/Discussion

An important issue that came to light as a result of these simulations is the issue of mismatch between optimum load conditions for peak output power and peak gain. For example, the conditions for peak amplifier gain occur at an output impedance of

$$R_{OUT} = \frac{1}{2\pi f_T C_{BC}} \tag{1}$$

which is approximately 6 ? for the emitter-up device, where f_T is the cut-off frequency and C_{BC} is the base-collector capacitance. The conditions for peak power, however, assuming 10W of output power and a maximum of 100V applied to the collector, occur at 125 ? . This mismatch leads to a large degradation of gain described by a mismatch factor, M_L, according to Eq (2),

$$M_L = \frac{4R_{OUT}R_L}{|Z_{OUT} + Z_L|^2} \tag{2}$$

where R_{OUT} and R_L are the output and load resistances, and Z_{OUT} and Z_L are the output impedances. Minimizing this mismatch requires careful design, including a reduction of the base-collector capacitance, and if necessary, a reduction of f_T. Collector-Up structures therefore become an important consideration as a result of a reduced base-collector capacitance, which increases the output impedance and offers the potential for higher gain and higher efficiency.

Microwave performance was evaluated at 1GHz in Class B mode, at an output power of 10W, for both emitter-up and collector-up HBT amplifiers. For these amplifiers, the load line was chosen such that a maximum voltage of approximately 70V is applied to the collector, and in Fig. 3, a dynamic load line computation illustrates the fact that the voltage and current swings are properly centered within the transistor I-V curves. From

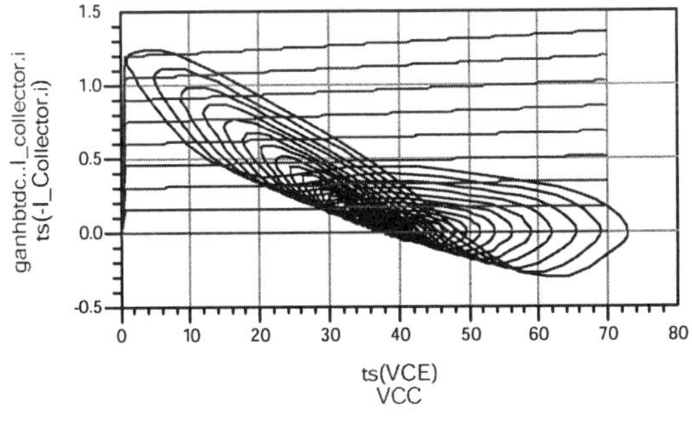

Fig. 3. Simulated dynamic load line for in Class B amplifier

harmonic balance simulations, shown in Fig. 4, it can be seen that the emitter-up amplifier achieves high power and high efficiency, with a maximum gain of approximately 18dB, 10W output power, and 65% power added efficiency. Though the

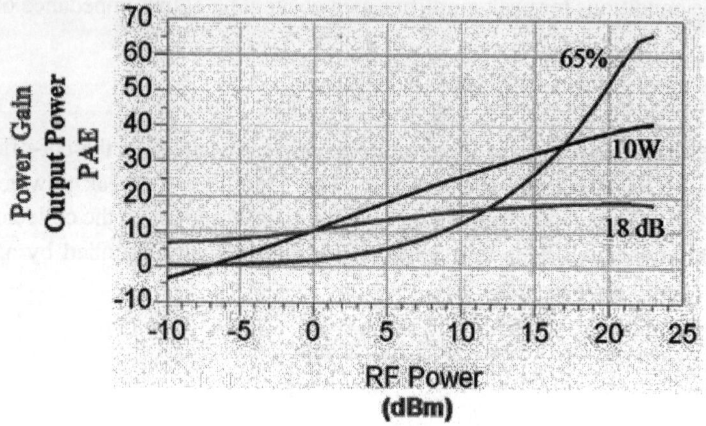

Fig. 4. Harmonic balance simulation results for emitter-up HBT operating at 1 GHz and 10W output power.

amplifier achieves 18 dB of gain, this is lower than what was expected, due to the large mismatch between the load conditions for peak gain and peak output power. As explained above, it is expected that a collector-up device, as a result of a reduced base-collector capacitance, should have higher gain and efficiency as compared to the emitter-up device. Simulation results for a collector-up amplifier are shown below, in Fig. 5. As

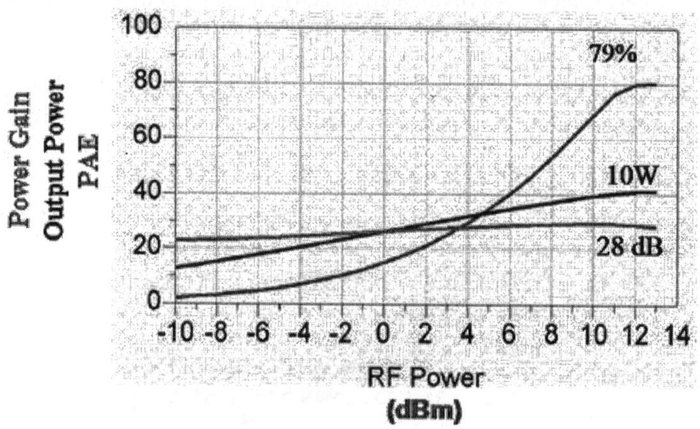

Fig. 5. Harmonic balance simulation results for C-up HBT operating at 1 GHz and 10W output power.

expected, this amplifier shows significant improvement, with a maximum transducer gain of 28 dB, 10W output power, and 79% power added efficiency. For the collector-up device, the output impedance is significantly increased, to approximately 15 ? , as compared to 6 ? for the emitter-up, which reduces the load mismatch and improves the power gain performance.

The distributed nature of the model implemented within ADS also allows us to evaluate the effect of emitter geometry on microwave performance. As a result of high base sheet resistance, significant current crowding is expected to occur, degrading amplifier performance. In order to offset the effect of current crowding, it is necessary to move to smaller emitter widths, thereby eliminating significant IR drops that accumulate within the intrinsic device. Fig. 6 details the current distribution across the width of a 1μm emitter for an emitter-up amplifier operating at 1GHz. Note that the current in this

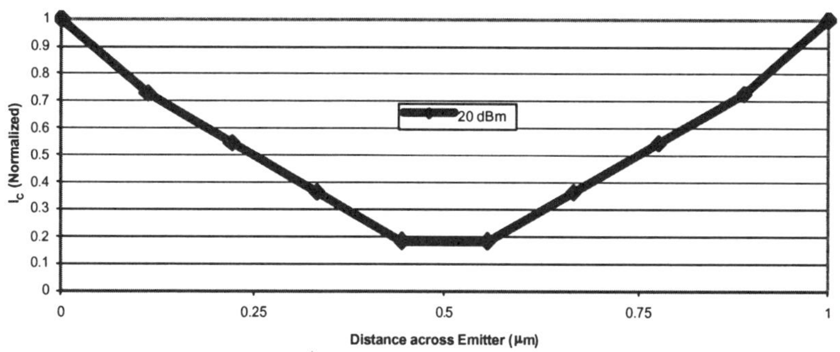

Fig. 6. Collector current distribution across the width of 1 um emitter at 1GHz.

figure is normalized to the peak current, and shows severe current crowding even for a 1μm emitter width. Aggressively scaling the emitter to 0.25 μm would certainly improve the situation, but current crowding is still expected to play a role, especially at higher frequencies. Ultimately, the base sheet resistance must be reduced to realize the full potential of these devices. Fig. 7 demonstrates how the emitter width influences the power gain performance at 1 GHz.

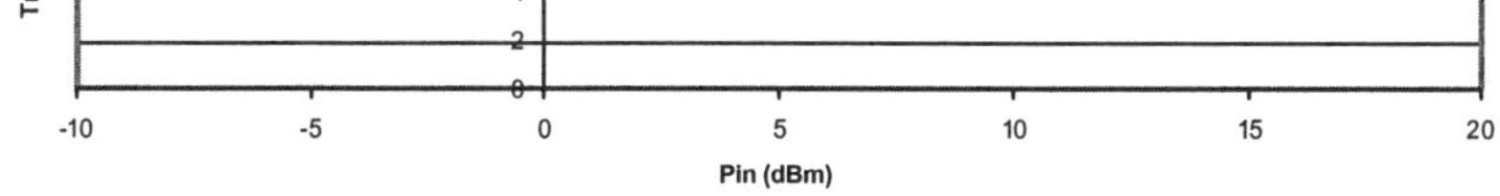

Fig. 7. Transducer Power Gain versus P_{IN} curves for various emitter widths.

Peak power gain for the 1 μm is 19 dBm, but drops to less than 16 dBm when the emitter is scaled to 3 μm. This further underscores the importance of scaling the emitter width to smaller dimensions.

3. Conclusions

Using a distributed HBT model, microwave simulations of Class B amplifiers based on emitter-up and collector-up device structures were investigated. The simulations showed that careful device design is critical for optimizing performance of microwave amplifiers, in particular, minimizing the mismatch between conditions for peak gain and peak power. Furthermore, current crowding plays a significant role due to the high base sheet resistance, and requires aggressive contact scaling. When properly accounted for, however, simulations demonstrate high power and high efficiency performance, promising for use in microwave circuits.

Acknowledgements

The authors would like to acknowledge partial support from DARPA.

References

1. Makimoto T, Kumakura K, Kobayashi N., "High current gain (>2000) of GaN/InGaN double heterojunction bipolar transistors using base regrowth of p-InGaN," Applied Physics Letters, Vol.83, No.5, 1035-7, 4 Aug. 2003.
2. H. Kroemer, "Heterostructure Bipolar Transistors and integrated circuits," *Proc. IEEE*, vol. 70, pp. 13-25, 1982.

International Journal of High Speed Electronics and Systems
Vol. 14, No. 3 (2004) 837–846

World Scientific
www.worldscientific.com

LEAKY SURFACE ACOUSTIC WAVES IN SINGLE-CRYSTAL AlN SUBSTRATE

GANG BU, DAUMANTAS CIPLYS[a], and MICHAEL S. SHUR

Department of Electrical, Computer, and Systems Engineering, Rensselaer Polytechnic Institute, Troy, New York 12180, USA

LEO J. SCHOWALTER and SANDRA B. SCHUJMAN

Crystal IS, Inc., Latham, New York 12110, USA

REMIS GASKA

Sensor Electronic Technology, Inc., Columbia, South Carolina 29209, USA

We report on the velocity V and the electromechanical coupling coefficient K^2 of the first and the second leaky surface acoustic waves in various propagation directions in the a-plane AlN single-crystal. For c-propagation direction, the second leaky wave exhibited the velocity of 11016 m/s and K^2 of 0.45%. For this direction, the temperature coefficient of frequency was found to be -30 ppm/°C. A near match of the velocities of the plane and leaky waves in the a-plane AlN allowed us to establish analytical relationships between the piezoelectric and elastic constants. A full set of elastic and piezoelectric constants of AlN has been evaluated by fitting the measured and calculated dependencies of velocities and electromechanical coupling coefficients on the propagation direction for both Rayleigh and leaky waves.

1 Introduction

AlN is a promising material for the surface acoustic wave (SAW) applications due to the high SAW velocity, good temperature stability, and relatively large electromechanical coupling coefficient, K^2. Although a lot of results have been published on Rayleigh type SAWs, little is known about the leaky surface acoustic waves (LSAWs) in AlN. Compared to the SAWs, the LSAWs have higher velocities. Also, in conventional piezoelectric materials like $LiNbO_3$, $LiTaO_3$ and quartz, the leaky waves exhibit larger K^2 values and better temperature stability for certain orientations.[1-3] Finally, investigations of leaky waves yield additional information about materials piezoelectric and elastic constants.

There are two types of leaky waves: the wave with the velocity between those of the fast and slow shear plane waves (called the first leaky wave or pseudo-SAW), and the wave with the velocity close to the longitudinal plane wave velocity (called the second leaky wave or high-velocity pseudo-SAW). At certain orientations, the first and the second leaky waves have the

[a] On leave from: Department of Radiophysics, Vilnius University, Vilnius 2040, Lithuania.

velocities almost equal to those of the horizontal shear and longitudinal bulk waves. Since the velocities of bulk waves can be analytically expressed in terms of elastic and piezoelectric constants, material parameters can be extracted if the plane wave velocities known.

In this paper, we report for the first time the elastic and piezoelectric constants obtained from measuring K^2 and V of Rayleigh and leaky waves in bulk single-crystal AlN samples.

2 Theory of leaky waves

The coordinate system of a semi-space substrate is shown in Fig.1, with x_1 being the wave propagation direction and x_3 being the surface normal direction. The propagation of elastic wave in a piezoelectric film is described by the following set of equations: [1]

$$\rho\frac{\partial^2 u_j}{\partial t^2} - c_{ijkl}\frac{\partial^2 u_k}{\partial x_i \partial x_l} - e_{kij}\frac{\partial^2 \phi}{\partial x_i \partial x_k} = 0$$

$$e_{ikl}\frac{\partial^2 u_k}{\partial x_i \partial x_l} - \varepsilon_{ik}\frac{\partial^2 \phi}{\partial x_i \partial x_k} = 0, \qquad i,j,k,l = 1,2,3 \tag{1}$$

where ρ is the mass density of the material, c_{ijkl}, e_{ijk} and ε_{ij} are the elastic stiffness, the piezoelectric, and the dielectric permittivity tensors, respectively, u_i is the elastic displacement vector, and ϕ is the electric potential.

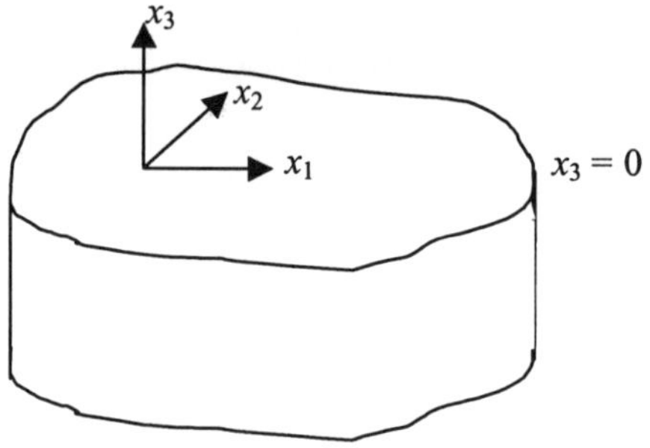

Figure 1. Semi-space substrate and coordinate system used in calculations. Surface acoustic wave propagates along x_1 axis.

The displacement and electric potential of the wave are sought in the form:

$$u_j = \{\sum_n C_n \alpha_j^{(n)} \exp(ikb^{(n)}x_3)\}\exp[ik(lx_1 - vt)],$$
$$\phi = \{\sum_n C_n \alpha_4^{(n)} \exp(ikb^{(n)}x_3)\}\exp[ik(lx_1 - vt)]. \qquad n=1\ldots4. \tag{2}$$

Here, k is the wave number, α_j (j=1,2,3) is the surface displacement of each sub-wave in three directions, and α_4 is the surface potential. The coefficients C_n are the weight factors of each sub-wave. The value l equals to $1+i\delta$, with δ being decay constant per wavelength in the wave propagation direction. For Rayleigh waves, δ is 0, and the decay constants $ikb^{(n)}$ in the depth direction for each sub-wave are positive. However, for the leaky wave, δ is not equal to 0 and one (or more than one) decay constant is negative, i.e. the relevant sub-wave grows in the depth direction.

The boundary conditions are:

(1) vanishing stress, $T_{3i}=0$, at the free surface, $x_3=0$;

(2) continuity of the normal component of the electrical displacement, $D_3=k\varepsilon\phi$, for the free surface (the potential above the top surface must satisfy the Laplace's equation in the free space), and

(3) vanishing potential, $\phi=0$, for the surface covered with an infinitely thin perfect conductor.

The phase velocity V and the decay constant l can be found using the above boundary conditions.[1]

The electromechanical coupling coefficient is obtained from the relation

$$\frac{K^2}{2}=\frac{V_f-V_c}{V_f}, \tag{3}$$

where V_f and V_c are the SAW velocities on free and conducting surfaces, respectively.

3 Results

Experiment

The c-plane and a-plane oriented bulk AlN samples have been grown using the sublimation-recondensation technique.[5] Aluminum interdigital transducers (IDTs) of 16 μm and 24 μm finger pair periods (Λ) have been fabricated on the samples using E-beam evaporation and wet etch. In the c-plane sample, the orientation of IDT was arbitrary because of the SAW isotropy in this plane. Because of the a-plane anisotropy, the IDT on the a-sample was oriented in various directions to facilitate the estimation of material parameters. To obtain the LSAW velocity and K^2, the S_{11}-parameters of IDT were measured and fitted to the IDT equivalent circuit. The phase velocity V and K^2 were determined using the following relationships:[6]

$$V=f_c\Lambda, \tag{4}$$

$$K^2=G_a/(8f_cC_TN), \tag{5}$$

where f_c is the IDT center frequency, G_a is the radiation conductance, C_T is the static capacitance, and N is the number of digit pairs of IDT, respectively. The aluminum mass loading effect was estimated by considering the aluminum-on-AlN layered structure, and its effect on the velocity was found to be less than 20 m/s for 1000 Å aluminum for both SAW and LSAW on a-plane AlN at 24 μm wavelength.

Fig. 2. shows the measured S_{11}-parameters of two IDT (Λ=24μm) on the a-plane for the c- and c+20° propagation directions. In the c-direction, there are two circles corresponding to the second leaky and Rayleigh waves, while in the c+20° direction there are three circles corresponding to the first leaky, second leaky and Rayleigh waves, respectively.

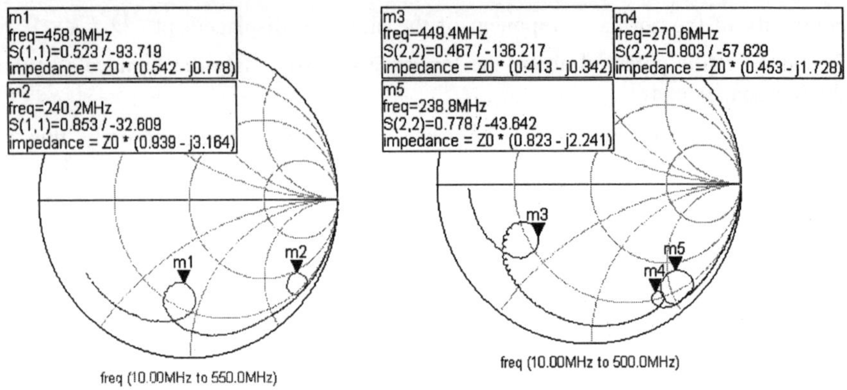

(a) c-direction (b) c+20° direction

Figure 2. S_{11}-parameter of IDT (Λ=24μm, aperture width L=0.5mm, number of electrode pairs=70) on a-plane AlN bulk sample corresponding to different acoustic propagation directions.

Material Parameter Extraction

If the bulk wave velocities are known, the AlN elastic constants can be expressed in terms of three piezoelectric constants. By performing the simulation with material parameters published in literature, we found that in a-plane AlN, the first and the second leaky waves have free surface velocities very close to those of horizontal shear and longitudinal bulk waves in the same direction (differences are in the order of 1 m/s). This enables us, using the analytical expressions for bulk velocities, to relate the leaky wave velocities to the elastic constants of the hexagonal crystal, except the constant c_{12}, which can not be determined from our experiment since it influences only the vertical shear bulk wave velocity. Fortunately, the a-plane Rayleigh wave velocity does depend on c_{12}. Thus, three piezoelectric constants e_{15}, e_{31}, e_{33}, and the elastic constant c_{12} were the independent variables used to fit the Rayleigh wave velocities and K^2 of both the a- and c-planes and K^2 of leaky waves in a-plane. The extracted piezoelectric and elastic constants corresponding to the best fit of our data are listed in Table 1, together with the density and the relative dielectric constant values used in the simulations.

Table 1. AlN elastic constants (10^9N/m^2), mass density (kg/m^3), piezoelectric constants (C/m^2) and relative permittivity

c_{11}	c_{12}	c_{13}	c_{33}	c_{44}	ρ^7
390	140	106	362	123	3260

e_{15}	e_{31}	e_{33}	$\varepsilon_{11}/\varepsilon_0{}^8$	$\varepsilon_{33}/\varepsilon_0{}^8$	
-0.16	-1.06	1.65	8.5	8.5	

The fitted velocities perfectly match the measured values as shown in Fig. 3. For the Rayleigh waves, on both the *a*-plane and *c*-plane the simulated velocity deviations from the measured values were less than 0.3%. For the leaky waves, the differences were less than 0.6%.

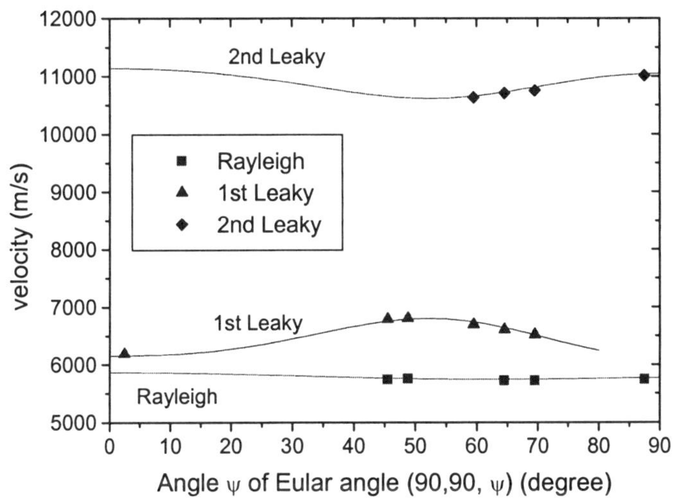

Figure 3. Simulated and measured SAW and LSAW velocities as functions of propagation direction on a-plane of AlN single crystal; Euler angles (90°, 90°,ψ), with ψ =90° corresponding to the c-propagation.

The fitted and the measured K^2 values are shown in Fig. 4. The fitted K^2 values of Rayleigh waves agree well with the measured values in both planes (for *c*-plane, they do not depend on propagation direction). For the leaky waves, the measured non-zero K^2 values span over a larger angle range than the simulated curve. However, the measured peak values are lower than the simulated ones. This is the result of a compromise made in the fitting process. During the fitting process, we could better fit the peak values with a narrower angle range, or have a wider range with much larger peak values. The reason of this relative large deviation of the leaky wave K^2 values might be attributed to the simultaneous existence of the surface skimming bulk wave when the LSAW were excited. However, this needs further verifications.

Fig. 5 shows the comparison of the AlN a-plane longitudinal and horizontal shear bulk wave velocities simulated using the material parameters reported by various authors.[7,9-12] The simulation using our set of elastic and piezoelectric constants agrees very well with the simulation using the elastic constants measured by McNeil *et al.* using Brillouin spectroscopy.[9] It should be noted that their elastic constants were obtained without separating the contribution of piezoelectricity, thus they are piezoelectrically stiffened elastic constants (in the simulation we set the piezoelectric constants to zero to prevent accounting for the piezoelectricity twice). With e_{33} being around 1.5 C/m^2 and $\varepsilon = 8.5\varepsilon_0$, the influence of piezoelectricity to elastic constant c_{33} is estimated to be about 7%, which is not negligible.

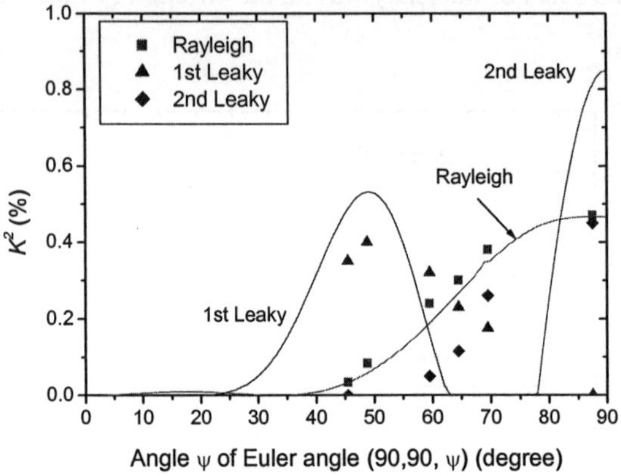

Figure 4. Simulated and measured SAW and LSAW K^2 as functions of propagation direction on a-plane of AlN single crystal; Euler angles (90°, 90°, ψ).

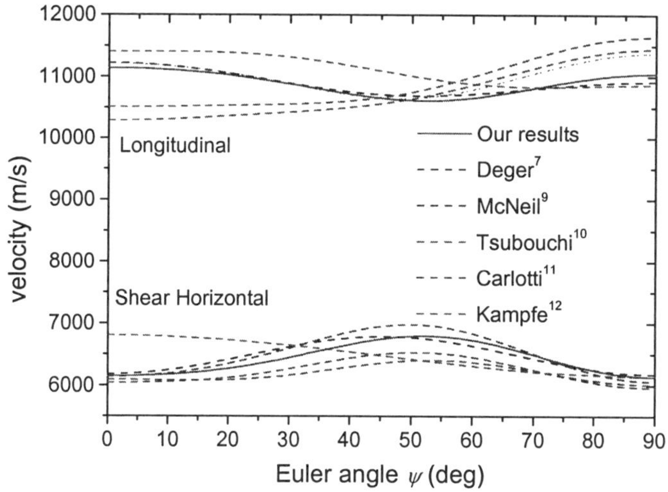

Figure 5. Simulated longitudinal and shear horizontal bulk wave velocity as functions of propagation direction polarized in a-plane of AlN single crystal; Euler angles, (90°, 90°, ψ), with ψ =90° corresponding to the c-propagation.

Leaky Wave Attenuation

With the decay constant δ known, the propagation attenuation per wavelength Λ can be found as:

$$A\,(\mathrm{dB}) = 20\log(e^{k\delta\Lambda}) = 20\log(e^{2\pi\delta}) = 54.55\delta \quad \mathrm{dB}. \tag{6}$$

The simulated attenuations of the first and the second LSAW in the a-plane as functions of the propagation direction are shown in Fig. 6. From the figure, we can see that the wave attenuation on the metalized surface is larger than the free surface attenuation. This is not always true for other materials. For the first LSAW, the direction that has the highest attenuation ($\psi \sim 55°$) is close to the direction of highest K^2 ($\psi \sim 50°$), whereas for the second LSAW, the highest K^2 ($\psi = 90°$) direction has the lowest free-surface attenuation rate. The attenuation magnitude of the second LSAW is one to two orders of magnitude larger than that of the first LSAW. The minimum attenuation value of the former, 0.24 dB/wavelength in c-direction is still a considerably large value not favorable for applications such as for delay lines and filtering.

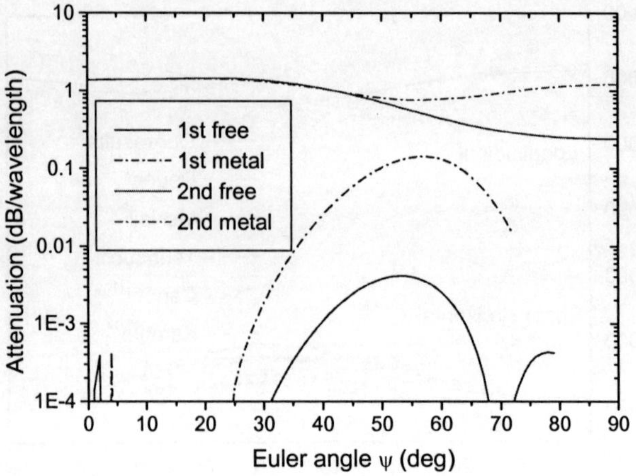

Figure 6. Simulated LSAW attenuation as functions of propagation direction on a-plane of AlN single crystal; Euler angles (90°, 90°,ψ).

Temperature Coefficient of Frequency

The temperature stability can be characterized by the temperature coefficient of frequency (TCF):

$$TCF = \frac{1}{f_c}\frac{df_c}{dT},$$

where T is temperature and f_c is the IDT center frequency.

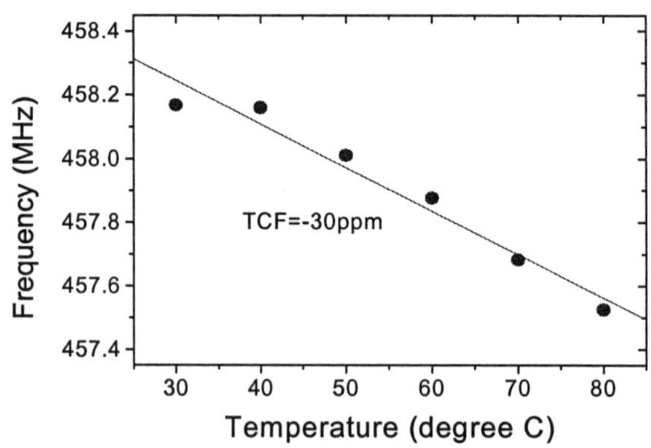

Figure 7. Measured AlN *a*-plane *c*-propagation second leaky wave TCF.

For the *a*-plane *c*-propagation, the measured TCF of the second LSAW is -30 ppm/deg as shown in Fig. 7. In the same orientation, the TCF value of the Rayleigh wave was measured as -16 ppm/deg. These values are lower than 75-94 ppm/deg in $LiNbO_3$, 120 ppm/deg in $Bi_{12}GeO_{20}$, 35 ppm/deg in $LiTaO_3$ and GaAs.[6]

Table 2. Summary of results for SAW and LSAW on bulk AlN c- and a-planes.

Euler Angles	Wave Type	V_m (m/s)	V_s (m/s)	K^2_m (%)	K^2_s (%)	TCF (ppm/°C)	A (dB/Λ)
(0 0 0)	Rayleigh	5755	5743	0.11	0.11	-19	0
(90 90 90)	Rayleigh	5760	5765	0.47	0.47	-16	0
(90 90 90)	Second leaky	11016	11047	0.45	0.85	-30	0.24(free) 1.21(metal)
(90 90 0)	First leaky	6186	6151	0.02	0	-41	~0 (both)

Table 2 presents a summary of the simulated and measured SAW results for the bulk AlN *c*-plane and *a*-plane *c*-propagation directions. In columns V_m, V_s, K^2_m and K^2_s, the subscripts *m* and *s* stand for measured and simulated values, respectively.

4 Conclusion

The leaky surface acoustic waves in bulk AlN have been studied for the first time. The LSAW velocities have been measured on *c*- and *a*- planes of the crystal using S_{11}-parameter method. The simulation shows that the first and second LSAW velocities on the free *a*-plane AlN are approximately equal to the velocities of horizontal shear and longitudinal plane waves in the same direction. A full set of elastic and piezoelectric constants of AlN has been extracted by fitting the velocity and K^2 results of both Rayleigh and leaky waves. The temperature coefficients of frequency for the *a*-plane *c*-propagation Rayleigh and second leaky waves have been measured as -16 and -30 ppm/°C, respectively. Low TCF, high velocity (~11000m/s for the second leaky wave) and reasonably large K^2 (~ 0.5%) indicate that bulk AlN is a promising material for SAW applications.

References

1. A. Takayanagi, K. Yamanouchi, and K. Shibayama, "Piezoelectric leaky surface wave in $LiNbO_3$", *Appl. Phys. Lett.*, vol. 17, no. 5, pp. 225-227 (1970).
2. Kazuhiko Yamanouchi and Masao Takeuchi, "Applications for piezoelectric leaky surface waves", *1990 Ultrasonics Symposium*, pp.11-18 (1990).
3. Shigetaka Tonami, Atsuhiro Nishikata and Yasutaka Shimizu, "Characteristics of leaky surface acoustic waves propagating on $LiNbO_3$ and $LiTaO_3$ substrates", *Jpn. J. Appl. Phys.*, Part 1, vol. 34, no. 5B, pp.2664-2667 (1995).

4. Daniel Royer and Eugene Dieulesaint, *Elastic Waves in Solids I: Free and Guided Propagation*, Chapter 4, Springer, 2000.
5. L. J. Schowalter, G. A. Slack, J. B. Whitlock, K. Morgan, S. B. Schujman, B. Raghothamachar, M. Dudley, and K. R. Evans, "Fabrication of native single-crystal AlN substrate", *Phys. Status Solidi C*, pp. 1997–2000 (2003) / DOI 10.1002/pssc.200303462.
6. Colin K. Campbell, *Surface acoustic wave devices for mobile and wireless communications*, Academic Press (1998).
7. C. Deger, E. Born, H. Angerer, O. Ambacher, M. Stutzmann, J. Hormsteiner, E. Riha, and G. Fischerauer, "Sound velocity of $Al_xGa_{1-x}N$ thin films obtained by surface acoustic-wave measurements", *Appl. Phys. Lett.*, vol. 72, no. 19, pp. 2400-2402, May (1998).
8. J. Noreika, M. H. Francombe, and S. A. Zeitman, "Dielectric properties of reactively sputtered films of aluminum nitride", *J. Vac. Sci. and Technol.*, vol. 6, no. 1, pp. 194-197, Jan. (1969).
9. L. E. McNeil, M. Grimsditch, and R. H. French, "Vibrational spectroscopy of aluminum nitride", *J. Am. Ceram. Soc.*, vol. 76, no. 5, pp.1132-1136, (1993).
10. K. Tsubouchi, K. Sugai, and N. Mikoshiba, "AlN material constants evaluation and SAW properties on AlN/Al_2O_3 and AlN/Si", *IEEE Ultrasonics Symposium Proc.*, pp.375-380 (1981); K. Tsubouchi and N. Mikoshiba, "Zero-temperature-coefficient SAW devices on AlN epitaxial films", *IEEE Trans. Sonics and Ultrason.*, vol. SU-32, no. 5, pp. 634-644 (1985).
11. G. Carlotti, F. S. Hickernell, H. M. Liaw, L. Palmieri, G. Socino, and E. Verona, *IEEE Ultrasonic Symposium Proc.*, pp. 353 (1995).
12. A. Kampfe, B. Eigenmann, O. Vohringer, and D. Lohe, "X-ray and single crystal values of the elastic constants and determination of matching residual stresses in AlN", *High temperature material processes*, vol. 2, pp. 309-326 (1998).

International Journal of High Speed Electronics and Systems
Vol. 14, No. 3 (2004) 847–852

World Scientific
www.worldscientific.com

HIGH LINEARITY GaN HEMT POWER AMPLIFIER WITH PRE-LINEARIZATION GATE DIODE*

SHOUXUAN XIE, VAMSI PAIDI, STEN HEIKMAN, LIKUN SHEN, ALESSANDRO CHINI, UMESH K. MISHRA, MARK J. W. RODWELL, STEPHEN I. LONG

Department of ECE, University of California, Santa Barbara
Santa Barbara, CA 93106, USA

A high linearity MMIC RF power amplifier is reported in the AlGaN/GaN HEMT technology. In order to obtain high linearity, a pre-linearization gate diode is added at the input to compensate for the nonlinear input capacitance C_{gs} of the GaN HEMT device. Another single-ended Class B power amplifier without the gate diode is also designed for comparison. The circuit with the pre-linearization gate diode demonstrates at least 4dB improvement on 3rd order intermodulation distortion (IMD_3) performance over the one without the diode over the useful power range in two-tone measurement.

1 Introduction

Our previously described single-ended Class B power amplifier design using GaN HEMT technology has shown that high linearity and high efficiency can be obtained simultaneously if the circuit is biased at exactly the pinch off point (Class B configuration) [1].

In order to further improve the linearity performance, linearization techniques, such as feedback, feed forward, or predistortion, could be used. Feedback is good for audio power amplifiers, but for RF power amplifiers the stability and efficiency can be compromised. System-level feed forward or predistortion can be useful for RF frequencies, but the overall efficiency will be decreased because extra components have to be added. This will also make it difficult for MMIC implementation.

In this work, three major sources of AlGaN/GaN HEMT device nonlinearity are described: nonlinear I-V characteristic (transfer function), nonlinear G_{ds} and nonlinear C_{gs}. All the above nonlinear sources in the device will create the 3rd order intermodulation distortion (IMD_3) either at the output or at the input of the circuit.

2 Sources of GaN HEMT Device Nonlinearity

The nonlinear I-V characteristic of the device will generate intermodulation products when two or more signals are present. It is the dominant nonlinear source when the input power exceeds the compression point. But at moderate input power level, as we discussed in [2], best linearity performance can be obtained by biasing at Class B.

For GaN HEMT devices, the output conductance G_{ds} is a nonlinear function of the input and output signal. Moreover, substantial drain voltage dependent leakage current is

* This work is supported by the Office of Naval Research under N00014-00-1-0653.

observed in 0.3μm single gate devices. This results in a negative threshold voltage shift with increasing drain voltage. Because linearity in Class B operation is sensitive to gate bias, the threshold shift effectively increases IMD_3. However, by using a dual-gate device, the nonlinear G_{ds} effect can be minimized. The upper device of the cascode reduces the drain voltage variation on the lower device. Thus output conductance is decreased, and the threshold voltage dependence on V_{DS} is greatly reduced.

Finally, the input capacitance C_{gs} vs. V_{gs} characteristic of the AlGaN/GaN HEMT device is not linear, and it can be proved that the nonlinear C_{gs} characteristic can generate intermodulation distortion at the input of the circuit [3]. The distorted signal at the gate is amplified by the active device, also affecting the linearity of the circuit. The effect due to nonlinear C_{gs} can be compensated by adding a pre-linearization gate diode [4].

3 Pre-linearization Gate Diode

The nonlinear C_{gs} vs. V_{gs} characteristic can be modeled as a hyperbolic tangent function, which is an anti-symmetric function around the center point, V_C (shown in figure 1). Linearity can be improved if the nonlinear C_{gs} is compensated with an opposite nonlinear capacitance. Suppose the C_{gs} vs. V_{gs} relation is modeled as $C_{gs} = C_1*tanh(V_{gs}-V_C) + C_2$, as shown in figure 2. An ideal nonlinear capacitor $C_{pd} = -C_1*tanh(V_{gs}-V_C) + C_2$ can be added in parallel with C_{gs} such that the total input capacitance of the circuit is constant (not a function of bias voltage V_{gs}). This idea can be easily implemented by adding another HEMT device at the input, whose source and drain are shorted, and the gate is biased at twice of the V_C (shown in figure 3).

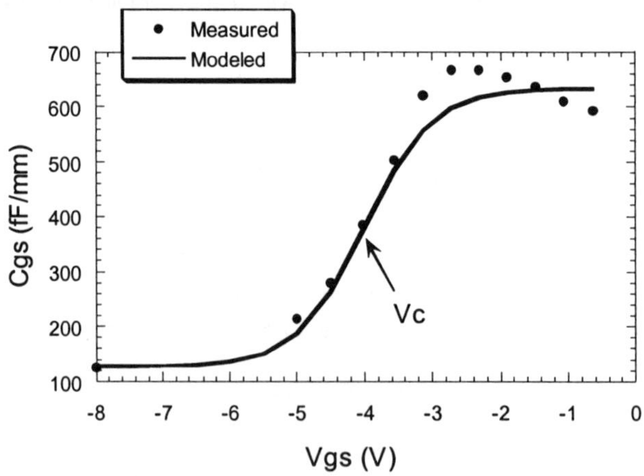

Figure 1. Nonlinear C_{gs} vs. V_{gs} characteristic of AlGaN/GaN HEMT.

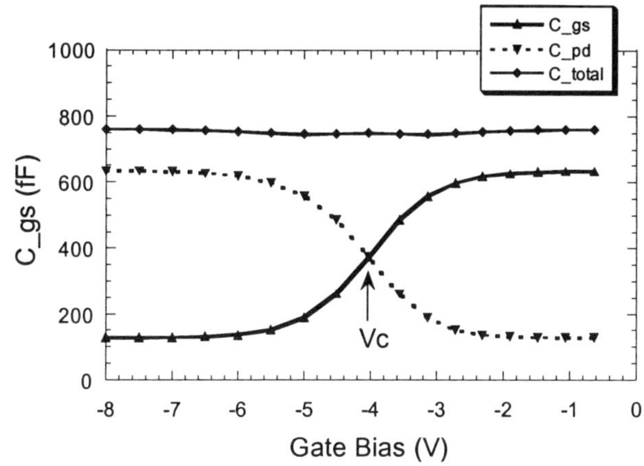

Figure 2. A pre-linearization gate diode compensates for the nonlinear C_{gs}.

4 Sources of GaN HEMT Device Nonlinearity

The dual gate Class B power amplifier with the prelinearization gate diode is designed at 10GHz. The Curtice cubic model (C_FET3) is used to model the GaN HEMTs device and the circuit is simulated using Agilent ADS. The circuit schematic and IC layout are shown in Figures 3 and 4. Both the input and output matching networks are on-chip. In the output matching network, the output capacitance of the device C_{out} is absorbed into the Pi low pass filter. A 10pF on-chip capacitor C_0 is needed as close to the second gate as possible to provide a good AC ground at the frequency of interest. The prelinearization gate diode is beside the main device, both of which have the same gate area. Another single-ended Class B power amplifier without the gate diode is also designed for comparison. According to simulation using a simplified model which only includes nonlinear C_{gs}, the IMD_3 performance can be improved by about 4dB with the pre-linearization gate diode at all power levels (shown in Figure 5).

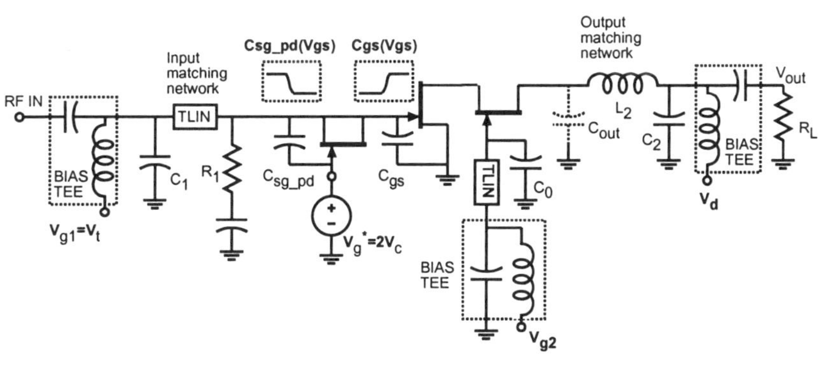

Figure 3. Circuit schematic of the dual-gate Class B PA with the pre-linearization diode.

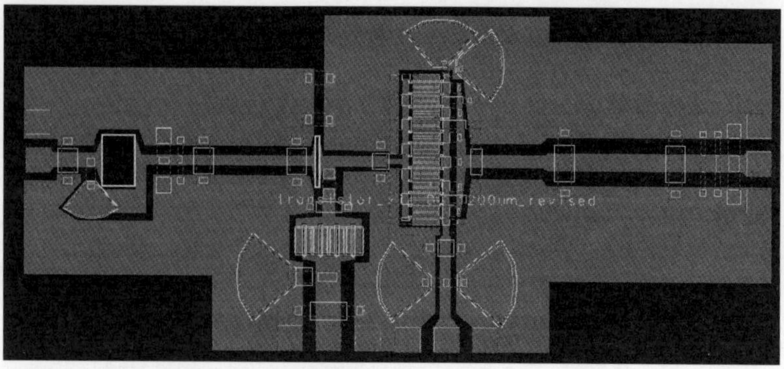

Figure 4. IC layout (dimensions 3mm x 1.5mm).

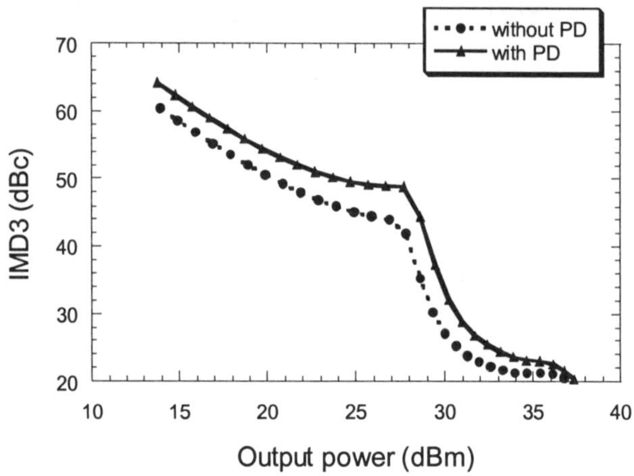

Figure 5. Simulation result on IMD_3 improvement with the pre-linearization diode.

5 Measurement Results

All the MMIC Class B power amplifiers are fabricated on a SiC substrate in the standard AlGaN/GaN HEMT technology [5] [6] and then tested. The 1.2mm dual gate GaN HEMT gives 1A/mm of I_{dss} and a breakdown voltage of approximately 55V. The measured f_T for the 0.3μm L_g dual gate device is 50GHz.

The single tone and two tone measurement results of the two circuits are shown in figures 6, 7 and 8. For single tone measurement, the circuit without the gate diode demonstrates 30dBm output power at 1dB gain compression point (P_{1dB}) with 9.5dB gain from 9GHz to 10GHz, while the circuit with the gate diode demonstrates 29.1dBm of P_{1dB} output power with 8.4 dB gain at 10GHz. The difference in gain and power between the two amplifiers is due to non-optimum input matching for the amplifier with diode.

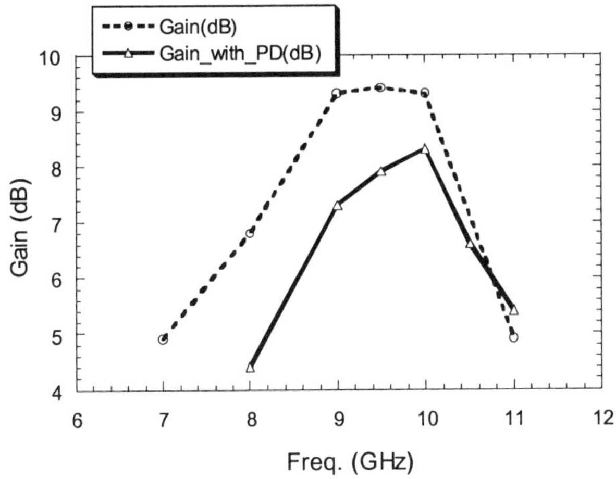

Figure 6. Measurement result: Gain vs. Frequency.

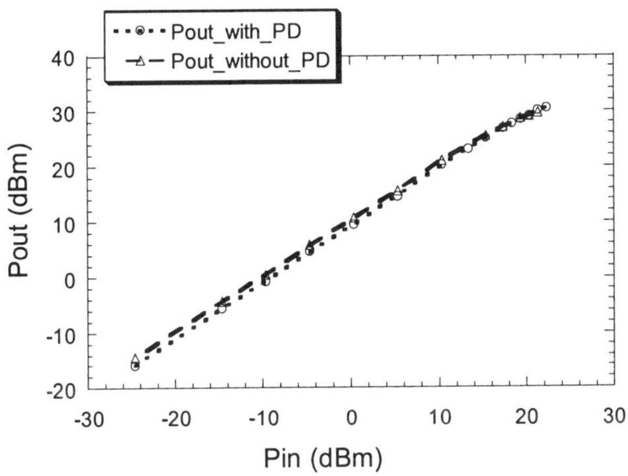

Figure 7. Single tone measurement result: Pout vs. Pin.

Two-tone measurement at 10GHz with 1MHz tone separation produced a peak of 51dBc IMD_3 with the gate diode MMIC at around 21dBm per tone output power. A 4dB improvement in IMD3 is obtained for the whole power range up to this point (as shown in figure 8).

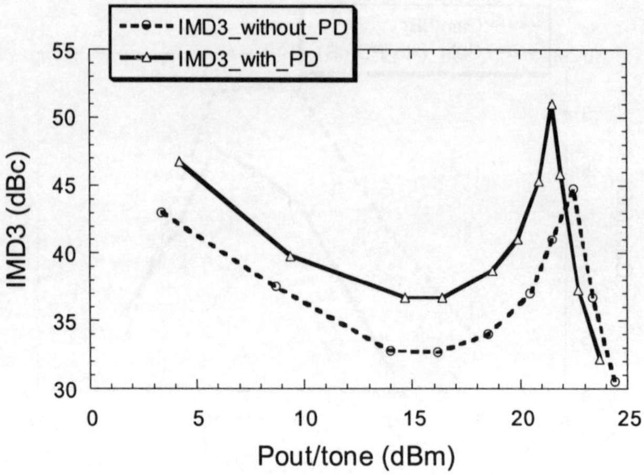

Figure 8. Two-tone measurement result IMD3 vs. Pout.

6 Conclusions

A high linearity MMIC RF power amplifier is reported in the AlGaN/GaN HEMT technology. Three major nonlinear sources of the GaN HEMT device are identified and investigated. To further improve linearity, a pre-linearization gate diode is added at the input to compensate the nonlinear effect of C_{gs}. The circuit with the pre-linearization gate diode demonstrates at least 4dB improvement on IMD_3 performance over the useful power range in two-tone measurement.

Acknowledgments

This work is supported by the Office of Naval Research under N00014-00-1-0653. And thanks to Dr. Walter Curtice, who provided us the Curtice C_FET3 model for the simulation.

References

1. Shouxuan Xie, *et al.*, *Microwave and Wireless Components Lett.*,**V 13 I 7**, 284 (2003).
2. Vamsi Paidi, *et al.*, *IEEE Transactions on MTT,* **V 51 I 2**, 643, (2003).
3. Joel Vuolevi, *et al.*, *IEEE Transactions on MTT*, **V 49 I 8**, 1383, (2001).
4. Chengzhou Wang, *et al.*, *RFIC Symposium*, *Digest of Papers*. *IEEE,* (2001).
5. Ching-Hui Chen, *et al.*, *Electron Device Letters, IEEE,* **V 21 I 12**, 549, (2000).
6. B.M. Green, *et al.*, *IEEE Transactions on MTT*, **V 49 I 12**, 2486, (2001).

International Journal of High Speed Electronics and Systems
Vol. 14, No. 3 (2004) 853–859

World Scientific
www.worldscientific.com

SPICE MODEL OF AlGaN/GaN HEMTs AND SIMULATION OF VCO AND POWER AMPLIFIER

SYED S. ISLAM

Department of Electrical Engineering, Rochester Institute of Technology
79 Lomb Memorial Drive, Rochester, NY 14623, United States

A.F.M. ANWAR

Department of Electrical and Computer Engineering, University of Connecticut
Storrs, CT 06269, United States

SPICE model parameters are extracted from reported experimental data. The model is implemented in the Cadence Affirma Analog Circuit Design Environment and Spectre simulator is used to simulate class-E power amplifier and ring voltage controlled oscillator (VCO) circuits. The availability of the SPICE model for GaN HEMTs ensures optimization of analog/RF circuits before an expensive cut-and-try method is employed.

1 Introduction

After a decade of extensive research in material growth and process development, GaN based devices are now used to design high power circuits operating at high temperature and frequency. GaN based HEMTs have been demonstrated to operate with power density five times higher than that of conventional GaAs/InP-based HEMTs and HBTs. As a result, GaN based circuits have reduced complexity, improved gain, efficiency and reliability. Green *et al.* [1] have demonstrated multioctave broadband MMIC power amplifier based on GaN/AlGaN HEMTs fabricated on high thermal conductivity SiC substrates with output power densities more than 5W/mm and PAE of 36% at 8GHz. Lee *et al.* [2] have demonstrated broadband push-pull microwave power amplifier using GaN/AlGaN HEMT operating at 4-9 GHz. Drozdovski *et al.* [3] have reported the application of GaN/AlGaN HEMTs as control components in high-power microwave and RF control devices such as switches, phase-shifters, *etc.* Shealy *et al.* [4] reported an AlGaN/GaN HEMT based 6GHz varactor-tuned voltage control oscillator (VCO). Chung *et al.* [5] demonstrated high power wideband AlGaN/GaN HEMT feedback amplifier module with drain and feedback loop inductances. The present authors have reported the design of a class-E power amplifier with AlGaN/GaN HEMTs using empirical models [6].

The current research trend clearly shows that with the combination of high power, high temperature, high frequency and low noise, GaN is becoming the future technology for both military (e.g., radar) and commercial (e.g., communication systems) applications. However, a circuit designer friendly, compact and scalable SPICE model based on BSIM equations [7], which is widely used to model the electrical behavior of silicon CMOS and

GaAs based HEMTs with different geometries, needs to be developed for GaN based HEMTs. Recently, Lee *et al.* [2] reported an artificial neural network (ANN) based modeling of AlGaN/GaN HEMT. Compared to ANN based models, SPICE models avoids complex algorithms and are user friendly. Besides, due to the inherent physics-based nature of BSIM equations, the SPICE models based on BSIM equations are scalable and compact [7]. Moreover, the SPICE models can be implemented in commercial and powerful Cadence integrated circuit design environment to simulate complex circuitries. The SPICE model for GaN HEMT can be used to analyze DC, AC and RF performances of GaN based integrated circuits.

In this paper, an SPICE model for GaN/AlGaN HEMT using simplified BSIM equations is reported. Model parameters are extracted from experimental data. The model is implemented in Cadence Affirma Analog Circuit Design Environment and Spectre simulator is used to design class-E power amplifier and ring voltage control oscillators.

2 Analysis

The following BSIM equations are used to model the electrical characteristics of an AlGaN/GaN HEMT [7]:

$$I_{DS} = \frac{I_{DS,0}}{1 + R_{DS}\dfrac{I_{DS,0}}{V_{DS,eff}}} \tag{1}$$

where

$$I_{DS,0} = \frac{W\mu_{eff}C_{AlGaN}V_{GST,eff}}{L\left(1+\dfrac{V_{DS,eff}}{E_{sat}L}\right)}\left(1-\frac{V_{DS,eff}}{2\left[V_{GST,eff}+2\dfrac{kT}{q}\right]}\right), \tag{2}$$

$$E_{sat} = \frac{2V_{SAT}}{\mu_{eff}}, \tag{3}$$

$$\mu_{eff} = \frac{U_0}{1+\dfrac{U_A V_{GST,eff}}{T_{AlGaN}}+\dfrac{U_B V_{GST,eff}^2}{T_{AlGaN}^2}}, \tag{4}$$

$$V_{DS,eff} = V_{DS,SAT} - \frac{1}{2}\left(ALPHA + \sqrt{ALPHA^2 + 4\cdot\Delta\cdot V_{DS,SAT}}\right), \tag{5}$$

$$ALPHA = V_{DS,SAT} - V_{DS} - \Delta, \tag{6}$$

$$V_{GST,eff} = \frac{2n(kT/q)\ln\left[1+\exp\left(\frac{V_{GS,eff}-V_T}{2n(kT/q)}\right)\right]}{1+2n\frac{C_{AlGaN}}{C_{dep0}}\left(-\frac{V_{GS,eff}-V_T-2V_{OFF}}{2n(kT/q)}\right)}, \tag{7}$$

$$n = 1 + N_{factor}\frac{C_{dep}}{C_{AlGaN}} + \frac{C_{IT}}{C_{AlGaN}}, \tag{8}$$

$$R_{DS} = R_{DSW}\left(1 + P_{RWG}V_{GST,eff}\right), \tag{9}$$

$$V_T = V_{TH0} + K_1\left(\sqrt{2\phi_f - V_{BS,eff}} - \sqrt{2\phi_f}\right) - K_2 \cdot V_{BS,eff}, \tag{10}$$

$$2\phi_f = 2\frac{kT}{q}\ln\left(\frac{N_{CH}}{n_i}\right), \tag{11}$$

$$C_{AlGaN} = \frac{\varepsilon_{AlGaN}}{T_{AlGaN}}, \tag{12}$$

$$C_{dep} = \frac{\varepsilon_{GaN}}{X_{dep}}, \tag{13}$$

$$C_{dep0} = \frac{\varepsilon_{GaN}}{X_{dep0}}, \tag{14}$$

$$X_{dep} = \sqrt{\frac{2\varepsilon_{GaN}\left(2\phi_f - V_{BS,eff}\right)}{qN_{CH}}} \text{ and} \tag{15}$$

$$X_{dep0} = \sqrt{\frac{2\varepsilon_{GaN}\cdot 2\phi_f}{qN_{CH}}}. \tag{16}$$

Here, W and L represent the channel width and length respectively, C_{AlGaN} is the AlGaN layer capacitance, C_{dep} is the depletion region capacitance, V_{SAT} is the saturation velocity, R_{DS} is the drain source resistance, μ_{eff} is the effective mobility, E_{sat} is the saturation electric field, V_T is the threshold voltage and $2\phi_f$ is the strong inversion surface potential. Also, $V_{DS,eff}$ and $V_{GST,eff}$ are smoothing functions that are used to fit the I-V characteristics of the device simultaneously in the linear and saturation regions and in the subthreshold and strong inversion regions respectively. AlGaN/GaN HEMT reported in [8] is used for modeling in this work using Eqs. (1)-(16). The AlGaN barrier layer thickness is 30nm, doping level is 1×10^{19} cm^{-3}, gate length is 0.12μm and channel width is 100μm. Parasitic capacitances for comparable feature size Si CMOS fabrication processes have been used in the absence of measured data for GaN HEMTs.

The extracted fitting parameters are used to simulate the device in Cadence Affirma Analog Circuit Design Environment using Spectre simulator. Fig.1 shows the simulated *I-V* characteristics using the developed SPICE model. The experimental data is also plotted on the same figure to show the agreement. The extracted parameter values are shown in Table 1.

Fig.2 shows the circuit diagram of a class-E power amplifier. The load network of the class-E amplifier is designed to ensure minimum switching loss. To consider the switching nonidealities, the load network parameters have been calculated assuming exponential decay of the drain current during fall time and a finite loaded Q-factor of the inductor. The complete analysis of the class-E power amplifier is reported by the present authors in [6].

An odd number of AlGaN/GaN HEMT based inverters are connected in ring configuration to form a single ended ring oscillator. A 3-stage ring oscillator can be converted to a single-ended VCO using the topology shown in Fig.3. Here the controlling transistor (M) is the variable delay element and can be used to control the frequency of oscillation by varying its gate voltage, $V_{control}$.

3 Results and Discussion

Fig.4 shows the calculated output power and power conversion efficiency as a function of frequency of a class-E power amplifier using the BSIM3v3 model of a 0.12μm $\times$100μm GaN/$Al_{0.20}Ga_{0.80}N$ HEMT [8]. At 1GHz, output power is 156mW with 10V supply voltage. Output power decreases to 120mW at 11GHz. The power conversion efficiencies are 52% and 43% at 1GHz and 11GHz respectively. The decrease in output power and efficiency is attributed to the increase in switching loss with increasing frequency.

The SPICE model of the AlGaN/GaN HEMT is used to simulate inverter based ring oscillators. The frequency of oscillation of the ring oscillator is 53GHz. Per-stage delay is calculated as 2.55ps which corresponds to an oscillation frequency of 65GHz. The difference is due to the loading effect of the following inverter stages. Fig.5 shows the output voltage waveform of the 3-stage ring VCO of Fig.3. The supply voltage V_{DD} = 10V, and $V_{control}$ = 0V and the aspect ratio of the control transistor is 800μm/0.12μm.

In Fig.6, the oscillation frequency of the 3-stage ring VCO is plotted as a function of the *W/L* ratio of the FETs. The above results are obtained by optimizing the VCO for maximum frequency of oscillation for a given value of the aspect ratio of the controlling transistor (M). As observed the frequency of oscillation initially increases and eventually saturating at a higher control voltage. This is because when the controlling transistor is operating in the active region the delay becomes fixed which results in a constant frequency of oscillation. With increasing aspect ratio the delay decreases resulting in an increase in the frequency of oscillation.

4 Conclusions

An SPICE model of AlGaN/GaN HEMT based on simplified BSIM equations is reported. The SPICE parameters have been extracted from experimental device characteristics. The model has been embedded in the SPECTRE circuit simulator to facilitate the design of power amplifier and voltage controlled oscillator.

Table 1. Extracted BSIM3v3 parameters for $Al_{0.20}Ga_{0.80}N$/GaN HEMT [8]

Parameter	Symbol	Value	Units
Zero field mobility at 300K	U_0	1100	$cm^2/V.s$
Channel doping concentration	N_{CH}	1×10^{19}	cm^{-3}
Electron saturation velocity at 300K	V_{SAT}	1×10^{8}	cm/s
AlGaN layer thickness	T_{AlGaN}	30	nm
Threshold voltage at zero V_{BS}	V_{TH0}	-5.83	V
First-order mobility coefficient	U_A	0.37×10^{-7}	m/V
Second-order mobility coefficient	U_B	0.40×10^{-15}	m^2/V^2
First-order body effect factor	K_1	3.73×10^{-5}	$V^{0.5}$
Second-order body effect factor	K_2	-1.87×10^{-6}	unitless
Drain source resistance parameter	R_{DSW}	0.270	Ω
Gate bias coefficient of R_{DSW}	P_{RWG}	0	V^{-1}
Offset voltage in subthreshold region	V_{OFF}	0	V
Subthreshold turn on swing factor	N_{factor}	0	Unitless
Interface trap capacitance per unit area	C_{IT}	0	Unitless
Effective drain voltage smoothing parameter	Δ	1.095	Unitless

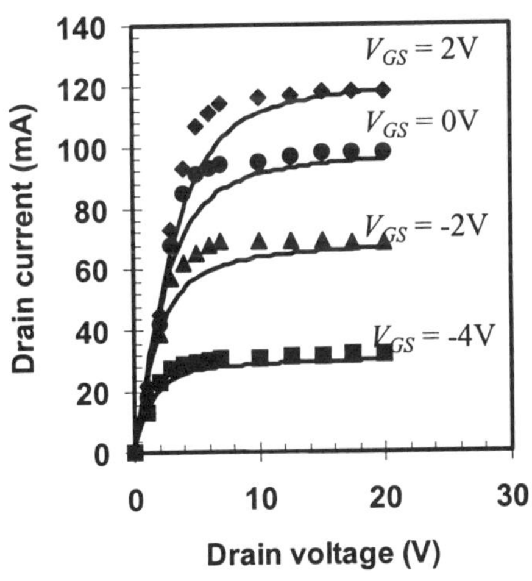

Fig.1. Simulated (solid lines) I-V characteristics of a 0.12μm×100μm GaN/ $Al_{0.20}Ga_{0.80}N$ HEMT [8]. Experimental results are shown by symbols.

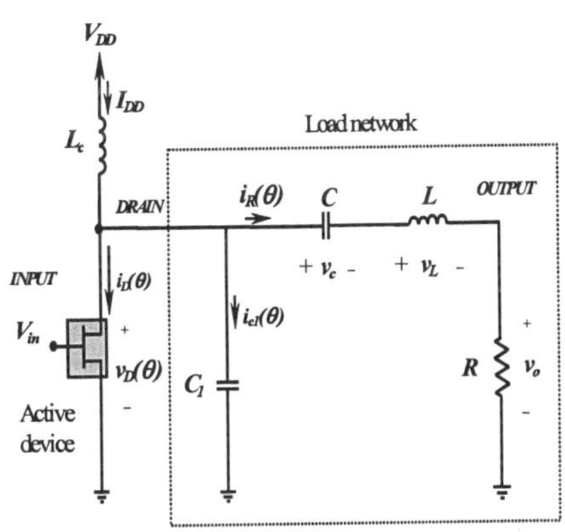

Fig.2. Basic circuit of the class-E power amplifier.

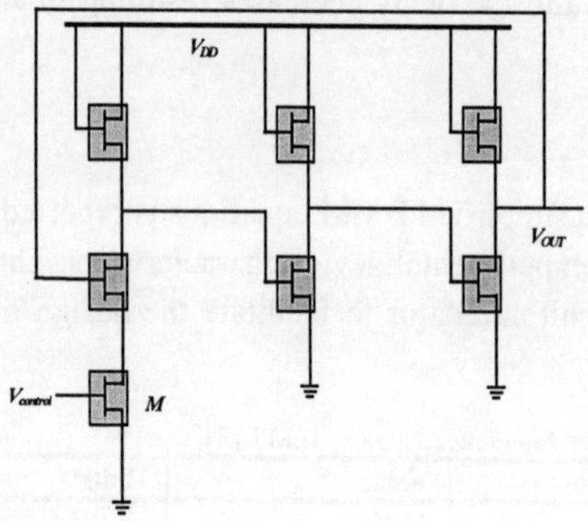

Fig. 3. Circuit diagram of a 3-stage ring VCO.

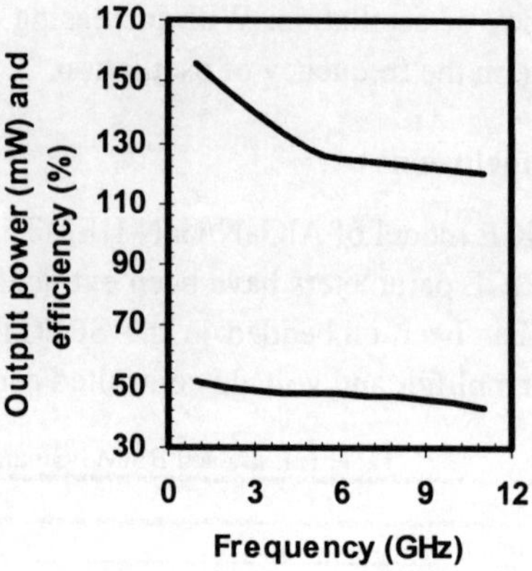

Fig.4. Calculated output power (top) and efficiency (bottom) of the class-E amplifier as a function of operating frequency. The active device is a 0.12μm×100μm GaN/ $Al_{0.20}Ga_{0.80}N$ HEMT [8].

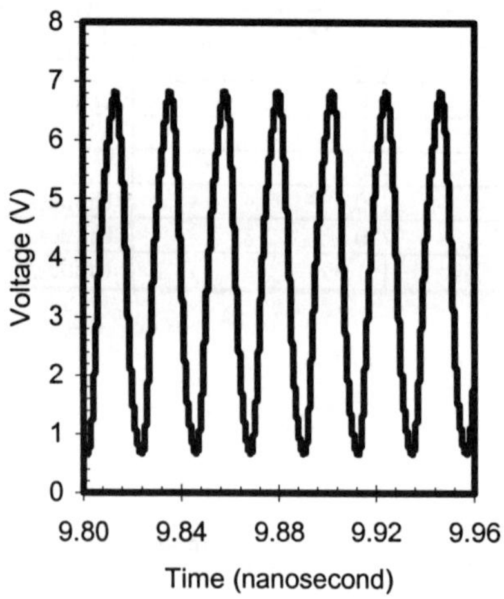

Fig.5. Output signal of a 3-stage ring VCO.

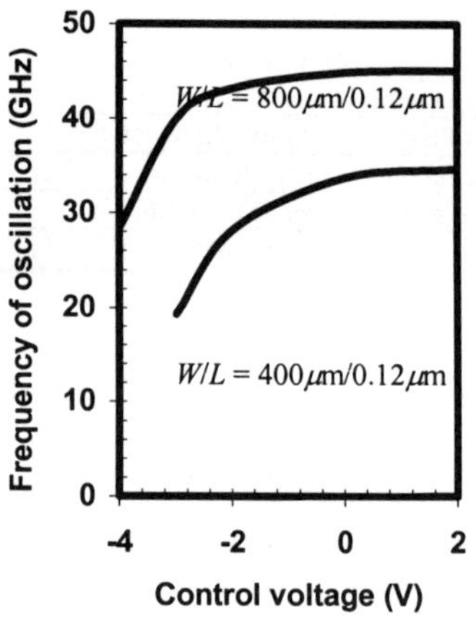

Fig.6. Oscillation frequency as a function of control voltage for a 3-stage ring VCO. W/L denotes the aspect ratio in (μm/μm) of the control transistor.

References

1. B.M. Green, V. Tilak, S. Lee, H. Kim, J.A. Smart, K.J. Webb, J.R. Shealy and L.F. Eastman, *IEEE Trans. Microwave Theory and Tech.*, **49**, 2486, (2001).
2. S.Y. Lee, B.A. Cetiner, H. Torpi, S.J. Cai, J. Li, K. Alt, Y.L. Chen, C.P. Wen, K.L. Kang, and T. Itoh, *IEEE Trans. Electron Dev.*, **48**, 495, (2001).

3. N.V. Drozdovski and R.H. Caverly, *IEEE Trans. Microwave Theory and Tech.*, **50**, 4, (2002).
4. J.B. Shealy, J.A. Smart and J.R. Shealy, *IEEE Microwave and Wireless Components Lett.*, **11**, 244, (2001).
5. Y. Chung, S. Cai, W. Lee, Y. Lin, C.P. Wen, K.L. Wang and T. Itoh, *Electron Lett.*, **37**, 1199, (2001).
6. S. S. Islam and A.F.M. Anwar, **46**, 1621, (2002).
7. W. Liu, *MOSFET Models for SPICE Simulation including BSIM3v3 and BSIM4*, John Wiley and Sons, (2001).
8. W. Lu, J. Yang, M.A. Khan and I. Adesida, *IEEE Trans. on Electron Dev.*, **48**, 581, (2001).

International Journal of High Speed Electronics and Systems
Vol. 14, No. 3 (2004) 860–864

World Scientific
www.worldscientific.com

HIGH POWER, DRIFT-FREE 4H-SiC PIN DIODES

MRINAL K. DAS, JOSEPH J. SUMAKERIS, BRETT A. HULL,
JIM RICHMOND, SUMI KRISHNASWAMI AND ADRIAN R. POWELL

Cree, Inc., 4600 Silicon Drive, Durham, NC, 27703, USA

The path to commercializing a 4H-SiC power PiN diode has faced many difficult challenges. In this work, we report a 50 A, 10 kV 4H-SiC PiN diode technology where good crystalline quality and high carrier lifetime of the material has enabled a high yielding process with V_F as low as 3.9 V @ 100 A/cm^2. These 10 kV diodes demonstrate pulsed current handling capability up to 328 A which represents over 3 MW of pulsed power. Furthermore, incorporation of two independent basal plane dislocation reduction processes (LBPD 1 and LBPD 2) have produced a large number of devices that exhibit a high degree of forward voltage stability. The more benign LBPD 2 process yields wafers with better reverse blocking capability resulting in a total yield (forward, 10 kV blocking, and drift) of >20% for 8.7 mm x 8.7 mm power PiN diode chips—the largest SiC chip reported to date.

1 Introduction

Rapid advancement of 4H-SiC PiN diode technology has enabled record high blocking voltage (up to 20 kV) with relatively low forward voltage drops (V_F).[1] This is a direct consequence of improved crystal growth (reduced micropipe and defect density) and epitaxy (reduced epi defects and high minority carrier lifetimes) which permits full utilization of the large critical field and enhanced conductivity modulation. Despite such progress, the PiN diodes may suffer from significant forward voltage instability due to stacking faults spawned from basal plane dislocations (BPD).[2] This problem is exacerbated with thick epitaxy because the faulted area scales with the drift layer thickness. Clearly, the reduction of BPDs is vital in developing a stable SiC PiN diode. Innovations in epitaxy have produced 100 μm drift layers with BPD densities of 10 cm^{-2} and 20 cm^{-2} for independent processes LBPD 1 and LBPD 2, respectively.[3] PiN diodes fabricated with LBPD 1 have already shown excellent V_F stability across the wafer, but poor surface morphology reduced the reverse blocking yield to just 2%.[4] Recent optimization of the LBPD 1 process has improved the surface morphology while maintaining the low BPD density. In this contribution, we report the effect of incorporating standard (continuous growth of the n$^-$ drift and p$^+$ injector layers), improved LBPD 1, and LBPD 2 epitaxy in the 50 A, 10 kV 4H-SiC PiN diode process.

2 Experimental

High quality epitaxial layers (100 μm drift layer, N_D = 2E14 cm^{-3}, and 2 μm injection layer, N_A = 8E18 cm^{-3}) are continuously grown on 0.6 cm^{-2} micropipe density 8° off-axis (0001) 4HN conducting substrates. The devices are mesa isolated by etching down to the n- drift layer in the field areas. Next, an aluminum implanted guard ring-based termination is formed around the device periphery while a nitrogen implanted channel stop region grounds the field area away from the devices. The implants are activated with a 1600°C anneal in Ar. The surface is passivated with a silicon dioxide layer. Sintered Ti-Al and Ni contacts are made to the anode and cathode, respectively, before depositing thick Au overlayers. The 8.7 mm x 8.7 mm chip contains a single 50 A PiN diode (0.50 cm^2 active area), while a drop-in chip contains a 20 A PiN diode (0.20 cm^2 active area) and five small PiN diodes (0.0078 cm^2 active areas).

3 Results

All three epitaxial processes yield PiN diodes with good blocking and low, stable V_F. Figure 1 shows the reverse blocking capability of a typical large diode. Reverse blocking is measured out to 9 kV where the leakage current exceeds the current compliance limit of the measurement apparatus. In terms of leakage current density, the 0.5 mA/cm^2 corresponds to a minimal value that is well below the typical current densities allowed in reverse operation. Hence, a 10 kV blocking is implied for the large devices blocking >7 kV. A low forward voltage drop of 3.9 V is observed at 50 A (100 A/cm^2) indicating a high degree of conductivity modulation (Fig. 2). PiN diodes are capable of handling very high current densities and we have measured this large device out to a pulsed current of 328 A (656 A/cm^2) at a low forward drop of 5.9 V. For a single chip, this demonstrates over 3 MW of pulsed power! These devices also show excellent forward characteristics with respect to temperature (Fig. 3). The V_F at low current densities decreases slightly due to temperature enhanced carrier lifetime which is a concern for thermal runaway. However, at high current densities, the behavior reverses to a positive temperature coefficient with a zero temperature coefficient observed at 400 A/cm^2, thereby making these devices attractive for use in a parallel configuration where the current must be properly shared.

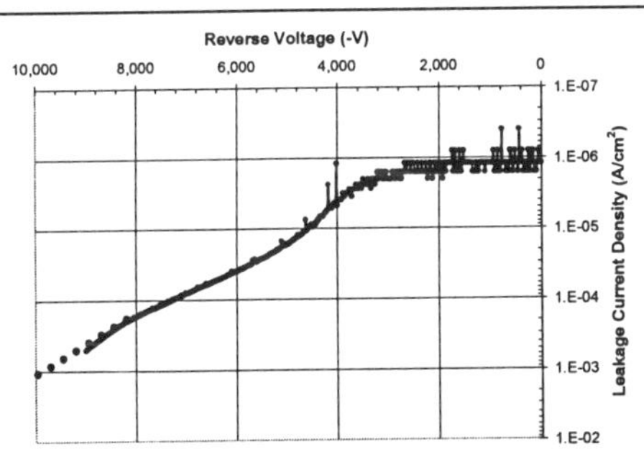

Figure 1. Reverse J-V characteristic of a 50 A diode (8.7 mm x 8.7 mm) showing >9 kV blocking (solid curve) that is current compliance limited. The dashed line extrapolates the reverse blocking to 10 kV.

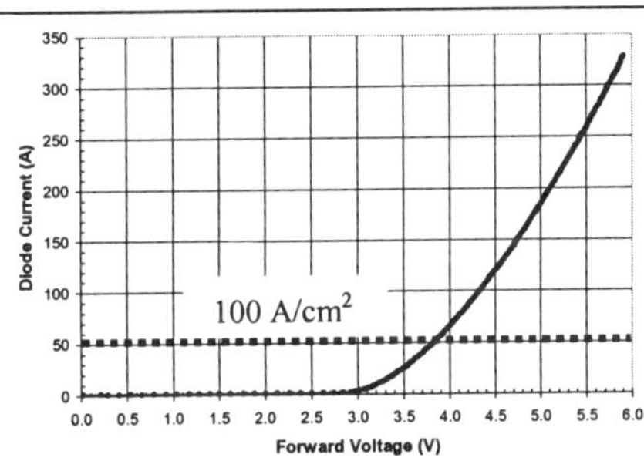

Figure 2. Pulsed forward I-V characteristic of a large 10 kV diode with a 3.9 V forward drop at 50 A and 5.9 V forward drop at 328 A indicating a high level of conductivity modulation.

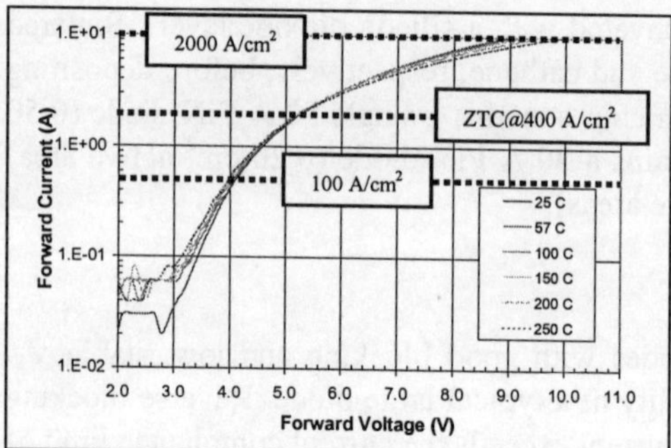

Figure 3. Forward I-V curve measured at temperatures up to 250°C on a small 10 kV device where a positive temperature coefficient (observed at high current densities) facilitates device paralleling.

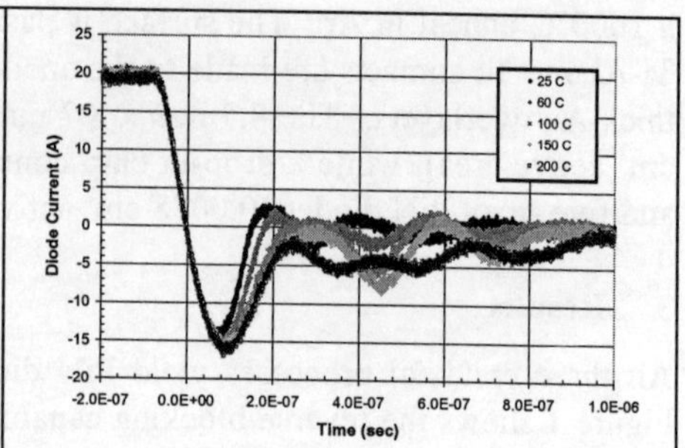

Figure 4. Large dI/dt (330 A/μsec) reverse recovery transient showing fast turn-off with very little reverse recovery charge for a 20 A, 10 kV device at temperatures up to 200°C..

The low forward voltage drops obtained on these devices are derived from high carrier lifetimes that increase the conductivity of the nominally high resistance drift layer. One potential drawback is that the switching ability may suffer, especially the turn-off transient, due to the extra time needed for carrier removal in the drift layer. The reverse recovery of the 20 A, 10 kV diodes (Fig. 4) shows a fast ~200 nsec recovery with very little reverse recovery charge despite an aggressive dI/dt of 330 A/μsec.

To be a commercially viable product, good device performance must be demonstrated across the entire wafer. Figure 5 illustrates the 10 kV blocking wafermaps for the three epitaxial splits (please recall the aforementioned discussion regarding the implied 10 kV blocking for devices leaking 300 μA at >7 kV). As expected, the standard epitaxy and the relatively benign LBPD 2 processes give the best yields of 37% and 35%, respectively. Although the more aggressive LBPD 1 process shows a lower yield of 27%, this is still a substantial improvement over our previous attempt where the blocking yield was a mere 2%. Key yield limiters appear to be particles and defects generated in the epigrowth as well as the ability to fully resolve the guard ring photolithography across the wafer. Overcoming these limitations should increase the reverse blocking yield to well over 50%.

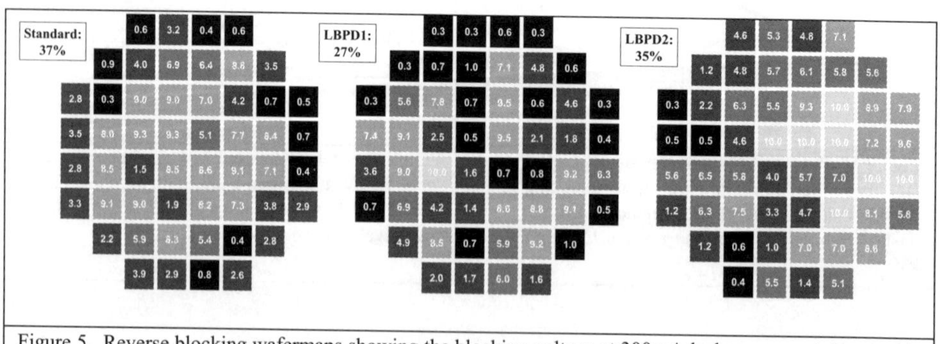

Figure 5. Reverse blocking wafermaps showing the blocking voltage at 300 μA leakage current. Due to this low current compliance, we are unable to measure the devices to their full blocking capability, hence 10 kV blocking is implied for >7 kV devices.

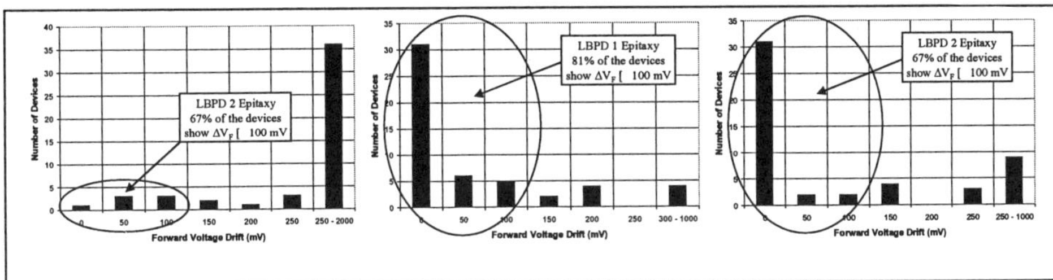

Figure 6. Forward voltage drift histograms for the three epitaxial processes. Both LBPD processes show dramatic improvement in the 100 mV drift yield.

In the forward operation, the initial V_F measured across the wafer is very uniform. The more discriminating test is to measure the stability of the forward voltage. With the advent of continuous growth epitaxy, the diodes have become more well behaved such that they may be screened with a simple on-wafer stress measurement: 1.) measure pulsed initial V_F @ 50 A/cm^2 at room temperature; 2.) constant current stress at 50 A/cm^2 for 30 min which also heats the device; 3.) cool down to room temperature; 4.) measure pulsed final V_F @ 50 A/cm^2; 5.) move to next die on wafer and repeat until the entire wafer is tested. Histograms in Fig. 6 illustrate the number of stable devices by binning the devices according to the amount of forward voltage increase. Using 100 mV as an acceptable amount of V_F drift, we observe V_F drift yields of 17%, 81%, and 67% for the standard, LBPD 1, and LBPD 2, respectively. These numbers are consistent with the observed occurrence of BPDs in the epitaxy and demonstrate that a large number of stable devices can be manufactured per wafer.

Table 1. Improved device yields corresponding to historical improvements in thick epitaxy

Epitaxy	BPD (cm^{-2})	Device	Drift Yield	10 kV Yield	Total Yield
2 Step Growth	300	7.5 A	0 %	10 %	0 %
Standard 1 Growth	180	20 A	15 %	40 %	6 %
Original LBPD 1	10	20 A	86 %	2 %	2 %
Original LBPD 2	20	20 A	67 %	39 %	26 %
Standard 1 Growth	**180**	**50 A**	**17 %**	**37 %**	**6 %**
Improved LBPD 1	**10**	**50 A**	**81 %**	**27 %**	**22 %**
Original LBPD 2	**20**	**50 A**	**67 %**	**35 %**	**23 %**

4 Summary

Isolated hero devices have been historically reported by many groups. Our emphasis, however, is to take this one step further and make the hero devices the norm, rather than the exception, on the wafer. Table 1 shows our progress toward this goal. From the 20 A experiment, we demonstrated two processes that dramatically increased the drift yield. However, the LBPD 1 process had poor blocking yield due to surface morphology issues. In the ensuing 50 A experiment, we optimized the LBPD 1 process to improve the surface morphology, thereby giving us two independent 4H-SiC PiN diode processes capable of yielding >20% of the 8.7 mm x 8.7 mm chips (the largest SiC chip reported to date). This trend of making the large hero devices more common on the wafer is a significant step in the evolution of a commerically viable SiC PiN diode technology expected to meet the demands of high voltage, high frequency power conversion applications.

Acknowledgements

This work was supported in part by DARPA and Dr. John Zolper through ONR contract number N00014-02-C-0302, monitored by Dr. Harry Dietrich.

References

1. M.K. Das: Latest Advances in 4H-SiC PiN and MOS Power Devices (International Semiconductor Device Research Symposium, Washington, DC, December 2003).
2. H. Lendenmann, F. Dahlquist, J.P. Bergman, H. Bleichner and C. Hallin, Mater. Sci. Forum, Vol. 389-393 (2002) p. 1259.
3. J.J. Sumakeris, M.K. Das, H.M. Hobgood, S.G. Mueller, M.J. Paisley, J.W. Palmour, and C.H. Carter, Jr.: Mater. Sci. Forum, Vol. 457-460 (2004) p. 1113.
4. B.A. Hull, M.K. Das, J.J. Sumakeris, J. Richmond, and S. Krishnaswami: Drift Free, 10 kV, 20 A, 4H-SiC PiN Diodes (Electronic Materials Conference, South Bend, IN, June 2004).

International Journal of High Speed Electronics and Systems
Vol. 14, No. 3 (2004) 865–871

World Scientific
www.worldscientific.com

DESIGN OF HIGH VOLTAGE 4H-SiC SUPERJUNCTION SCHOTTKY RECTIFIERS

Lin Zhu, Peter Losee and T. Paul Chow

Center for Integrated Electronics, Rensselaer Polytechnic Institute, Troy, NY-12180 USA

This paper presents a novel Schottky rectifier structure based on the superjunction approach, which is utilizes 2- and 3-D field shaping to increase the avalanche breakdown voltage. Device forward and blocking characteristics are analyzed with numerical simulations and compared with conventional 4H-SiC Schottky rectifiers. Optimal design tradeoffs between breakdown voltage and specific on-resistance are obtained for high voltage 4H-SiC superjunction Schottky rectifiers. The results show that the new structure can provide a 20 × lower $R_{on,sp}$ than conventional Schottky rectifier for 6kV device. In addition, device termination and possible fabrication steps for the superjunction devices are also presented.

1 Introduction

Unipolar devices are widely used in power electronic circuits because of their fast switching speed. The devices are preferred to have small on resistance (R_{on}) to reduce power loss and have high blocking voltage (BV) to sustain reverse voltage. From 1D analysis of unipolar devices, such as MOSFET and Schottky rectifier, the specific on-resistance ($R_{on,sp}$) is proportional to $BV^{2.5}$[1], which means that increasing BV can result in a significant increase in $R_{on,sp}$.

A new device concept, called superjunction (SJ) device, which utilizes 2- and 3-D non-planar pn junction for field shaping, has been demonstrated in silicon [2-4], that can be used to overcome the trade-off between blocking voltage and on state losses of power MOSFETs. The drift region of the new device is composed by alternatively stacked, heavily doped N and P pillars. Fig.1 shows a schematic cross section of a superjunction Schottky structure. The on state current flows through the N pillar. The P pillar does not contribute to on state conduction, but is essential in achieving high blocking voltage. The superjunction structure can achieve high blocking voltage based on the charge balance between the two pillars. In the off state, a space charged region across the vertical p-n junction builds up, leading to depletion of N and P pillars at relatively low reverse bias. The field distribution in the device is now two-dimensional, which allows higher blocking voltage with the same epi thickness than the conventional 1-D breakdown voltage. In the superjunction structure, the doping concentration of the drift region is limited by the pillar width, not the breakdown voltage. For the same breakdown voltage, the doping concentration of the N pillar can be made much higher than the drift region of a conventional unipolar device.

The advantages of superjunction structure have been experimentally demonstrated in many silicon devices. In this work, 4H SiC Schottky rectifiers based on the superjunction technique will be analyzed using numerical simulations. Termination technique and fabrication steps will also be presented.

This work was supported by National Science Foundation under Award Number EEC-9731677.

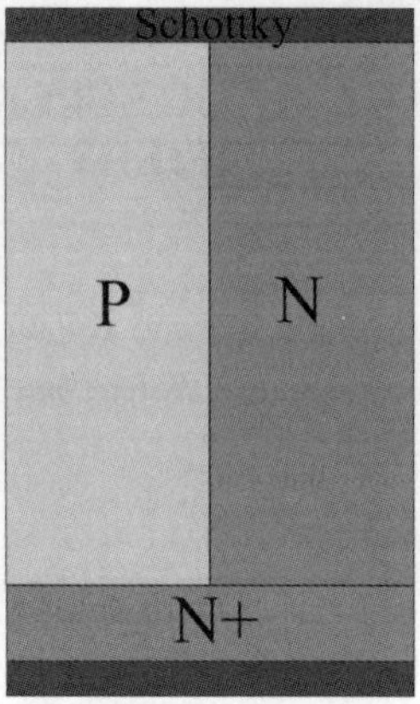

Figure 1. Schematic cross-section of a superjunction Schottky rectifier.

2 Forward and Blocking Simulation

The forward and blocking characteristics are modeled by numerical simulations to determine the design parameters with MEDICI^{TMA}. In Fig.1, we show the simulated unit cell structure of the 4H-SiC superjunction Schottky (SJS) rectifier. The key design parameters that affect the performance of SJS include epi thickness, epi doping concentration and p and n pillar width.

In the on state, current flow from the Schottky contacts, through the drift region (the n-type side of the superjunction), to the substrate. Thus only half of the drift region is conducting forward current. To achieve the same on-resistance as the conventional Schottky rectifier, the doping concentration in the n-pillar has to be more than two times higher than that of the conventional structure. To minimize the on-resistance, the doping concentration of the drift region should be as high as possible. For conventional Schottky rectifier, the doping concentration of the drift region is only related to the blocking voltage [1]. But for SJS rectifier, the doping concentration is limited by the pillar width, which further explained with the blocking results below. Fig. 2 shows the forward simulation results for a SJS rectifier with 35μm thick drift region, comparing with a conventional Schottky rectifier with the same epi thickness. The doping concentration for the conventional Schottky rectifier is $3.5\times10^{15}cm^{-3}$. The SJS rectifier simulated has a pillar width of 2μm. The curve for SJS with $1\times10^{15}cm^{-3}$ n-pillar doping concentration, has a turn on knee because the n-pillar is pinched off by the built-in potential between the n and p pillar at zero bias. With higher doping concentration, the forward drop decreases, eventually lower than that of the conventional Schottky rectifiers.

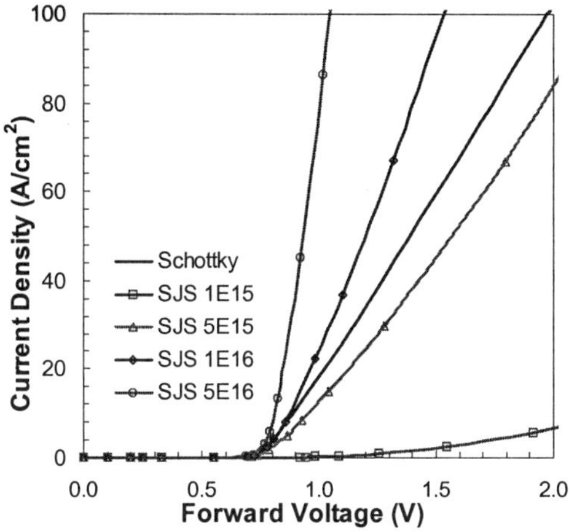

Figure 2. Forward characteristics of a SJS rectifier comparing with conventional Schottky rectifier.

During the reverse blocking, the reverse voltage appears as a reverse bias between the n and p pillars. The depletion region will form along the p-n junction, and at relatively small reverse bias, the drift region is fully depleted. This requires that the 2 pillars have the same amount of charge. Subsequently, the drift region behavior is similar to that of an intrinsic layer. With a rectangular 1-D electrical field, as shown in Fig. 3, versus the triangular electrical field in the conventional Schottky, the new structure can support more reverse voltage by the drift region with the same thickness. Fig. 3 shows the ideal case, which means the pillar doping is low, so the electric field is very uniform across the drift region. If the pillar doping concentration is very high, the electric field will be much less uniform across the pillar, resulting in a much lower breakdown voltage than that of the low doping concentration case, as shown in Fig. 4. Another factor is that the critical electric field increases with the doping concentration. As a combined effect of those two factors, the breakdown voltage slightly increases with pillar doping concentration at the beginning. Then the breakdown voltage decreases dramatically as the pillar doping concentration increases after it reaches the maxim point.

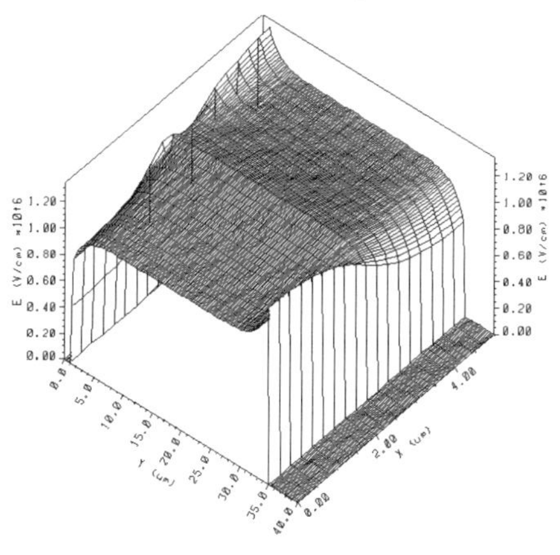

Figure 3. Ideal electric field for SJS rectifier at breakdown voltage.

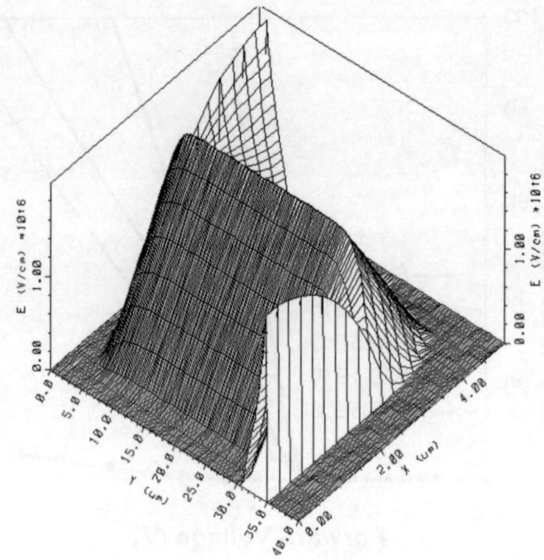

Figure 4. Electric field for SJS rectifier with high pillar doping concentration at breakdown voltage.

The breakdown voltage and specific on-resistance as a function of pillar doping concentration is shown in Fig. 5. There is a trade-off between breakdown voltage and specific on-resistance. The optimal concentration occurs when the charge dosage of the p and n pillar, (which is equal to the doping concentration multiplied by the pillar width), is around 1.5×10^{13}cm^{-2}. The advantage of the superjunction structure is that the doping concentrations of the n and p pillars are only a function of the pillar width, not breakdown voltage. In theory, when the pillar width is fixed, the optimal doping concentration is fixed. Then the specific on-resistance is almost linear function of breakdown voltage, which is a much better trade-off than that of the conventional structure as shown in Fig. 6. SiC superjunction devices can achieve better trade-off when the breakdown voltage more than 1kV. With 35μm epi thickness and 2μm pillar width, the optimized 4H-SiC superjunction Schottky rectifier can block 6kV while the on-resistance is 20 times smaller than that of the conventional Schottky rectifier. According to the figure, the advantage of superjunction devices will be more significant with scale-down of the pillar width.

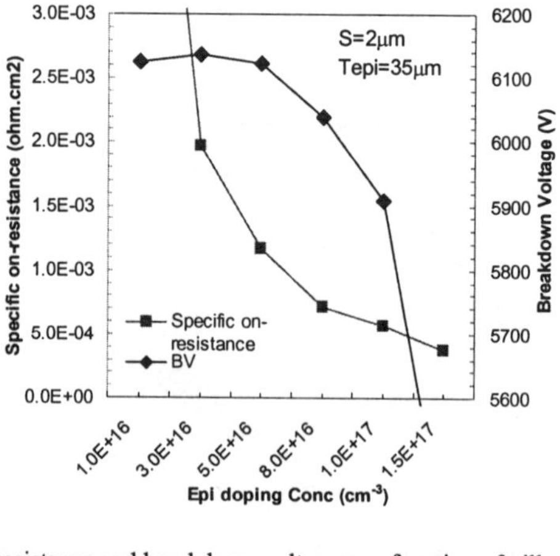

Figure 5. Specific on-resistance and breakdown voltage as a function of pillar doping concentration.

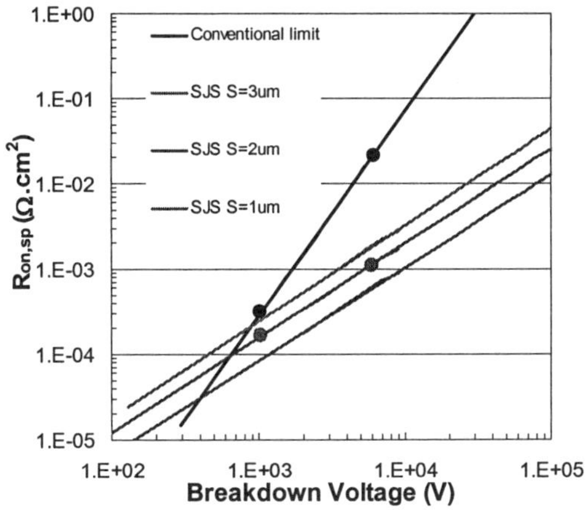

Figure 6. Trade-off between specific on-resistance and breakdown voltage of 4H-SiC SJS devices comparing with conventional structure.

The superjunction structure is based on charge balance theory. Perfect charge balance between the n and p pillars is required to completely utilize the benefits of the superjunction structure, which is very difficult in reality. The charge imbalance will decrease the blocking capability as shown in Fig. 7. Also the higher the doping is, more sensitive the breakdown voltage is to the charge imbalance [5].

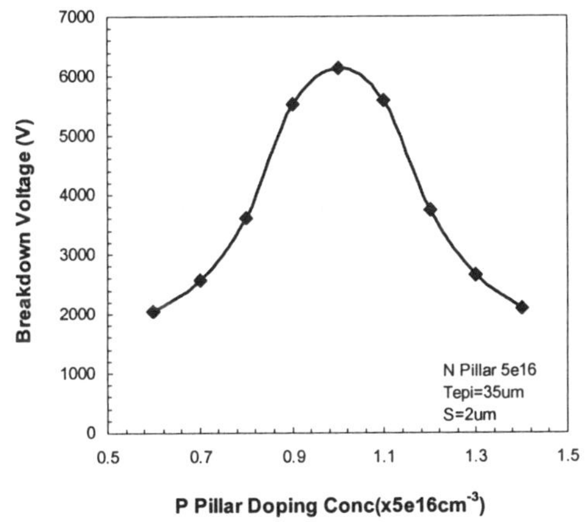

Figure 7. Sensitivity of BV to the charge imbalance.

3 Implementation in SiC

In silicon, superjunction devices are manufactured by multiple deposition of epitaxial layers and subsequent implantation and diffusion steps on a highly doped n+ substrate. For SiC, dopant diffusion is extremely slow even at high temperature and the fabrication using multiple epitaxial regrowth technology is not mature. For those reasons, an alternative way to form the pillar is necessary. A new structure using only one epitaxial regrowth and oxide trench refill is shown in Fig. 8. This structure requires a deep trench etching through the drift region, which is less complicated than multiple epitaxial regrowth. After etching, a single thin, p-type SiC layeris regrown. The rest of the trench can be refilled with oxide. For the superjunction structure to work, the charge in the p and n-pillar must be balanced. If the width of the p-pillar is *n* times smaller than the width of n-pillar, the doping concentration in p pillar should be *n* times higher than that of n-pillar to keep the total charge the same. Simulation of the new structures exhibits similar performance as the conventional superjunction structure.

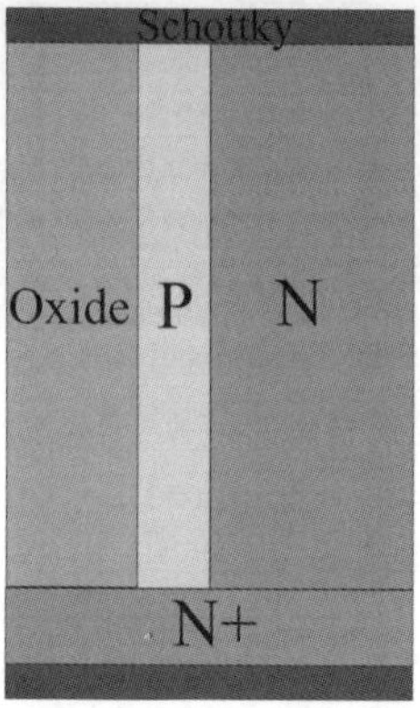

Figure 8. Schematic cross-section of the proposed superjunction Schottky rectifier.

A novel termination scheme that is compatible with the SJS SiO_2 refill process, has been simulated (Fig.9). This termination is a combination of superjunction, JTE and field plate. The critical design parameter is the mesa width in termination region. Proper design can ensure that potential drops evenly across mesa rings. A breakdown voltage close to 90% of that of the ideal SJS rectifier was achieved.

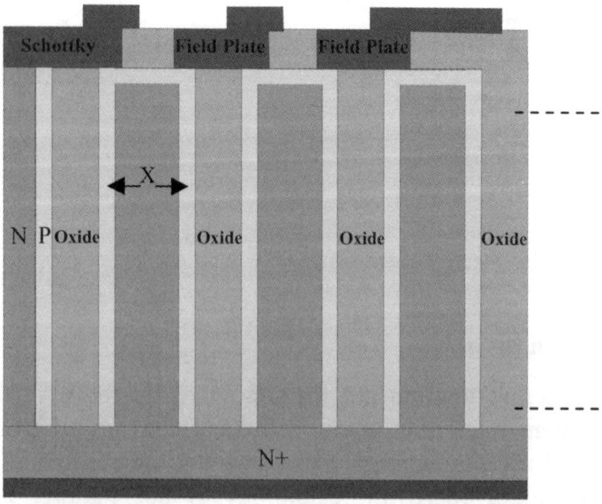

Figure 9. Schematic cross-section of the newly proposed termination for 4H-SiC SJS rectifier.

4 Summary

In summary, 4H SiC Schottky using superjunction structure is presented. The simulation results show significant improvement of trade-off between specific on-resistance and breakdown voltage. A novel device structure for SiC technology is proposed. Novel device termination technique for the new structure is also presented.

Acknowledgment

This work made use of ERC Shared Facilities supported by the National Science Foundation under Award Number EEC-9731677.

References

1. B.J. Baliga, *Modern Power Devices*. New York: Wiley, 1987.
2. L. Lorenz, G. Deboy, A. Knapp, M. Marz, "CoolMOS- a new milestone in high voltage Power MOSFET", *Proceedings of International Symposium on Power Semiconductor Devices and ICs*, pp. 3-10, (1999).
3. T. Fujihira, and Y. Miyasaka, "Simulated Superior Performances of Semiconductor Superjunction Devices", *Proceedings of International Symposium on Power Semiconductor Devices and ICs*, pp. 423-426, (1998).
4. E. Napoli, A. G.M. Stollo, "Design consideration of 1000V merged PiN Schottky diode using superjunction sustaining layer", *Proceedings of International Symposium on Power Semiconductor Devices and ICs*, pp. 339-342, (2001).
5. P. M. Shenoy, A. Bhalla and G.M. Dolny, "Analysis of the effect of charge imbalance on the static and dynamic characteristics of the super junction MOSFET", *Proceedings of International Symposium on Power Semiconductor Devices and ICs*, pp. 99-102, (1999).

International Journal of High Speed Electronics and Systems
Vol. 14, No. 3 (2004) 872–878

World Scientific
www.worldscientific.com

HIGH-VOLTAGE DIAMOND SCHOTTKY RECTIFIERS

W. HUANG, T.P. CHOW
Rensselaer Polytechnic Institute, Troy, NY 12180, U.S.A.

J. YANG
NOVA research Inc., Alexandria, VA 22308, U.S.A

J.E. BUTLER
Naval Research Laboratory, Washington, DC 20375, U.S.A.

In this paper, we simulate and fabricate diamond schottky rectifiers. The growth rate of pure diamond single crystal epitaxial is from 0.5 up to 100μm/hr with boron doping concentration around 1×1014 cm-3 to 1×1016 cm-3. A "liftoff" technology is used to provide the wafer. Theoretical calculation indicates that the diamond shottky rectifier has a significant lower voltage drop than SiC schottky rectifier and comparable with SiC PiN diode with the blocking voltage higher than 10 kV. A maximum 50 kHz operating frequency at switching voltage 25 kV is shown based on thermal limit. Vertical structure devices with 70μm epi layer achieve 18 A/cm2 at 250°C at 7 V forward drop as shown with a breakdown voltage of only 600 V. A breakdown voltage of 8 kV at 100μm distance is recorded for lateral structure devices without ohmic contact (back to back Schottky diodes), 12.4 kV at 300μm distance.

1 Introduction

Diamond is one of the wide bandgap semiconductors that is projected to be very useful for high-voltage power electronic devices.[1] Diamond-based semiconductor devices offer the potential of operation at high voltages, power levels, and temperatures and under extreme radiation conditions due to its high breakdown field ($5\text{-}10\times10^6 V\cdot cm^{-1}$), high saturated current velocity, high thermal conductivity.

There are only few stable dopants for diamond due to its highest atomic density. Boron is a p-type dopant with an activation energy of 0.37 eV, and nitrogen is a deep n-type dopant with activation energy about 1.7 eV. In this paper, we have successfully synthesized high quality undoped homoepitaxial diamond films with flat surface on atomic scale using microwave-plasma chemical vapor deposition (CVD) technique. The grow rate was from 0.5 to nearly 100μm/hr, liftoff process was followed to cut the epitaxial layer. C-V measurement indicated that the low boron concentration is less than $10^{16}\ cm^{-3}$.

Various devices including Bipolar Transister, Schottky rectifier, MESFET, MISFET have been widely studied. Schottky rectifier and MESFET have attracted most of the attention due to the simple processing, the high speed switching of unipolar devices and also the difficulty of a stable, high quality gate material for MISFET.

In this work, we went through the theoretical calculation of diamond devices and the comparison with other semiconductor materials, where diamond shows significant advantage at high voltage and high temperature. We also fabricated both vertical Schottky diodes and lateral Schottky diodes. Breakdown voltage of 8 kV and 12.4 kV were recorded for the lateral Schottky didoes. This is the highest breakdown voltage for diamond Schottky rectifiers as known to the authors.

2 Epitaxial diamond growth

Single crystal epitaxial layers were grown at NRL in a conventional 1.5 kW Astex HPMM microwave plasma chemical vapour deposition (MPCVD) system. The substrates were 3.5×3.5×1.5mm (100)-type Ib high-pressure high-temperature synthetic diamonds. The grow rate was from 0.5 to nearly 100μm/hr. The epitaxial layers were removed from the substrate using a pregrowth ion implant, followed by a postgrowth electrochemical etch, or by laser cutting. An example of the stages of this 'liftoff' technology is shown in Figure 1.

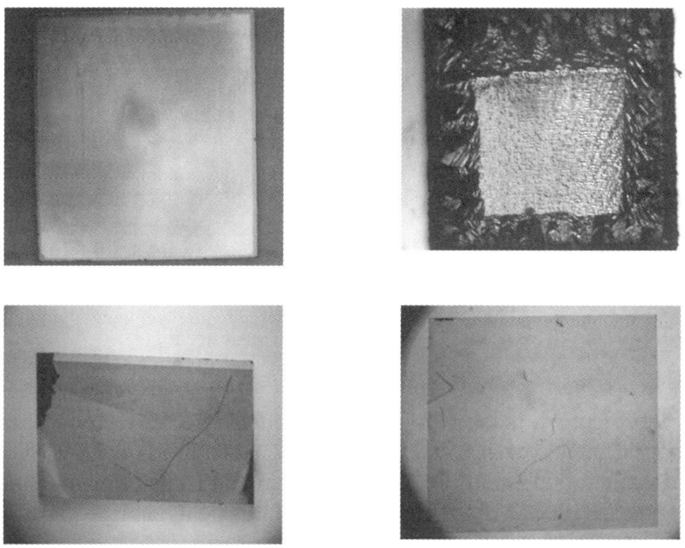

Figure 1: a) an optical image (reflection) of the implanted seed surface, b) an optical image of the as grown surface, c) a optical image of the seed and epilayer from the side view after mechanically polishing all surfaces, and d) a top view of the released epilayer (90 microns thick).

3 Theoretical calculation

As shown in chapter 3 in Baliga's *Power Semiconductor Device*[3], for a certain doping concentration, we are able to find the corresponding breakdown voltage with the assumption of critical electrical field. Using space charge neutrality equation, we are able to find the corresponding fermi level and ionised doping concentration for certain doping concentration. As shown in Figure 2, with the assumption of mobility[3] of 3800 $cm^2/V{\cdot}S$, Diamond Schottky rectifiers with different dopants were compared with SiC Schottky rectifier and SiC PiN diode (a typical 1μs lifetime is assumed). For a p-type boron doped schottky diode, the forward drop is significantly less than that of SiC Schottky diode and comparable with SiC pin diode. For the potentially n-type (with 0.2 eV activation energy) doped Schottky diode, diamond Schottky rectifier has less forward drop than SiC

Schottky or PiN at even 1000V breakdown voltage. Figure 3 shows the forward drop Vs. Breakdown Voltage at elevated temperatures, where diamond device shows even more advantage against SiC comparing to room temperature due to the diamond's deep dopant level and more ionized percentage at higher temperatures. This gives an addition to diamond high temperature application when conventional semiconductor cannot be used.

Switching loss calculation is based on depletion capacitance charge and discharge. C=ε*A/d=ε*A/ $\sqrt{2*\varepsilon*A/q/Na}$ =A* $\sqrt{q*Na*\varepsilon/2/V}$ (1), where C is depletion capacitance, A is device area, ε is dielectric constant, d is depletion width, V is reverse voltage, Na is semiconductor doping.

$q = 1.6\times10^{19}$ C, is constant.

Then Q=C*V=A* $\sqrt{q*Na*\varepsilon*V/2}$ (2), where Q is stored charge for capacitor.

I=dQ/dt=A*1/2* $\sqrt{q*Na*\varepsilon/2/V}$ *dV/dt (3), where I is the current.

Energy loss during turn off:

$$E=\int_0^V I*Vdt=\int_0^V A*1/2*\sqrt{q*Na*\varepsilon/2/V}*V*dV \quad (4)$$

$$=1/3*A*\sqrt{q*Na*\varepsilon/2}*V^{1.5} \quad (5)$$

Where V is the maximum reverse voltage, typically we use breakdown voltage.

E/A=1/3* $\sqrt{q*Na*\varepsilon/2}$ * $V^{1.5}$ (6), is the energy loss per unit area.

For switching frequency f, we can get power loss per unit area:

P/A=2*E/A*f=2/3*f* $\sqrt{q*Na*\varepsilon/2}$ * $V^{1.5}$ (7)

As shown in Figure 4, there is no significant difference between different materials due to the slight difference of dielectric constant. For the high switching voltage as high as 25 kV, the maximum operating frequency for diamond is 50 kHz based on maximum cooling packaging of 1000 W/cm, which is higher than SiC and Si due to its highest thermal conductivity among all the materials.

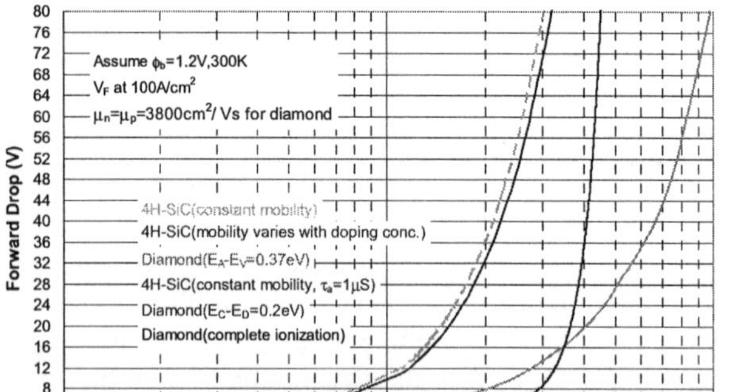

Figure 2: Schottky Barrier Diode Forward Drop versus Breakdown Voltage for 4H-SiC and Diamond of different dopant energy level (100A/cm^2)

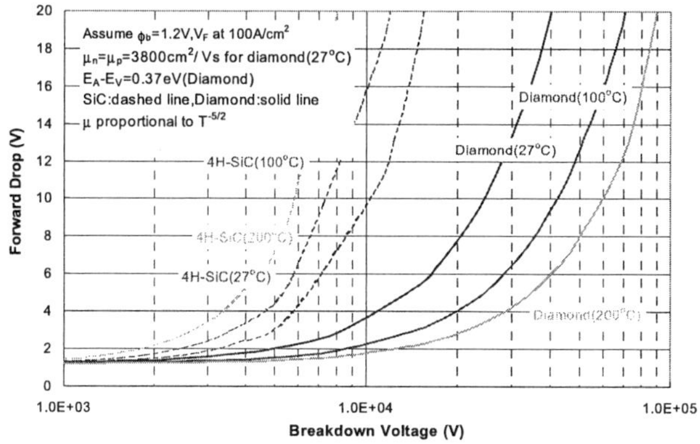

Figure 3: Forward Drop versus Breakdown Voltage for 4H-SiC and Diamond at different temperatures (100A/cm^2)

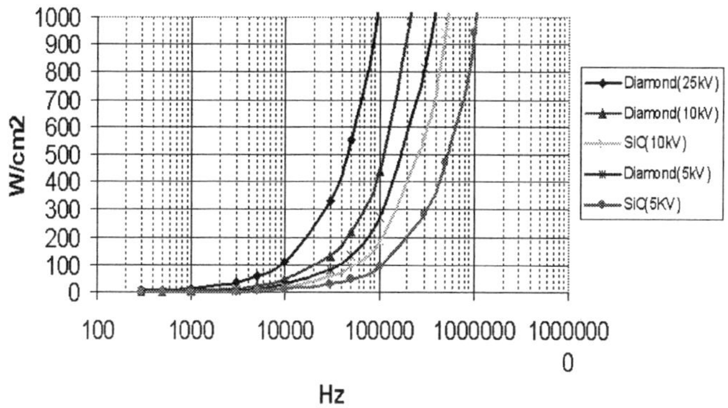

Figure 4: Switching loss vs. frequency for different material and voltages

4 Device fabrication and testing

As a critical step for diamond device fabrication, excellent ohmic contact formation has been done on natural diamond or chemical vapor deposition grown diamond.[4-7] M. Werner et al.[4] has achieved a specific on-resistance as low as 1×10^{-7} $\Omega\cdot cm^2$ on heavily boron doped polycrystalline diamond films using Al/Si metallization. Others[4-5] also are able to achieve the specific on-resistance value of the order of 10^{-5} $\Omega\cdot cm^2$.

In our experiment, Ohmic contacts have been formed on heavily boron doped as grown diamond films (with boron doping concentration at 1×10^{19} cm^{-3}) and also high boron ion-implanted diamond films. An ion implantation dose 5×10^{16} cm^{-2} at 200 °C has been done to achieve a surface peak boron doping concentration as high as 1.5×10^{21} cm^{-3}.

After 1200 °C post-implant anneal, and a subsequent chemical removal of the graphite layer, Ti/Mo/Au (10nm/50nm/100nm) trilayer was deposited on diamond surface. After using photolithography and TLM test pattern mask, a 850 °C anneal was done. Ti is believed to form carbide to achieve low contact resistance and Mo is believed to act as a barrier to prevent Ti outdiffusion into capping layer Au. TLM1 is the three ring circular transmission line model structure, which can avoid mesa isolation and simplify the processes.[8] TLM2 is another circular transmission line geometry[9]. The as grown heavy boron doped diamond samples achieve a specific contact resistance value of the order of 10^{-2} Ω·cm^2, where the more heavily boron doped ion-implanted sample yield a specific contact resistance value of the order of 10^{-5} Ω·cm^2. As described above, a heavy boron ion implantation and subsequent Ti/Mo/Au metallization can successfully form low contact resistance and was used in the device fabrication.

Figure 5a shows the structure of a vertical diamond Schottky diode with Junction Termination Extension (JTE). The starting material is 6mm×6mm single crystal CVD diamond from Apollo. First a high dose boron implant was done on the backside, with the same process step as previous, a Ti/Mo/Au trilayer metal is deposited to the backside and annealed to achieve ohmic contact On the front side, A H+ ion implantation was done to increase the resistivity in the JTE region and expected to achieve a higher breakdown voltage. Cr/Au (50nm/100nm) bilayer metallization was deposited to form Schottky contact.

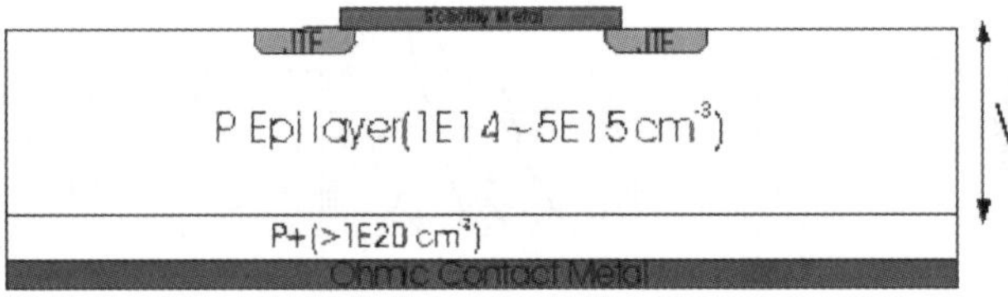

Figure 5a: Cross-section view of vertical rectifier

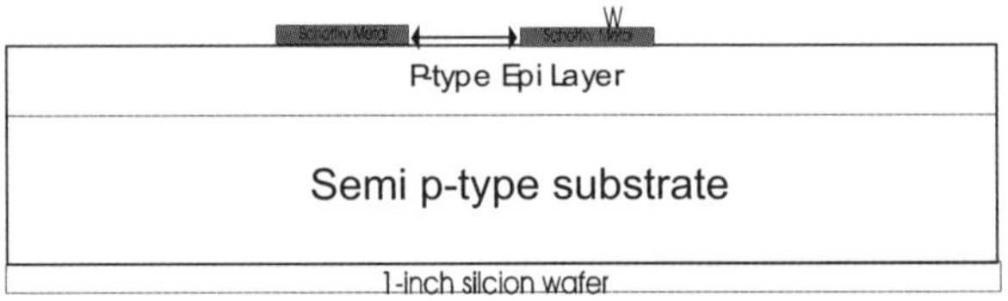

Figure 5b: Cross-section view of vertical rectifier

Due to the 70μm low doping (3×10^{15} cm^{-3} from C-V measurement) drift layer and deep activation energy level, the Schottky forward characteristics is resistive as expected. The forward current density Vs. Voltage at room temperature and higher temperatures were measured as shown in Figure 6. As the temperature increased, more boron get ionised and the forward resistance decreases. The measurement yields 18 A/cm^2 at 250°C with 7 V forward drop. Also the I-V data results in an effective mobility of 80 cm^2/V·S. The reverse breakdown voltage is only about 600 V. The 1200 °C post-implantation anneal may be responsible for the low breakdown voltage. The high temperature may break the carbon hydrogen bonds which were formed during the growth. Hydrogen diffuses out and leaves a defect path for the current.

Lateral diamond back to back Schottky pairs (without ohmic contact) were fabricated on 3mm×3mm undoped CVD epitaxial diamond films as shown in Figure 5b. Cr/Au (50nm/100nm) bilayer metallization was deposited to form Schottky contact. The metal pad is 100μm in diameter. The measurement was performed by probing the diodes in oil to minimize the failure by arcing at the surface. J E Butler et al. exhibited breakdown voltage as high as 6 kV, which is the highest breakdown voltage for diamond diode by then.[10] A breakdown voltage of 8 kV at 100μm distance is recorded as shown in Figure 7, and the highest breakdown voltage reported for any diamond diode. At 300μm distance, the breakdown voltage is as high as 12.4 kV. It is noticed that even with the oil, the diode sustained only surface damage when the breakdown occurred and metal usually got destroyed at this high voltage. Also oil breakdown can restrict the breakdown testing with very high voltage at very short distance.

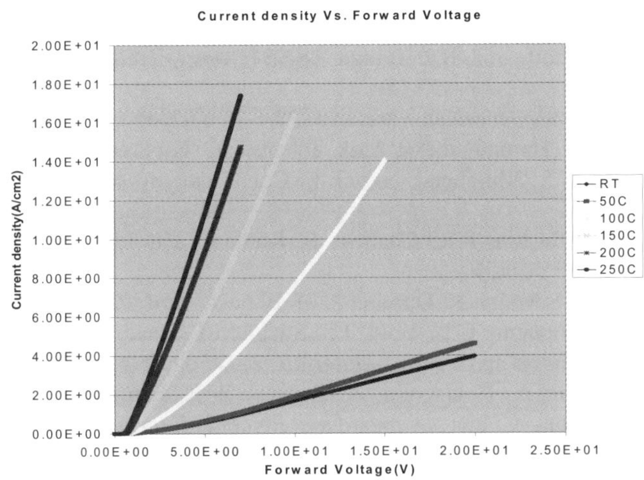

Figure 6: Forward I-V at different temperatures

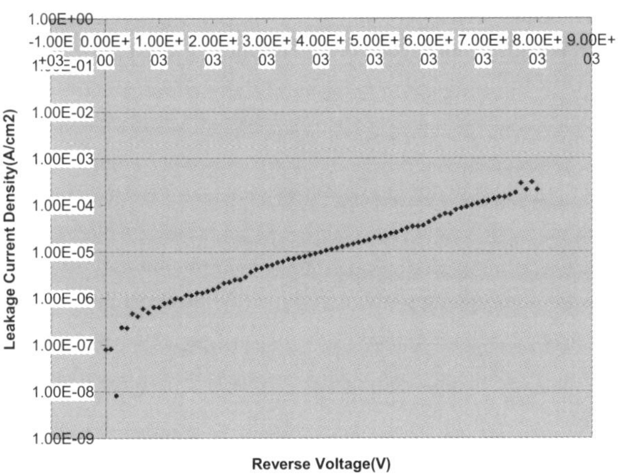

Figure 7: Reverse I-V characteristics

5 Summary

In summary, this paper reports on theoretical calculation, on low levels of impurities boron, nitrogen and hydrogen in epitaxial diamond film growth and on the vertical, lateral schottky diodes fabrication and testing. A breakdown voltage of 8kV at 100um and 12.4kV at 300um made on the epitaxial films were recorded.

Acknowledgments

The authors would like to thank the support of ONR and Darpa program monitored by Dr. J. Zolper for proving funding for the designing and fabrication of these high voltage diodes. They would also like to thank Apollo for providing the vertical diamond samples.

References

1. K. Shenai, R.S. Scott, and B.J. Baliga, *IEEE Trans. Electron Devices*, **36**, 1811 (1989).
2. B.J. Baliga, *Power Semiconductor Devices*, PWS Publishing Company (1996).
3. Jan Isberg, Johan Hammersberg, Erik Johansson, Tobias WikstrÖm, Daniel J. Twitchen, Andrew J. Whitehead, Steven E. Coe, Geoffrey A. Scarsbrook, *Science*, **297**, 1670 (2002).
4. M. Werner, C. Johnston, P.R. Chalker, S. Romani, and I.M. Buckley-Golder, *J. Appl.Phys.* **79**, 2535 (1995).
5. V. Vnkatesan, D. M. Malta, K. Das and A.M. Belu, *J. Appl. Phys.* **74**, 1179 (1993).
6. H.A. Hoff, G.L. Waytena, C.L. Vold, J.S. Suehle, I.P. Isaacson, M.L. Rebbert, D.I. Ma, K.Harris, *Diamond and Related Materials*, **5**, 1450 (1996).
7. M. Werner, R. Job, A. Denisenko, A. Zaitsev, W.R. Fahrner, C. Johnston, P.R. Chalker, I.M. Buckley-Golder, *Diamond and Related Materials*, **5**, 723 (1996).
8. G.K. Reeves, *Solid State Electronics*, **23**, 487 (1980).
9. Dieter K. Schroder, *semiconductor material and device characterization* (2nd edition), New York: Wiley, 155 (1998).
10. J.E. Butler, M.W. Geis, K.E. Krohn, J. Lawless Jr, S. Deneault, T.M. Lyszczarz, D. Flechtner and R. Wright, *Semicond. Sci. Technol.* **17**, 1 (2002).

International Journal of High Speed Electronics and Systems
Vol. 14, No. 3 (2004) 879–883

World Scientific
www.worldscientific.com

2 KV 4H-SiC DMOSFETS FOR LOW LOSS, HIGH FREQUENCY SWITCHING APPLICATIONS

SEI-HYUNG RYU, SUMI KRISHNASWAMI, MRINAL DAS,
JAMES RICHMOND, ANANT AGARWAL, AND JOHN PALMOUR

Cree, Inc., 4600 Silicon Drive, Durham, NC 27703, USA

JAMES SCOFIELD

Air Force Research Laboratory, Wright-Patterson Airforce Base, OH 45433, USA

Due to the high critical field in 4H-SiC, the drain charge and switching loss densities in a SiC power device are approximately 10X higher than that of a silicon device. However, for the same voltage and resistance ratings, the device area is much smaller for the 4H-SiC device. Therefore, the total drain charge and switching losses are much lower for the 4H-SiC power device. A 2.3 kV, 13.5 mΩ-cm2 4H-SiC power DMOSFET with a device area of 2.1 mm x 2.1 mm has been demonstrated. The device showed a stable avalanche at a drain bias of 2.3 kV, and an on-current of 5 A with a VGS of 20 V and a VDS of 2.6 V. Approximately an order of magnitude lower parasitic capacitance values, as compared to those of commercially available silicon power MOSFETs, were measured for the 4H-SiC power DMOSFET. This suggests that the 4H-SiC DMOSFET can provide an order of magnitude improvement in switching performance in high speed switching applications.

1 Introduction

The high critical field (E_c) of 4H-SiC enables Power MOSFETs in 4H-SiC using drift layers with 10% of the thickness and 100 times the doping concentration as compared to silicon devices with comparable blocking voltages (*BV*). Therefore, the drift layer charge per unit area for a 4H-SiC MOSFET is expected to be a factor of 10 greater than that of a silicon MOSFET with the same *BV*. However, the device area for a 4H-SiC MOSFET is expected to be a factor of 100 smaller than that of a silicon MOSFET with comparable voltage and on-resistance ratings due to the reduction in drift layer resistance. Therefore, the total drift layer charge for a 4H-SiC MOSFET is a factor of 10 smaller than that of a silicon device, and a 10X improvement in switching characteristics over silicon devices are expected for 4H-SiC devices [1]. In this paper, characteristics of our 2.3 kV 4H-SiC DMOSFETs are presented, and its internal capacitances, which are an indicator of switching speeds, are compared to those of commercially available silicon MOSFETs.

2 Switching Characteristics of Power Switches

Turn-on and turn-off transients of power switches can be viewed as charging and discharging operations of parasitic capacitances. Simplified diagrams of power MOSFET turn-on and turn-off transients are given in Fig. 1. During the turn-off transient, the drain capacitances, C_{GD} and C_{DS}, are charged by the load current. The charge stored in the drain capacitance can be obtained by calculating the depletion charge in the drift layer using critical field (E_C) and Gauss's law,

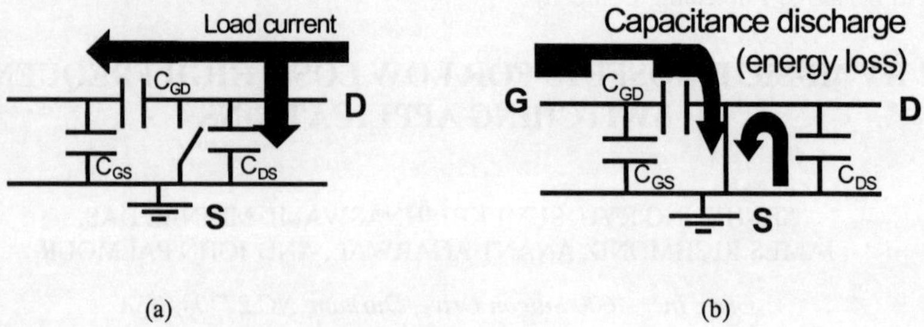

Figure 1. Simple diagrams of (a) turn-off and (b) turn-on transients of a power switch.

$$Q_D = A \cdot \varepsilon \cdot E_C . \tag{1}$$

where A is the device area, including the edge termination structure. The load current (I_D) determines the charging rate. Therefore, the voltage rise time (t_r) can be given as:

$$t_r = Q_D / I_D = A \cdot \varepsilon \cdot E_C / I_D \tag{2}$$

During the turn-on transient, the charge stored in the drain capacitance is discharged through the switch. The turn-on time, or voltage fall time (t_f), is determined by the transconductance of the switch, and does not depend on the load current. The t_f can be substantially reduced by using low resistance gate drives, i.e., gate drives with higher current capabilities. However, t_r is determined by the parasitic capacitances and external load current, and cannot be reduced by using stronger gate drives. The switching time is generally determined by t_r, which is proportional to drain charge.

The switching power loss can be estimated by multiplying the stored energy by switching frequency.

$$P_{sw} = A \cdot f \cdot \int_0^{V_{DD}} C \cdot V \cdot dV = A \cdot f \cdot (1/3) \cdot \varepsilon \cdot E_C \cdot V_{DD} \qquad (3)\ [2,3]$$

It is shown in Eqs. (1) and (3) that the drain charge and switching loss per unit area is directly proportional to E_C. Since E_C in 4H-SiC is approximately 10 times higher than that of silicon, the drain charge and switching loss per unit area are also 10 times higher in 4H-SiC. However, the drift layer of a 4H-SiC device is 10X thinner, and 100X higher doped than that of a silicon device with the same blocking voltage. The electron mobility in 4H-SiC is approximately 70% of that of silicon. Hence, the specific on-resistance of a 4H-SiC device can be theoretically calculated to be a factor of 700 smaller than that of a silicon device. The device area for the same blocking voltage and the same on-resistance will be a factor of 700 smaller for 4H-SiC devices. Therefore, from Eqs. (2) and (3), the drain charge and switching loss can be 70X smaller for the 4H-SiC device when compared to a silicon device with the same on-resistance and blocking voltage ratings.

3 Experimental

A simplified cross-section of a 4H-SiC power DMOSFET is shown in Fig. 2. The DMOSFET structure is preferred in 4H-SiC because it is an MOS controlled planar structure. Also, the implanted p-wells shield the MOS gate region from high electric fields.

A 15 μm thick, 5.5 x 10^{15} cm^{-3} doped n-type epilayer was grown on an n^+, 4H-SiC substrate as the drift layer for the device. P-wells, 0.7 μm deep, were formed via aluminum ion implantation, after which 0.3 μm deep n^+ source regions were formed by nitrogen implantation. The MOS gate length, defined by these two implants, was 0.5 μm. A heavy dose aluminum implantation formed p^+ contacts to the p-wells as well as the floating guard rings. A 1 μm thick oxide layer was then deposited and patterned as the field oxide. Approximately 600 Å thick gate oxide layer was thermally grown at 1200°C in dry O_2, then nitrided in N_2O. A degenerately doped polysilicon layer was deposited and patterned as the gate electrode. An LPCVD oxide layer (1 μm thick) was then deposited as an inter-metallic dielectric layer, and via holes were opened. The source and drain ohmic contacts were formed with alloyed Ni. Finally, an aluminum overlayer was deposited and etched to form electrodes.

Figure 3 shows the I_D-V_{GS} characteristics of a 150 μm / 150 μm test MOSFET, built on a p-well adjacent to the power DMOSFET. A V_{DS} of 50 mV was used for the measurement. A peak effective channel mobility of 17 cm^2/Vs was extracted from the first derivative of the IV curve, and a threshold voltage of 8 V was extracted from the linear portion of the IV curve.

Figure 4(a) shows the room temperature on-state characteristics of a 4H-SiC power DMOSFET. The active area of this device is 2.78×10^{-2} cm^2, and the total chip size is 2.1 mm x 2.1 mm. The threshold voltage extracted from saturation region is approximately 4 V, which is significantly lower than the value measured from the 150 μm long test MOSFET. With a gate bias of 20 V (E_{OX} = 3.3 MV/cm), the device showed an on-current of 5 A at a forward drop of 2.6 V. The specific on- resistance measured was 13.5 mΩ-cm^2. The off-state characteristics are shown in Fig. 4(b). The device is normally-off, and showed stable avalanche behavior at a V_{DS} of 2.3 kV. The theoretical parallel plate electric field is approximately 2.1 MV/cm at this voltage. It is shown that reducing the channel length down to 0.5 μm does not affect the breakdown voltage.

The internal capacitances of the 4H-SiC DMOSFET are compared to those of commercially available silicon power MOSFETs in Table 1. When compared to a conventional 1200 V silicon MOSFET [4], a factor of 9 reduction in all capacitance values (normalized in on-resistance only) is shown. The advantages in input and reverse transfer capacitances reduce when the comparison is made to the charge compensated MOSFETs [5]. However, there still is a factor of 8 difference in output capacitance, which determines the switching time and losses, between the 2.3 kV 4H-SiC DMOSFET and the 800 V CoolMOS transistor. This suggests that the 4H-SiC DMOSFETs can have a significant advantage over silicon devices in high speed power switching applications, regardless of the device structure.

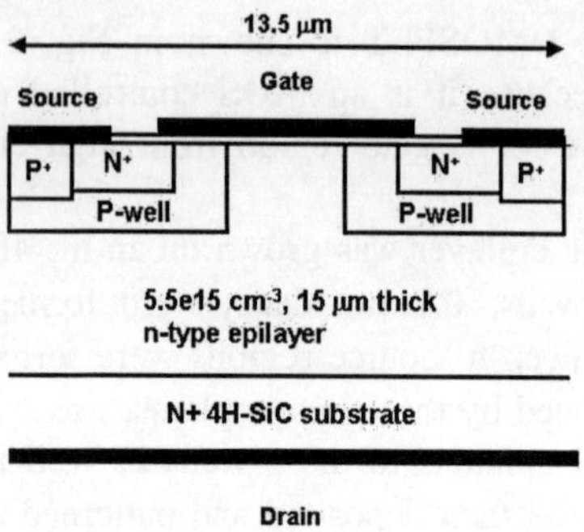

Fig. 2. Simplified 4H-SiC DMOSFET cross-section.

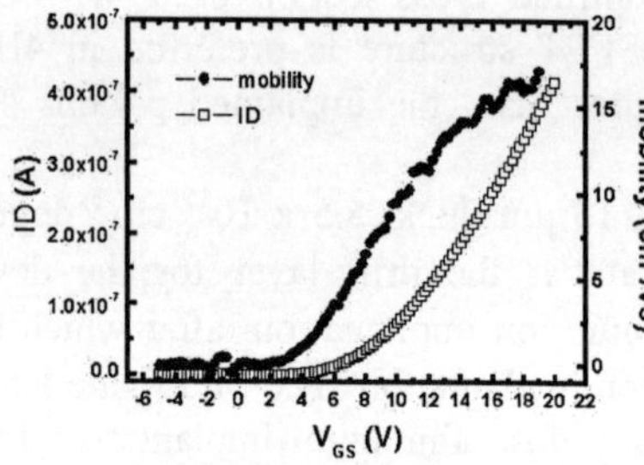

Fig. 3. I_D vs. V_{GS} characteristics measured from a 150 μm / 150 μm FATFET.

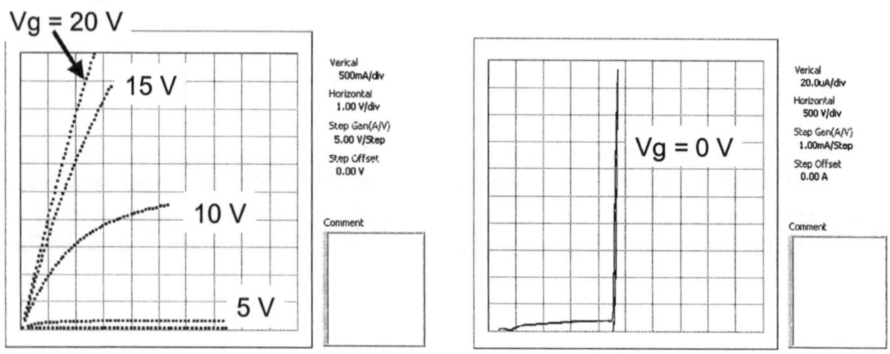

(a) (b)

Figure 4. IV characteristics of a 2.1 mm x 2.1 mm (active area: 2.78×10^{-2} cm^2) 4H-SiC power DMOSFET. (a) on-state, and (b) off-state.

4 Summary

4H-SiC power DMOSFETs are suggested for high speed power switching applications. The drain charge and switching loss per unit area of a 4H-SiC device are approximately 10 times higher than those of a comparable silicon device. However, the total drain charge and switching loss for the 4H-SiC device are expected to be much smaller due to the very small chip size, which, in theory, can be a factor or 700 smaller than that of the silicon device. A 2.3 kV, 13.5 mΩ-cm2 4H-SiC power DMOSFET is demonstrated. The internal capacitances of the 4H-SiC DMOSFETs were compared to those of commercial silicon power MOSFETs, a 9X improvement in all capacitance values were achieved when compared to a 1200 V conventionalsilicon MOSFET, and an 8X improvement in output capacitance value was observed when compared to a 800 V charge compensated silicon MOSFET. This suggests that 4H-SiC power DMOSFETs will result in at least an order of magnitude improvement over silicon devices in switching characteristics.

Table 1. Comparison of internal capacitances of Power MOSFETs in Si and SiC.

	APT Power MOS 7 [4] (APT1201R4BLL)	Infineon CoolMOS [5] (SPP11N80C3)	Cree SiC DMOSFET
Vdss	1200V	800V	2300V
Ron	1.4 Ω	0.45 Ω	0.48 Ω (Vg = 20V)
Input capacitance	2030 pF*	1600 pF*	377.64 pF*
Output Capacitance	309 pF*	800 pF*	100.2 pF*
Reverse Transfer capacitance	60 pF*	40 pF*	18.52 pF*

*measured at VGS = 0 V, VDS = 25 V, 1 MHz

Acknowledgement

This work is supported in part by AFRL TIA # FA8650-04-2-2410, monitored by Dr. J. Scofield.

References

1. S. Ryu et al.,: MRS Proc. Vol. 764 (2003), p.C2.7.1
2. K. Adachi et al.,: Mat. Sci. Forum Vols. 457 – 460 (2004), p. 1233.
3. Q. Huang.: IEEE EDL, Vol. 25, No. 5 (2004), p. 298.
4. http://www.advancedpower.com/
5. http://www.infineon.com/cmc_upload/documents/085/068/SPP_A11N80C3.pdf

International Journal of High Speed Electronics and Systems
Vol. 14, No. 3 (2004) 884–889

SIMULATION STUDY ON BREAKDOWN BEHAVIOR OF FIELD-PLATE SiC MESFETs

HO-YOUNG CHA, Y. C. CHOI, LESTER F. EASTMAN, AND MICHAEL G. SPENCER
School of Electrical and Computer Engineering
Cornell University, Ithaca, NY 14853, USA

The effects of field-plate structures on SiC MESFETs were investigated using two-dimensional simulations. The simulation results without a field-plate were in good agreement with the characteristics of fabricated SiC MESFETs. The breakdown voltage was increased by 80 % when the optimum field-plate was applied to the device with a 0.5 μm long gate and a 1 μm spacing between gate and drain; a breakdown voltage of 240 – 250 V was obtained from a 0.35 μm field-plate, while 140 V was obtained without a field-plate.

1 Introduction

Wide bandgap semiconductor devices have been developed for high power applications and have shown great potentials in comparison with Si or GaAs based devices. Due to unique material properties (i.e., high saturated electron velocity, high breakdown field, and high thermal conductivity), SiC MESFETs have emerged as a promising high power microwave device.

It has been reported that recessed structures increased the breakdown voltage of SiC MESFETs. Among different recessed structures, the channel-recessed SiC MESFETs demonstrated the highest breakdown voltage.[1] According to the careful investigation of the breakdown mechanism of SiC MESFETs in our previous work, it was suggested that the breakdown of SiC MESFETs was caused by the gate field emission due to the high electric field at the gate edge.[2] Though SiC MESFETs have high breakdown voltages compared to other semiconductor devices, it is still worth investigating potential ways to increase the breakdown voltage. Recently, tremendous improvement in the breakdown voltage of GaN HEMTs was experimentally demonstrated by employing a field-plate structure.[3-5] A theoretical study of the effects of the critical geometries of the field-plate on the breakdown voltage of GaN HEMTs was performed using two-dimensional simulations.[6-7] However, no study has been performed on SiC MESFETs yet. In this work, the effects of the field-plate on the breakdown voltage of SiC MESFETs are investigated and the geometrical optimization is proposed on the basis of two-dimensional simulation studies.

2 Simulations and discussion

The physical models used in the simulations are summarized in Table 1 along with the model parameters. The velocity-field characteristics were investigated using TLM measurements under very short pulsed conditions.[8] The model parameters were adjusted according to the experimental data. Fig. 1 shows the device structure used in the simulation, which is the same as the channel-recessed device but has a Si_3N_4 layer on top of the surface. The gate length (L_g) is 0.5 μm and the spacing between source and gate (L_{sg}) and gate and drain (L_{gd}) are 0.5 μm and 1.0 μm, respectively. The thickness of Si_3N_4

is fixed (100 nm), and the extension of the field-plate toward the drain side on top of the insulator (L_{ext}) is varied.

Table 1. Model parameters used for two-dimensional simulations.

Model	Parameters (Electron)	Parameters (Hole)
Bandgap energy $E_g(T)=E_g(0)-\frac{\alpha T^2}{T+\beta}$		$E_g(0)=3.23$ eV $\alpha=6.5\times10^{-4}$ eV/K $\beta=1300$ K
Concentration dependent mobility $\mu(N)=\mu_{\min}+\frac{\mu_{\max}}{1+\frac{N}{N_{cn}}}$	$\mu_{min}=88$ cm^2/V.s $\mu_{max}=900$ cm^2/V.s $N_{cn}=0.5\times10^{17}$/cm^3	$\mu_{min}=44$ cm^2/V.s $\mu_{max}=73$ cm^2/V.s $N_{cn}=0.5\times10^{17}$/cm^3
Field dependent mobility $\mu(E)=\frac{\mu_0}{\left[1+\left(\frac{\mu_0 E}{v_{sat}}\right)^{\beta}\right]^{1/\beta}}$	$v_{sat}=1.7\times10^7$ cm/s $\beta=1$	$v_{sat}=1.7\times10^7$ cm/s $\beta=1$
Recombination $R=\frac{pn-n_i^2}{\tau_p\left[n+n_i\exp\left(\frac{E_{trap}}{kT}\right)\right]+\tau_n\left[p+n_i\exp\left(\frac{-E_{trap}}{kT}\right)\right]}$		$E_{trap}=0$ eV $\tau_n=1\times10^7$ sec $\tau_p=1\times10^7$ sec
Impact ionization model $\alpha=a\cdot\exp\left[-\left(\frac{b}{E_0}\right)\right]$	$a=1.31\times10^6$ cm^{-1} $b=1.3\times10^7$ V/cm $E=4\times10^5$ V/cm	$a=2.98\times10^6$ cm^{-1} $b=1.3\times10^7$ V/cm $E=4\times10^5$ V/cm

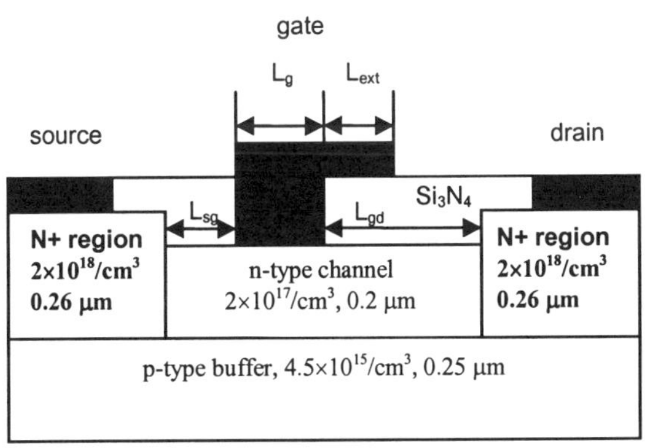

Figure 1. Field-plate SiC MESFET structure used in simulation.

Both DC I-V and breakdown characteristics were simulated for six different values of L_{ext} (L_{ext} = 0, 0.15, 0.25, 0.35, 0.5, and 0.6 μm). L_{ext} of 0 μm corresponds to the channel-recessed structure without a field-plate. The simulation results without the field-plate (L_{ext} = 0 μm) were in good agreement with measured data of fabricated devices.[1] Fig. 2 (a) and (b) show the drain current and breakdown voltage depending on L_{ext}. Fig. 3 shows the behavior of the gate leakage current versus the drain voltage as a function of L_{ext}. The saturation current decreases slightly as the extension of the field-plate increases. The channel region under the extension of the field-plate is modulated with the applied gate bias, increasing the effective channel length. Consequently, the saturation current decreases due to the increase in the effective channel length. While the saturation current decreases continuously as L_{ext} increases, the breakdown voltage shows a peak at L_{ext} = 0.35 μm. This phenomenon will be discussed by investigating the electric field distribution of different structures. The breakdown voltage of 240 – 250 V is obtained from L_{ext} = 0.35 μm, which is 100 V higher than that obtained without a field-plate.

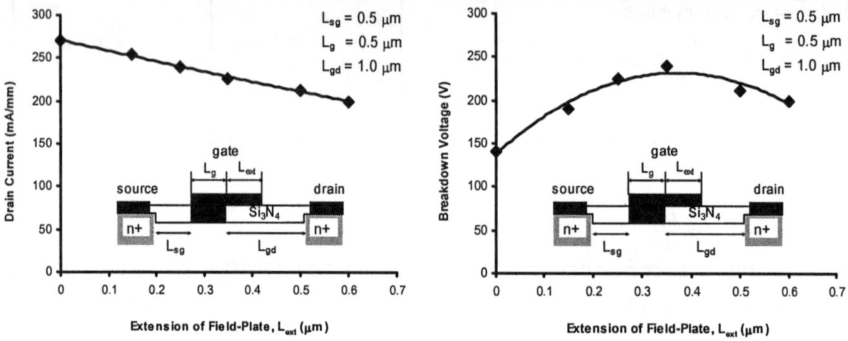

1 (b)

Figure 2. Change in (a) saturation current and (b) breakdown voltage versus Lext. The saturation current was sampled at Vds = 30 V and the breakdown voltage was defined when the gate leakage current exceeds 1mA/mm.

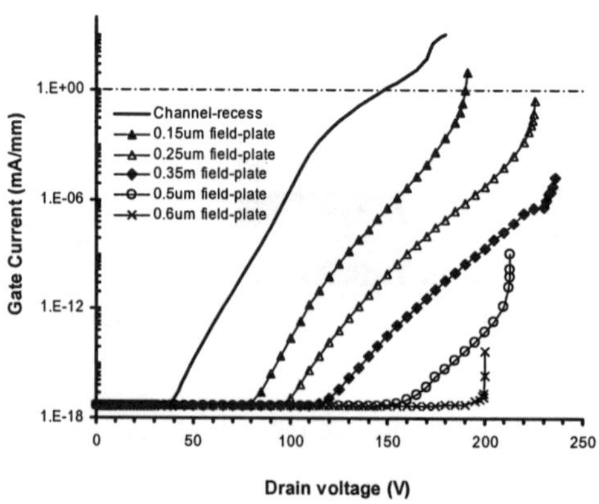

Figure 3. Behavior of the gate leakage current versus the drain voltage as a function of Lext. The simulation was performed at Vgs = -13 V (The pinch-off voltage is ~ -10 V). The highest breakdown voltage was obtained at Lext = 0.35 μm.

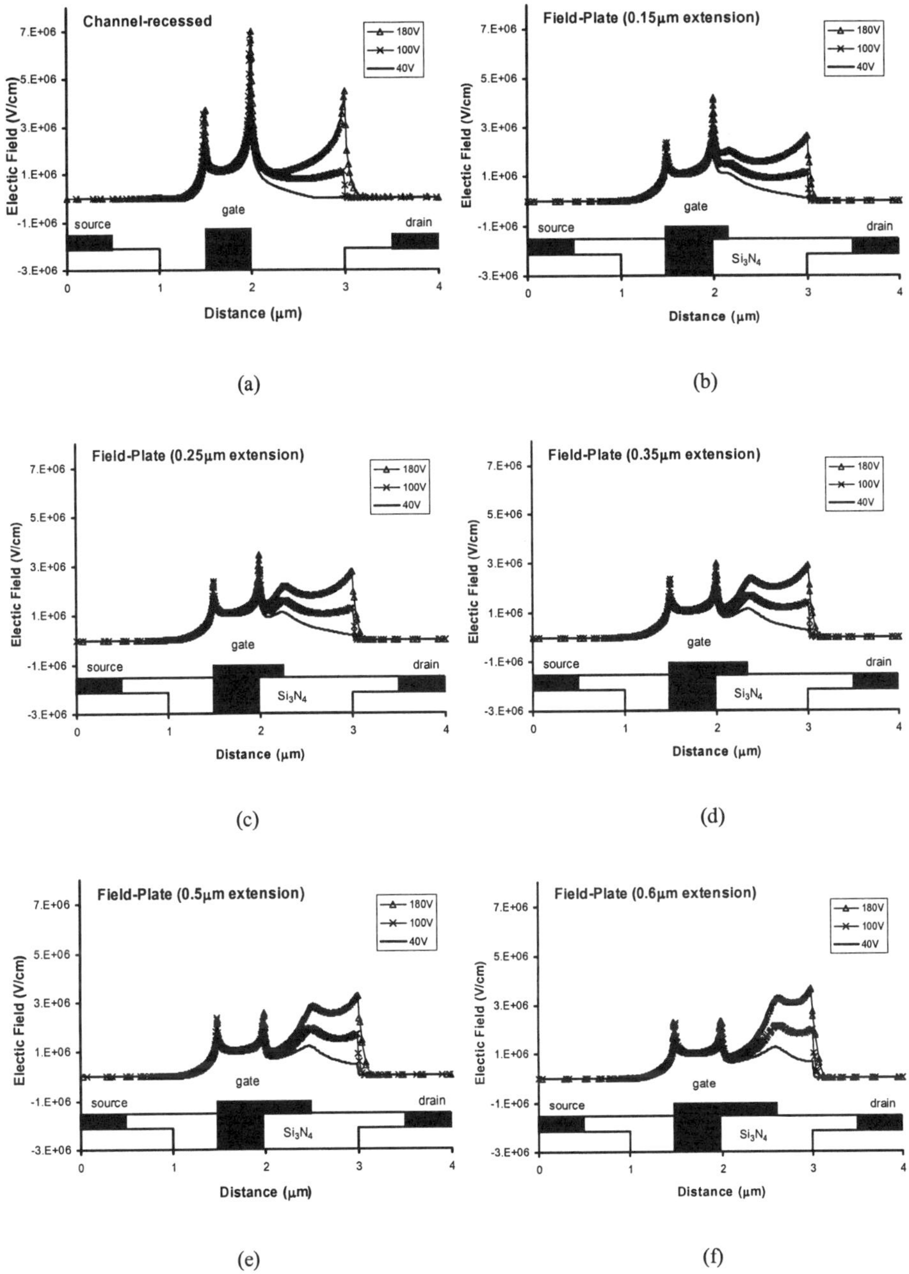

Figure 4. Electric field distribution near the surface. (a) Lext = 0 μm (Channel-recessed). (b) Lext = 0.15 μm. (c) Lext = 0.25 μm. (d) Lext = 0.35 μm. (e) Lext = 0.5 μm. (f) Lext = 0.6 μm.

The distribution of the electric field in the channel region was investigated carefully to analyze the breakdown mechanism of the field-plate SiC MESFETs. Fig. 4 shows the electric field distribution near the surface (50 nm below the surface) for different values of L_{ext}. The electric field distribution at three different drain voltages (i.e., 40 V, 100 V, and 180 V) was plotted for each. Regardless of the structures, the strength of the electric field at the drain edge increases as the drain voltage increases due to the dramatic change in the potential at the drain edge. It should be noted that the electric field has two important peaks, i.e., one at the drain side gate edge and the other one at the drain edge. Without a field-plate (L_{ext} = 0 μm), the electric field distribution shows the high, sharp peak at the drain side gate edge (See Fig. 4 (a)). This high electric field peak leads to the gate field emission induced breakdown of the device. On the other hand, the peak of the electric field at the bottom of the gate edge is suppressed dramatically by employing the field-plate structure: even with a small extension of the field-plate (L_{ext} = 0.15 μm), the electric field peak is reduced by 40 %, resulting in a gradual field distribution (See Fig. 4 (b)). The electric field peak at the bottom of the gate edge keeps decreasing as the extension of the field-plate (L_{ext}) increases. However, this peak is still higher than the peak at the drain edge until the field-plate extends up to 0.35 μm. For the further increase in the extension (L_{ext}), the peak at the drain edge becomes dominant at high drain voltages. Accordingly, it is suggested that the breakdown mechanism switches from the gate field emission induced breakdown to the avalanche impact ionization at the drain edge as the extension exceeds 0.35 μm. On the basis of the simulation results, the highest breakdown voltage is obtained when two electric field peaks are comparable at high drain voltages. i.e., the transition from the gate field emission induced breakdown to the avalanche impact ionization.

3 Conclusion

It was demonstrated based on simulations that the optimized field-plate structure produced breakdown voltages almost twice those of conventional structures. It was suggested that the breakdown mechanism changed from gate field emission to avalanche impact ionization as the field-plate extended toward the drain side, and the maximum breakdown voltage was obtained at the transition from one to another.

Acknowledgements

This work was supported under Contract N00014-04-1-0002 of the Office of Naval Research.

References

1. H.-Y. Cha, et al., “Reduced Trapping Effects and Improved Electrical Performance in Buried Gate 4H-SiC MESFETs”, IEEE Trans. Electron Devices, vol. 50, 1569-1574 (2003).
2. Ho-Young Cha, C. I. Thomas, Y. C. Choi, L. F. Eastman, and M. G. Spencer, “Gate field emission induced breakdown in power SiC MESFETs”, IEEE Electron Device Letters, vol. 24, 571 (2003).

3. N.-Q. Zhang, S. Keller, G. Parish, S. Heikman, S. P. Denbaars, and U. K. Mishra, "High breakdown in GaN HEMT with overlapping gate structure", IEEE Electron Device Lett., vol. 21, 421 (2000).
4. Y. Ando, Y. Okamoto, K. Hataya, T. Nakayama, H. Miyamoto, T. Inoue, M. Kuzuhara, "12 W/mm recessed gate AlGaN/GaN heterojunction field-plate FET", IEEE International Electron Devices Meeting, Washington D.C., 563 (2003).
5. W. Saito, Y. Takada, M. Kuraguchi, K. Tsuda, I. Omura, and T. Ogura, "600V AlGaN/GaN power-HEMT: design, fabrication, and demonstration on high voltage DC-DC converter", IEEE International Electron Devices Meeting, Washington D.C., 587 (2003).
6. S. Karmalkar and U. K. Mishra, "Enhancement of breakdown voltage in AlGaN/GaN high electron mobility transistors using a field plate", IEEE Trans. Electron Devices, vol. 48, 515 (2001).
7. S. Karmalkar and U. K. Mishra, "Very high voltage AlGaN/GaN high electron mobility transistors using a field plate deposited on a stepped insulator", Solid-State Electron., vol. 45, 1645 (2001).
8. Ho-Young Cha, Y. Choi, L. F. Eastman, M. G. Spencer, L. Ardaravicius, A. Matulionis, and O. Kiprijanovic, "Influence of low field mobility related issues on SiC MESFET performance", Electronic Materials Conference, Notre Dame, IN (2004).

International Journal of High Speed Electronics and Systems
Vol. 14, No. 3 (2004) 890–896

World Scientific
www.worldscientific.com

EFFECTS OF BUFFER LAYER THICKNESS AND DOPING CONCENTRATION ON SiC MESFET CHARACTERISTICS*

SANKHA S. MUKHERJEE and SYED S. ISLAM
Department of Electrical Engineering, Rochester Institute of Technology
79 Lomb Memorial Drive, Rochester, NY 14623, USA

Two-dimensional simulations have been carried out using the Atlas® device simulator to investigate the effects of the buffer layer thickness and doping concentration on the electrical characteristics of the SiC MESFET. The variations of transconductance, output resistance, gate-source capacitance, gate-drain capacitance and (cutoff frequency) f_T with respect to the change in buffer layer thickness and doping concentration have been investigated. It is observed that the performances of MESFET can be improved by reducing the leakage of channel carrier into the substrate at high drain bias, which is achieved by increasing buffer layer doping density and/or increasing buffer layer thickness. For a SiC MESFET with buffer layer thickness of 0.3μm and gate length of 1μm, drain current increases from 0.1A/ μm to above 0.45A/ μm as the buffer layer doping density is decreased from 1.9×10^{17} cm^{-3} to 1×10^{16} cm^{-3}. The simulations were carried out at a gate-source voltage of –1V and a drain-source voltage of 15V. Under similar conditions, the output resistance decreases from 1.2×10^6 Ω/μm to 1.0×10^5 Ω/μm, and the transconductance decreases from 5.9mS/ μm to 5.3mS/ μm, and f_T decreases from 11GHz to 8GHz.

Introduction

In recent years, SiC has emerged as the ideal material for high power, high frequency and high temperature applications [1]. SiC has a wider bandgap of 3.3 eV and a higher thermal conductivity of 4.5 W/cm-K compared to Si for which the respective quantities are 1.1eV and 1.5 W/cm-K [2, 3]. These allow SiC-based devices to operate at higher power levels. Besides, the lower relative dielectric constant of SiC (9.7) as compared to Si (11.8) and moderate electron mobility of SiC (900 cm^2/V-s) as compared to Si (1350 cm^2/V-s) allow SiC based devices to operate at higher frequencies [2, 3]. Moreover, better thermal stability of SiC allows the device to operate at higher temperature [2].

The performances of SiC-based devices are limited by the presence of traps located at the surface between gate and drain, and in the substrate. Recently, both surface and substrate trapping effects of SiC MESFETs have been reported [4-7]. The surface trapping effects can be limited by using a passivation layer of Si_3N_4 [8]. It has been shown by Noblanc *et al.* [9] that the buffer layer plays a critical role in the mitigation of the substrate trapping phenomena. Presently, the use of a buffer layer is a ubiquitous method for combating the substrate trapping effects in SiC MESFETs.

In this paper, the nature of the spatial dispersion of channel electrons and its effects on the electrical performance of SiC MESFETs have been investigated. A number of test

* This work is supported by Gleason Foundation of the Rochester Institute of Technology.

structures with varying buffer layer thickness and doping concentration have been simulated using the ATLAS® 2D device simulator [10]. Electrical properties in terms of output resistance, transconductance, capacitances, cutoff frequency *etc.* have been calculated and compared for various MESFET structures.

Analysis

In conventional MESFET devices, a conducting channel is grown over a semi-insulating substrate. However, during epitaxial channel growth a significant number of electron traps burgeon at the channel/substrate interface. These traps capture electrons and deteriorate the performances of the MESFET. To combat the effects of substrate traps, a buffer layer is introduced which is doped with opposite type dopant compared to the channel. The buffer layer shields the channel electrons from being captured by the substrate traps. The doping concentration and thickness of the buffer layer determine its effectiveness in combating trapping effects as well as the intrinsic electrical behavior of the MESFET as demonstrated in this paper.

For the purpose of simulation, the Atlas® 2D device simulator has been used [10] to analyze n-channel SiC-based MESFET. In the charge transport model, current flow due to minority carriers (holes) have been neglected. The electron mobility is implemented by using the Caughey-Thomas model, which is given by,

$$\mu_0 = \mu_{\min} + \frac{\mu_{\max} - \mu_{\min}}{1 + \left(N / N_{\mathrm{Ref}}\right)^{\alpha}} \tag{1}$$

where α, μ_{max}, μ_{min} and N_{Ref} depend on the material and for SiC these values are 0.5, 950 $cm^2/V.s$, 40 $cm^2/V.s$ and 2×10^{17} cm^{-3}, respectively [11]. The Caughy-Thomas model incorporates carrier-acoustic phonon and carrier-impurity scatterings. The electric field dependence of electron mobility have been incorporated by,

$$\mu = \frac{\mu_0}{\left\{1 + \left(\frac{\mu_0 |E|}{v_{sat}}\right)^{\beta}\right\}^{1/\beta}} \tag{2}$$

where β and v_{sat} are the characteristics of the material having values 1 and 2.07×10^7 cm/s [11]. (Shockley-Read-Hall) SRH and Auger recombination processes have also been incorporated in the simulation. The SRH parameters τ_p and τ_n are given by 6.0×10^{-7}s and 1.0×10^{-9}s respectively [2] (assuming that the trap energy level is at the center of the energy bandgap) and Auger recombination parameters C_n and C_p are given by 5×10^{-31} cm^6/s and 2×10^{-31} cm^6/s [2]. A gate workfunction of 1.2 eV has been used in the simulations [12].

Results and Discussion

In this analysis, three n-channel SiC-based MESFETs with different buffer layers have been considered as described in Fig. 1 and Table I.

Table I – Device structures in the analysis.

Structure	I	II	III
Buffer thickness (μm)	0.3	0.3	0.6
Buffer doping (cm^{-3})	1×10^{16}	1×10^{17}	5×10^{16}
Channel doping (cm^{-3})	2×10^{17}	2×10^{17}	2×10^{17}
Channel thickness (μm)	0.3	0.3	0.3

For comparison of the performance variations due to different buffer layers, same channel doping density and thickness of all three structures have been considered. These are 2×10^{17} cm^{-3} and 0.3 μm, respectively. Structure-I has a thin (0.3 μm) and lightly-doped (1×10^{16} cm^{-3}) buffer region, Structure-II has a thin (0.3 μm) but heavily-doped buffer region, and Structure-III has a thick (0.6 μm) and moderately-doped (5×10^{16} cm^{-3}) buffer layer. Fig. 1 depicts the channel current density and the vertical component of the electric field along the cross section of the devices. It can be seen from Fig. 1(a) that the electric field in Structure-I is very weak at the channel/buffer interface compared to that present in Structure-II and Structure-III. This is because the lower doping in the buffer layer of Structure-I would entail a much smaller built-in voltage at the channel/buffer interface and a correspondingly low electric field would result. However, a net electric field across the buffer exists as the buffer layer is fully depleted. As a result, the channel electrons in Structure-I are not completely confined within the channel and leaks into substrate region through the buffer layer and are subsequently captured by the traps. The dispersion of electrons causes the drain current in the saturation region to become linearly dependent upon the drain voltage. Thus, the output resistance of the channel is significantly reduced as shown in the *I-V* characteristics of Structure-I (see Fig. 2, curves shown by □). In Structure-II and Structure-III, the buffer layer is not fully depleted due to increase in doping density and/or buffer layer thickness. As a result, the carriers are confined inside the channel due to the absence of a net electric field across the buffer. Higher output resistance is observed in the I-V characteristics of Structures II and III as shown in Fig. 2. However, since the concentration of the buffer layer of Structure-II is double the concentration of the buffer layer of Structure-III, spatial dispersion of channel electrons is significantly lower and so is the output resistance. One consequence of improper confinement of the channel current density resulting from a low-doped buffer region is that the drain current is higher due to the effective increase in channel thickness.

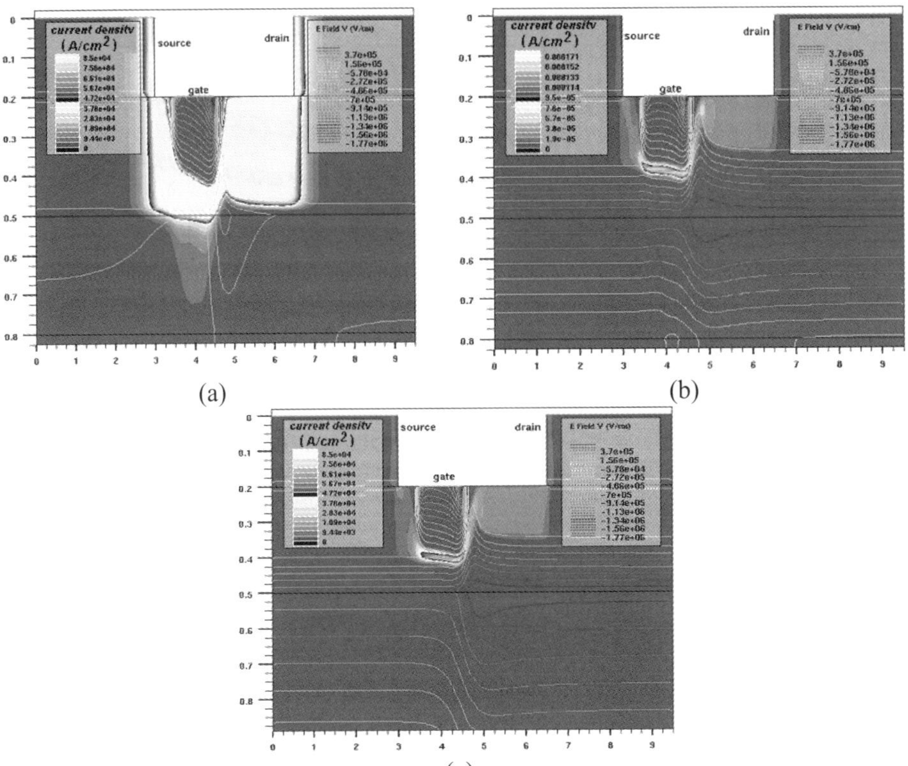

Fig. 1. The current density and the vertical component of electric field lines are plotted across the device cross-sections under similar biasing conditions (V_{GS} = -5V and V_{DS} = 15V). It may be observed that the spatial dispersion of free carriers is the highest in Structure-I (Fig. 1a) - where the buffer layer is totally depleted, and is lowest in the Structure-II (Fig. 1b) - where the buffer is least depleted. Structure-III (Fig. 1c) exhibits a greater degree of spatial dispersion of carrier density compared to Structure-II as the depletion region width in the buffer layer is increased.

The drain current varies significantly with the buffer doping concentration and only slightly with the buffer thickness, especially in buffers that are highly doped. The thick and heavily doped buffers have the least current while thin and lightly doped buffers have the maximum current. From the previous discussion, it may be inferred that the buffers that are heavily doped shall exhibit the greatest output resistances. This can be seen from Fig. 3(b), which shows the variation of the output resistance as a function of drain voltage with the buffer doping concentration as a parameter, at a gate bias of –1V. As expected, the output resistance steadily increases with increasing drain voltage and buffer layer concentration. The variation of the gate-source capacitance (C_{GS}) with respect to the buffer thickness and buffer doping concentration is shown in Fig. 4(a). The gate-source capacitance magnitude is of the order of 5 fF/μm and is nearly independent of the thickness of the channel. However, it increases linearly with decreasing doping concentration. The variation of the gate-drain capacitance (C_{GD}) with respect to the buffer thickness and doping concentration is shown in Fig. 4(b).

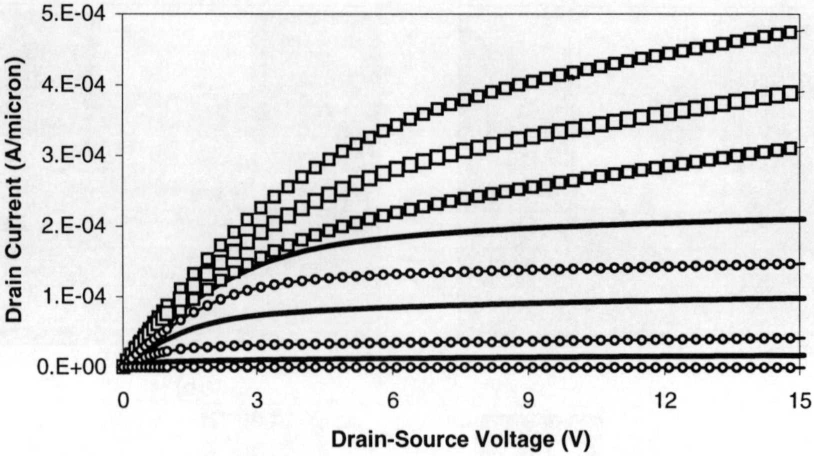

Fig. 2. *I-V* characteristics of Structure-I, -II and -III. The *I-V* characteristics of Structure-I (□) exhibits the highest saturation drain current, however, its output resistance is the lowest. Structure-II (○) has the lowest saturation drain current and the highest output resistance. In Structure-III (—), an increase in saturation drain current is obtained at the expense of reduced output resistance. Gate voltages are -1V (top), -3V (middle) and -5V (bottom).

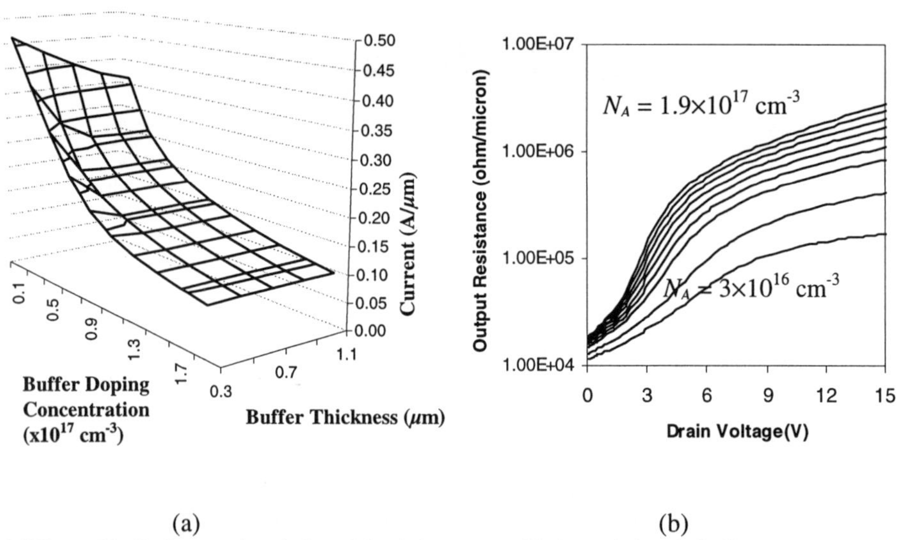

(a) (b)

Fig. 3 Effects of buffer layer - (a) variation of the drain current with the variations in buffer concentration and thickness at V_{GS}=-1V and V_{DS}=15V, and (b) variation of the drain resistance with the variations in drain voltage and buffer layer concentration at V_{GS}=-1V.

The gate-drain capacitance is nearly independent of the buffer thickness and increases almost linearly with decreasing buffer doping concentration. The transconductance (G_m) as a function of buffer concentration and buffer thickness is shown in Fig. 4(c). The transconductance is nearly independent of both the doping and the buffer thickness. However, for low values of buffer concentration, the transconductance decreases at low channel thickness.

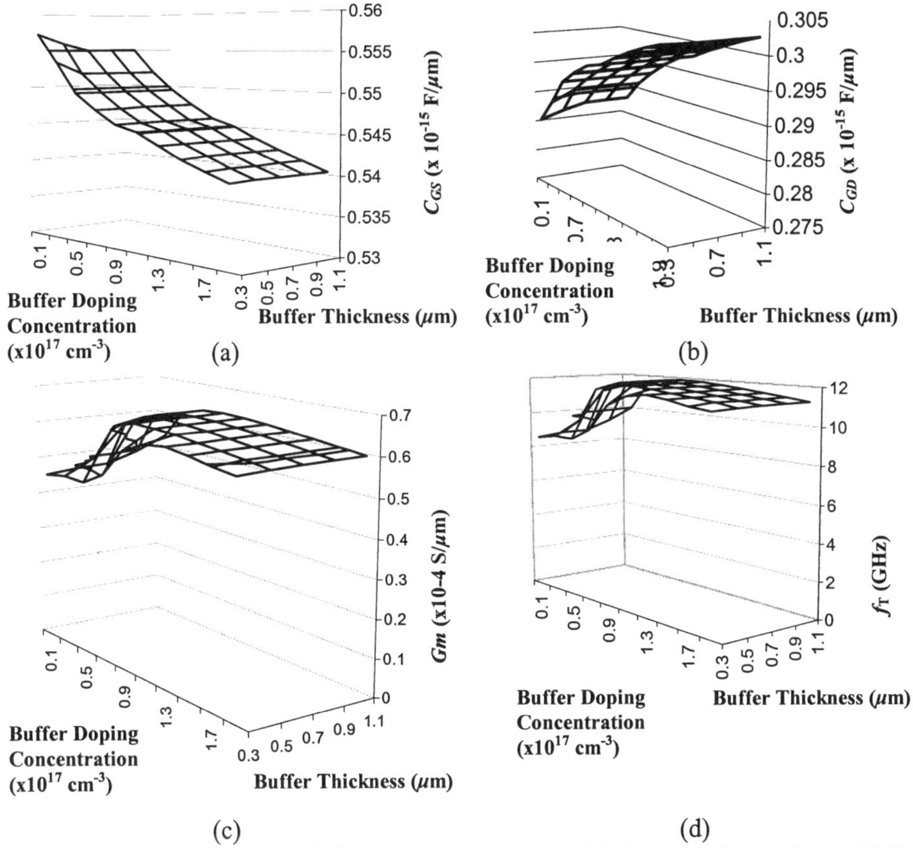

Fig. 4. Small-signal parameters- (a) the gate-source capacitance, (b) the gate-drain capacitance, (c) the transconductance and (d) f_T, as a function of the buffer concentration and buffer thickness.

Finally, f_T for the transistor is calculated using [13]:

$$f_T = G_m / 2\pi \left(C_{GS} + C_{GD} \right) \tag{3}$$

A plot of f_T is shown in Fig. 4(d). As can be seen from the figure, f_T is also nearly independent of both the buffer thickness and buffer concentration. For low values of buffer concentration, a sharp drop in f_T can be observed. This decrease in the value of f_T is seen to mimic the decrease in transconductance shown in Fig. 4(c). C_{GS} and C_{GD} exhibit complementary behavior toward doping concentration and the summation of C_{GD} and C_{GS} is almost independent of both buffer doping concentration and thickness.

Conclusions

This paper investigates the effects of the doping concentration and thickness of the buffer layer on the SiC-based MESFET characteristics. It was found that high values of doping concentration of the buffer layer results in higher output resistance and higher f_T, which are important for the devices to be used in RF power amplifiers. As the maximum drain current in the devices with higher doping concentration is significantly reduced, devices with larger channel widths are required to compensate for the decreasing current. A highly doped buffer prohibits channel electrons from reaching the electron traps located at the buffer-substrate interface and thus would result in the mitigation of substrate trapping effects.

References

1. R. J. Trew, *Proc. of the IEEE* **90**, 6, 1032 (2002).
2. M. E. Levinshtein, S. L. Rumyantsev and M. S. Shur, John Wiley & Sons, Inc., ISBN 0-471-35827-4.
3. M. E. Levinshtein, S. L. Rumyantsev and M. S. Shur, vol. 1, World Scientific, ISBN 981-02-2934-8.
4. S. Mitra, M. V. Rao, K. Jones, N. Papanicolaou and S. Wilson, *Solid-State Elect.* **47**, 193 (2003).
5. H.-Y. Cha, C. I. Thomas, G. Koley, L. F. Eastman and M. G. Spencer, *IEEE Trans. Elect. Dev.*, **50**, 7, 1569 (2003).
6. N. Sghaier, J.-M. Bluet, A. Souifi, G. Guillot, E. Morvan and C. Brylinski, *IEEE Trans. Elect. Dev.* **50**, 2, 297 (2003).
7. S. S. Mukherjee, S. S. Islam and R. J. Bowman, *Solid-State Electronics*, **48**, 10-11, 1709, (2004).
8. P. Tobias, B. Golding, and R. N. Ghosh, *IEEE Sensors Journal*, **2**, 5, 543, (2003).
9. O. Noblanc, C. Arnodo, C. Dua, E. Chartier and C. Brylinski, *Mater. Sci. Engg.*, **B61-62**, 339 (1999).
10. Atlas® Reference Manual, Silvaco International®, http://www.silvaco.com/
11. M. Roschke and F. Schwierz, *IEEE Trans. on Elect. Dev.*, **48**, 7, 1442 (2001).
12. J. H. Zhao, P. Alexandrov and X. Li, *IEEE Elect. Dev. Lett.*, **24**, 6, 402 (2003).
13. Singh J., John Wiley & Sons, Inc., ISBN 0-471-36245-X.

International Journal of High Speed Electronics and Systems
Vol. 14, No. 3 (2004) 897–905

World Scientific
www.worldscientific.com

ANALYTICAL MODEL FOR NON-SELF ALIGNED BURIED P-LAYER SiC MESFET

R.S GUPTA, SANDEEP KUMAR AGGARWAL, RITESH GUPTA, AND MRIDULA GUPTA
Semiconductor Device Research Laboratory, Department of Electronic Science, University of Delhi, South Campus, Benito Juarez Road, New Delhi 110021, India

SUBHASIS HALDAR
Department of Physics, Motilal Nehru College, University of Delhi, Benito Juarez Road, New Delhi 110021, India

A new analytical model for saturation region is proposed for buried p-layer 4H-SiC MESFET considering the effect of B.P. layer. This model provides the static characteristics, small signal parameters and is also extended to predict capacitance-voltage characteristics of the device. The results so obtained are in excellent agreement with experimental data confirming the validity of this model.

1 Introduction

Silicon carbide due to its unique thermal and electrical properties, specially band gap (2.97eV-3.26eV), high electron velocity ($2x10^7$cm/s), high critical field (2MV/cm, five times of GaAs), high thermal conductivity (4.75W/cmK, seven times of GaAs), large avalanche breakdown, is a potential material for high-power and high-temperature operation. However, 4H-SiC gain much attention among its poly-types as the electron mobility in 4H-SiC is about twice as large as of 6H-SiC that makes 4H-SiC MESFET more promising candidate for high power and high temperature application.

SiC MESFET's are usually fabricated on semi-insulating substrate at shorter gate length formation of adjacent source/drain n+ layers can lead to finite substrate current that degrades the device performance. More direct approach to reduce this current is to use a buried p-layer (B.P. layer) between the active layer and the substrate, so as to isolate n+ layers from the substrate.

In the present paper an analytical model has been proposed to calculate the I_d-V_d characteristics of B.P.layer 4H-SiC MESFET. The use of this layer has been emphasized by comparing the device characteristics without B.P.layer. The results obtained from the non-self aligned 0.7μm gate length SiC MESFET at 300^0 K is in close agreement with the experimental results.

1.1. *Model Formulation*

The schematic structure of 4H-SiC MESFET as shown in Figure 1 consists of 4H-SiC n-type active layer with doping N_d followed by a buried-p layer of doping N_a .In n-type active layer two kinds of depletion regions are formed;(i): due to metal semiconductor contact and (ii): due to buried p-layer. Before saturation both the regions are independent but at saturation these two regions will overlap and penetrate each other.

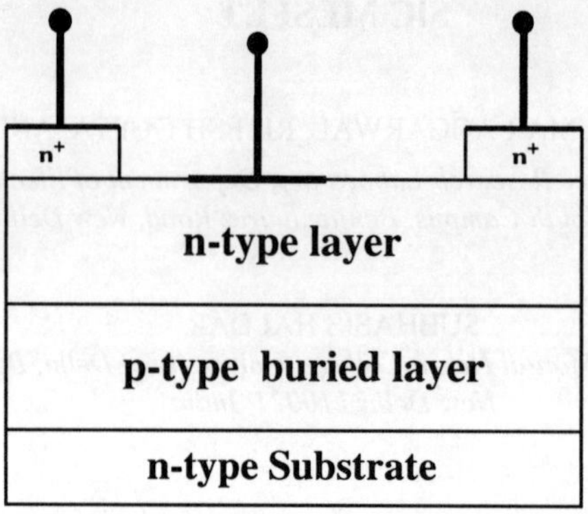

Figure 1 Schematic structure of 4H-SiC MESFET

The depletion region is also influenced by gate voltage (Vg), and the width of depletion region is given by **[1]**

$$h_2 = \frac{N_a}{N_d} \cdot \left[\left(A^2 + \frac{2\varepsilon}{qN_a} \left(V_g + V_o - \Phi_{bi} - V_{bs} + V_p - V_c(x) \right) \right)^{1/2} - A \right] \tag{1}$$

where A is channel depth, $V_c(x)$ is the channel potential, V_p is pinch off voltage, Φ_{bi} is Schottky-barrier height, V_o is diffusion potential other symbols are reported in [1]. Due to additional depletion region formed at the bottom of the active channel the charge is decreased and the drain current is modified as,

$$I_{dsat} = \frac{W.q.N_d.\mu_{eff}}{L - L_2} \cdot \left(A \left(V_{dsat} - I_{dsat}(R_s + R_d) - \frac{2.B_1}{3\sqrt{V_p}} \right) \right.$$

$$\left. - \frac{q.N_a.B_2}{3.\varepsilon} - A.\left(V_{dsat} - I_{dsat}(R_s + R_d) \right) \right) \tag{2}$$

Where

$$B_1 = \left(V_{dsat} + V_{bi} - V_{gs} - I_{dsat} R_d \right)^{\frac{3}{2}} - \left(V_{bi} - V_{gs} - I_{dsat} R_s \right)^{\frac{3}{2}}$$

$$B_2 = \left(A^2 + \frac{2\varepsilon}{qN_a}\left(V_g + V_o - \Phi_{bi} - V_{bs} + V_p - I_{dsat}R_s\right)\right)^{\frac{3}{2}}$$

$$-\left(A^2 + \frac{2\varepsilon}{qN_a}\left(V_g + V_o - \Phi_{bi} - V_{bs} + V_p - V_{dsat} + I_{dsat}R_d\right)\right)^{\frac{3}{2}}$$

For without B.P. layer the drain current is given by

$$I_{dsat} = \frac{W.A.q.N_d.\mu_{eff}}{L - L_2}.\left(V_{dsat} - I_{dsat}(R_s + R_d) - \frac{2}{3\sqrt{V_p}}\left(\left(V_{dsat} + V_{bi} - V_{gs} - I_{dsat}R_d\right)^{\frac{3}{2}}\right.\right.$$

$$\left.\left. - \left(V_{bi} - V_{gs} - I_{dsat}R_s\right)^{\frac{3}{2}}\right)\right) \tag{3}$$

where L_2 is the reduction in the channel length caused by the channel length modulation, (Appendix 1), and is given as:

$$L_2 = \frac{2.H}{\pi}.Sinh^{-1}\left[\pi.(L - L_2).\frac{V_d - V_{dsat}}{2.H.V_{dsat}}\right] \tag{4}$$

where H assumes A for analysis without B.P. layer and h_1 for analysis with B.P. layer.

Trans-conductance and Drain-conductance

Trans-conductance for saturation region is obtained as

$$g_{msat} = g_{msatL}\Big|_{L-L_2} + I_{dsat}\Big|_{(L-L_2)^2} \cdot \frac{\partial L_2}{\partial V_g}\Bigg|_{V_d} \tag{5}$$

where

$$\left.\frac{\partial L_2}{\partial V_g}\right|_{V_d} = -\frac{\left.\frac{\partial L_2}{\partial V_g}\right|_{V_d}.\frac{V_d - V_{dsat}}{V_{dsat}} + \frac{\pi.(L-L_2)}{V_{dsat}}.\left(\left.\frac{\partial V_{dsat}}{\partial V_g}\right|_{V_d} + \frac{V_d - V_{dsat}}{H}.\left.\frac{\partial H}{\partial V_g}\right|_{V_d} + \frac{V_d - V_{dsat}}{V_{dsat}}.\left.\frac{\partial V_{dsat}}{\partial V_g}\right|_{V_d}\right)}{1+\left(\pi.(L-L_2).\frac{V_d - V_{dsat}}{2.H.V_{dsat}}\right)^2}$$

$$+\frac{2}{\pi}.\left.\frac{\partial H}{\partial V_g}\right|_{V_d} Sinh^{-1}\left[\pi.(L-L_2).\frac{V_d - V_{dsat}}{2.H.V_{dsat}}\right]$$

The Drain conductance for saturation region, is obtained as

$$g_{dsat} = g_{dsatL}\big|_{L-L_2} + I_{dsat}\big|_{(L-L_2)^2}.\left.\frac{\partial L_2}{\partial V_d}\right|_{V_g} \quad (6)$$

where

$$\left.\frac{\partial L_2}{\partial V_d}\right|_{V_g} = \frac{-\left.\frac{\partial L_2}{\partial V_d}\right|_{V_g}.\frac{V_d - V_{dsat}}{V_{dsat}} + \frac{\pi.(L-L_2)}{V_{dsat}}}{1+\left(\pi.(L-L_2).\frac{V_d - V_{dsat}}{2.H.V_{dsat}}\right)^2}$$

Capacitance – Voltage Model

Charge under the gate will be the sum of the charges in the linear region and the saturation region i.e.

$$Q = Q_L + q.N_d.W.L_2.H \quad (7)$$

The gate-source capacitance is given as

$$C_{gs} = C_{gsL}\big|_{V_d=V_{dsat}} + q.N_d.W.H.\left.\frac{\partial L_2}{\partial V_g}\right|_{V_d-V_g} \quad (8)$$

where $\left.\frac{\partial L_2}{\partial V_g}\right|_{V_d-V_g} = \frac{2}{\pi}.\left.\frac{\partial H}{\partial V_g}\right|_{V_d-V_g}.Sinh^{-1}\left[\pi.(L-L_2).\frac{V_d - V_{dsat}}{2.H.V_{dsat}}\right]$

$$+\frac{-\left.\frac{\partial L_2}{\partial V_g}\right|_{V_d-V_g}\cdot\frac{V_d-V_{dsat}}{V_{dsat}}+\frac{(L-L_2)}{V_{dsat}}}{1+\left(\pi.(L-L_2).\frac{V_d-V_{dsat}}{2.H.V_{dsat}}\right)^2}\cdot\left(1-\left.\frac{\partial V_{dsat}}{\partial V_g}\right|_{V_d-V_g}-\left.\frac{V_d-V_{dsat}}{H}\cdot\frac{\partial H}{\partial V_g}\right|_{V_d-V_g}\right.$$

$$\left.-\left.\frac{V_d-V_{dsat}}{V_{dsat}}\cdot\frac{\partial V_{dsat}}{\partial V_g}\right|_{V_d-V_g}\right)$$

The gate-drain capacitance can be found by differentiating the above with respect to V_d keeping V_g constant and is found to be

$$C_{ds}=C_{dsL}\big|_{V_d=V_{dsat}}+q.N_d.W.H.\left.\frac{\partial L_2}{\partial V_d}\right|_{V_g} \tag{9}$$

1.2. *Result and Discussion*

Fig.2 shows the variation of drain current with drain voltage taking gate voltage as parameter for models with and without B.P. layer for a gate length of 0.7μm. The comparison of characteristics with and without B.P. layer shows that introduction of B.P. layer leads to a decrease in carrier concentration in the active layer resulting in decrease in the drain current. It can be seen from the figure that the effect of B.P. layer is prominent as there is significant difference in output characteristics especially in saturation region in which two dimensional analysis is to be done. Furthermore the results obtained with B.P. layer shows excellent match with the experimental results[2] that shows the importance of B.P. layer.

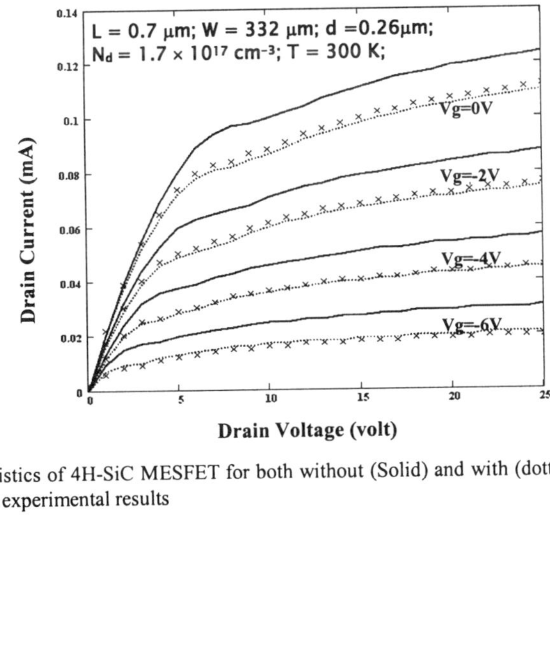

Figure 2 I_d-V_d Characteristics of 4H-SiC MESFET for both without (Solid) and with (dotted) the effect of B.P layer with crosses shows experimental results

Figure 3 shows the variation of drainconductance with drain voltage for various values of gate voltages. It is clear from the figure that due to B.P. layers the maximum value of drainconductance decreases from 19mS to 17mS at zero gate bias. Moreover B.P. layer in the MESFET leads to decrease in drainconductance of the device although in saturation region this difference reduces.

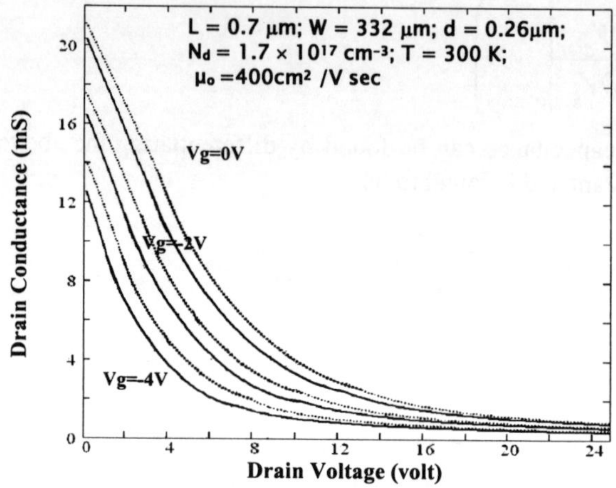

Figure 3 Variation of Drain Conductance with drain voltage at different gate voltages for both with (Solid) and without (dotted) the effect of B.P layer

Figure 4 shows the variation of transconductance with gate voltage for different drain voltages. The maximum transconductance obtained with and without the effect of B.P. layer are 15.5 mS and 17 mS respectively, showing that the introduction of B.P. layer in the MESFET leads to decrease in transconductance.

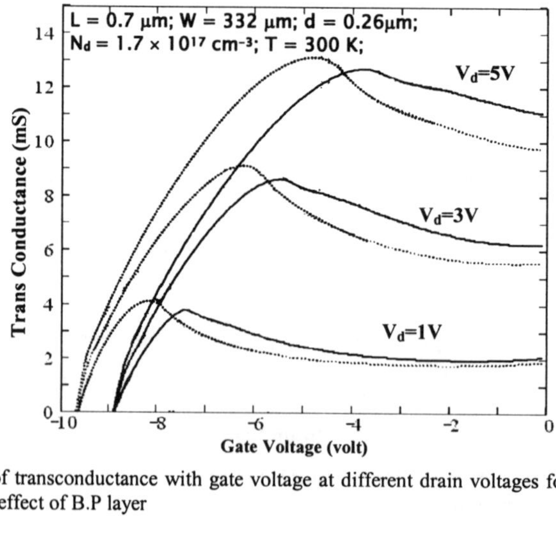

Figure 4 Variation of transconductance with gate voltage at different drain voltages for both with (Solid) and without (dotted) the effect of B.P layer

This is due the fact that introduction of B.P.layer provides an additional depletion region at, the bottom of the active layer which increases the channel resistance of the device. This leads to decrease in drain current and so is the transconductance. B.P.layer also provides isolation between active layer and the substrate thus, reducing the substrate current of the device. The reduction in substrate current is much more in comparison to the device increases with B.P. layer. This facto is incorporated experimentally Yamasaki et. al.[3]

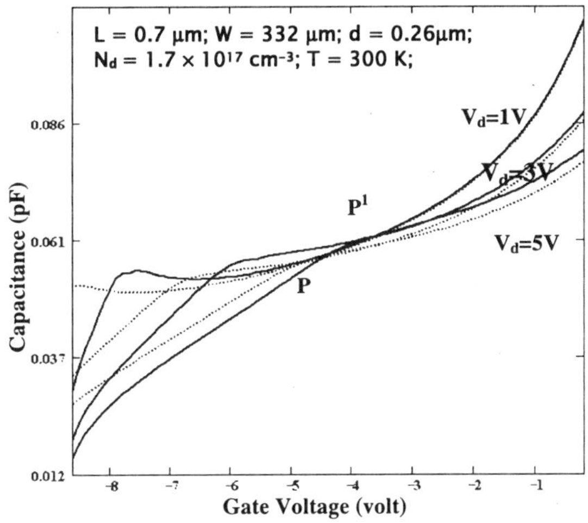

Figure 5 Variation of Gate Source capacitance with gate voltage at different drain voltages for both with (Solid) and without (dotted) the effect of B.P layer

Figure 5 shows the variation of gate-source capacitance with gate-source voltage at various drain voltages. From the graph pinch-off point can be obtain as a point where all the curves meet .This is true as in this condition charge in depletion region is same in all cases. Since saturation voltage is different for buried p-layer the pinch-off (**P**) point differs from the condition when B.P. layer is not taken (**P^1**). It is evident from the figure that pinch-off will occur at lesser gate voltage.

1.3. *Conclusions*

The analytical model for the MESFET proposed in this paper is based on two dimensional Poisson's equation. Present paper proposes a compact and accurate model to calculate the I_d-V_d characteristics of non-self aligned buried layered 4H-SiC MESFET considering the effect of B.P. layer. Various device characteristics (with and without B.P. layer) show that the introduction of B.P. layer leads to a considerable decrease in the drain current, transconductance and drainconductance of the device. But with the introduction of B.P.layer substrate current reduces considerably. It is useful to suppress trap induced phenomena like drain lag and frequency distortion necessary to stabilize operation of SiC MESFETs thus enhancing the performance of the device.[3]

Acknowledgments

The authors are thankful to DRDO, Ministry of Defence, Government of India, for providing necessary financial assistance to carry out the research work.

References

1. S. K. Aggarwal, Subhasis Haldar, Mridula Gupta and R. S. Gupta, "Analytical Model of Non-Self aligned 4H-SiC MESFET", IWPSD 2003,channei.
2. C. E. Weitzel, J. W. Palmour, C. H. Carter and K. J. Nordquist, "4H-SiC MESFET with 2.8W/mm power density at 1.8 GHz", IEEE-EDL 15 (1994) 369-371.
3. Kimiyoshi Yamasaki, Naoki Kato, and Masahiro Hirayama, "Buried p-layer SAINT for very High-Speed GaAs LSI's with Sub Micrometer Gate Length" IEEE-ED 32 no. 11 (1985) 2420-2425.
4. Kazuaki Kunihiro, Yasuo Ohno, "Design considerations for buried P-layers to suppress substrate-trapping effects in GaAs MESFET's," Solid State Electron. 45 (2001) 1763-1771.
5. A.B.Grebene and S.K.Gandhi, "General theory for pinched operation of the junction-gate FET, "Solid-State Electron 1969; Vol 12, pp 573.
6. M.S.Shur, "Analytical models of GaAs FETs," IEEE ED 32 no.1 Jan (1985).

Appendix 1

At higher drain-source voltage, the electric field in the conducting channel increases which leads to velocity saturation. In the non-self-aligned structure, there is substantial extension of the depletion region beyond the gate forming high field region at the drain side of the channel, now horizontal electric field comparable to vertical electric field, and one dimensional analysis is not suitable, so two-dimensional analysis is required. The solution of the 2-D potential in this region is evaluated as Grebene [5] in the high field region,

$$\Psi(x^1, y) = \frac{2.H.E_s}{(-1)^{n-1}(2n-1)\pi}.\sinh\frac{(2n-1)\pi.x^1}{2.h}.\sin\frac{(2n-1)\pi.y}{2.h} \tag{A-1}$$

Where h is maximum depletion width due Metal –Schottky contact at pinch-off condition, the boundary conditions to be satisfied by $\psi(x^1, y)$ are

$$\Psi(x^1,0) = 0, \Psi(0,y) = 0, \frac{\partial\Psi(x^1,H)}{\partial y} = 0, \frac{\partial\Psi(0,H)}{\partial x} = E_s \tag{A-2}$$

Where E_s is saturation electric field, x^1=x-L_2 and L_2=L-L_1 is the length of the velocity saturation region.
Now only the first term of the series is taken as the higher terms decreases, is given as [5], here we calculate general solution for both BP and without BP is

$$V_{sr} = \frac{2.H.E_s}{\pi}.\sinh\frac{\pi.x^1}{2.H} \quad \textbf{(A-3)}$$

V_{sr} can be written as [6]

$$V_{sr} = V_d - V_{dsat} \quad \textbf{(A-4)}$$

Now by equating (A-4) and (A-5) effective channel length with out BP layer

$$L_2 = \frac{2.H}{\pi} Sinh^{-1}\left[\pi.(L-L_2).\frac{V_d - V_{dsat}}{2.H.V_{dsat}}\right] \quad \textbf{(A-5)}$$

International Journal of High Speed Electronics and Systems
Vol. 14, No. 3 (2004) 906–908

World Scientific
www.worldscientific.com

ION IMPLANTED SiC STATIC INDUCTION TRANSISTOR WITH 107 W OUTPUT POWER AND 59% POWER ADDED EFFICIENCY UNDER CW OPERATION AT 750 MHZ

G.C. DE SALVO, P.M. ESKER, T.A. FLINT, J.A. OSTOP, E.J. STEWART, T.J. KNIGHT, K.J. PETROSKY, S.D. VAN CAMPEN, R.C. CLARKE, AND G.M. BATES

Northrop Grumman, Electronic Systems,
Advanced Materials and Semiconductor Device Technology Center,
Baltimore, Maryland 21203, USA

There are many commercial applications which require high RF CW power in the kilowatt to megawatt range. To date, these high RF power requirements can only be accomplished by incorporating traveling wave tubes (TWT) or inductive output tubes (IOT). However, recent advances in the output power capability of SiC power devices for radar systems now suggest that high power SiC static induction transistors (SIT) can be used to develop a 10 kW solid state RF power amplifier. From analysis of existing high power RF amplifiers, development of a transistor cell that can produce 10 W CW output power would provide the necessary building block for assembly of a 10 kW power amplifier. As a first step in this development, existing ion implanted SiC SIT device cells fabricated at Northrop Grumman were packaged and tested for CW operation at 750 MHz [1]. Identical transistor die from a single wafer of ion implanted SIT devices were selected, each having a layout of 10 device cells per die, with each cell having a gate periphery equal to 1.30 cm, and an active area of 2.92×10^{-4} cm^2. To minimize heating under CW operation, only 10 cells were bonded into a transistor package. This was accomplished by attaching only 2 die to a package, and wire bonding every other cell in each die. In this manner, only 5 cells from each die were connected together, and the remaining 5 cells were not used, thus allowing for improved heat spreading, as shown in Figure 1.

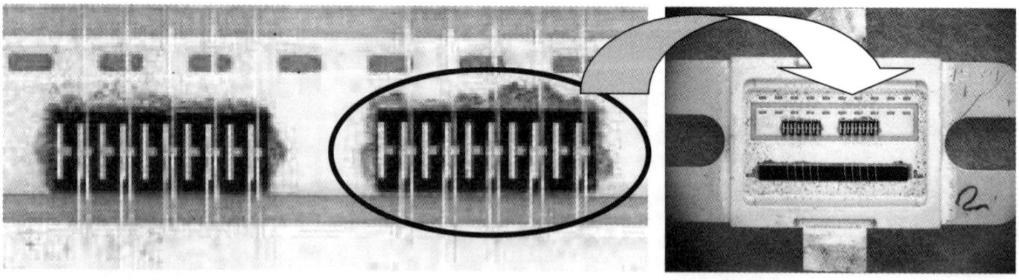

Figure 1. The transistor package used for the CW measurements consisted of attaching two die into a package, and wire bonding every other cell from both die, providing for a total of 10 active cells with improved heat spreading for CW operation.

The SiC SIT cells were connected in common-gate circuit configuration, to provide higher output power. A CW RF test fixture was made for the transistor package, which contained wider RF transmission lines to handle the higher power dissipation. Electrical DC characterization for one of the 10 cell ion implanted SITs is given in Figure 2, and shows classic SIT transistor response, providing a voltage gain ($\Delta V_{ds}/\Delta V_{gs}$) = 30 and blocking operation beyond 225 V, while providing a maximum current of I_{max} = 10 A at V_{ds} = 15 V, and a pinch-off voltage around V_{gs} = –2.5 V.

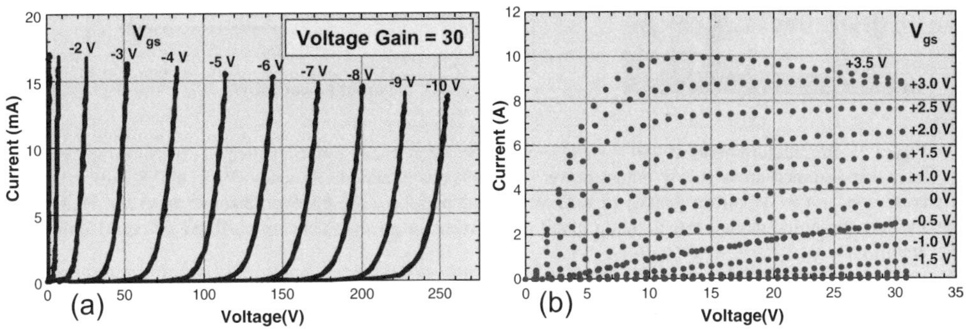

Figure 2. DC measurements of the (a) static induction transistor curve and the (b) FET family curve of a 10 cell ion implanted SIT transistor package shows good transistor response with a voltage gain (ΔVds/ΔVgs) equal to 30 and blocking operation beyond 225 V, while providing a maximum current of Imax = 10 A at Vds = 15 V, and a pinch-off voltage around –2.5 V.

Initial RF measurements taken at 750 MHz under CW operation were very encouraging, as output power levels above 100 W were obtained from several transistor packages, with the best result equaling 107 W output power with 59% power added efficiency (PAE) and 8.7 dB gain, as shown in Figure 3. These results were accomplished at the DC bias point of V_{ds} = 81.8 V and I_{ds} = 1.87 A. Due to the high PAE of the transistor, only air cooling was needed (by blowing ambient air from a fan onto the package). Higher output power may be possible by operating at a higher bias (V_{ds}) or incorporating more device cells, but further analysis of the transistor package characteristics will need to be considered before implementing higher bias operation.

The measurement of 107 W RF output power with 59% PAE provides an excellent starting point towards development of a 10 kW RF solid state driver amplifier for future commercial applications. For this 10 cell device, the 107 W output power corresponds to an average power of 10.7 W/cell (36.5 kW/cm^2), which meets the initial device cell requirements as the building block to provide a 10 kW RF solid state driver amplifier. Further work needs to be done to optimize the transistor assembly process for thermal management under high RF power CW operation, including cell layout, die attachment, wire bonding, and packaging, as well as their interactions which limit the available output power.

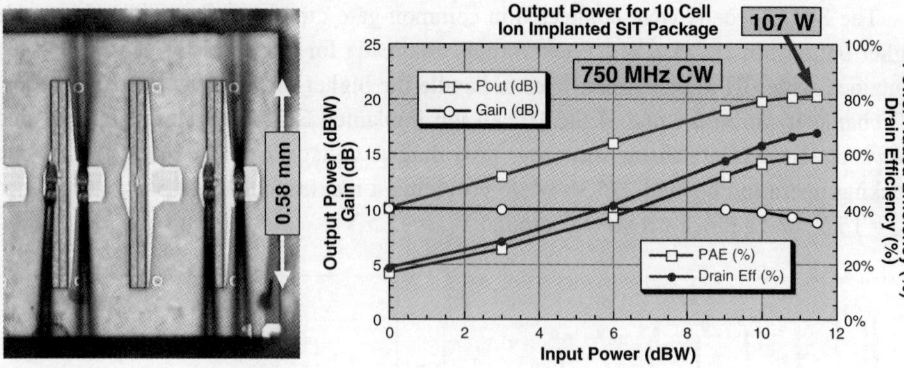

Figure 3. RF measurements at 750 MHz CW operation for a 10 cell ion implanted SIT at Northrop Grumman have demonstrated up to 107 W output power with 59% power added efficiency (PAE) and 8.7 dB gain. This corresponds to a CW power density of 8.23 W/cm gate periphery (36.5 kW/cm2 active area) and 10.7 W/cell, meeting the initial device cell building block requirements needed to develop a 10 kW RF solid state driver amplifier.

References

1. T.J. Knight, R.C. Clarke, R.R. Barron, J.A. Ostop, B.A. Morick, J.R. Gigante, W.J. Malkowski, A.W. Morse, G.C. DeSalvo, K.J. Petrosky, W.R. Curtice, Silicon Carbide Self Aligned and Ion Implanted Static Induction Transistor (SAI-SIT) for 150 Watt S-Band Operation, 60th Annual Device Research Conference, 179 (June 2002).

International Journal of High Speed Electronics and Systems
Vol. 14, No. 3 (2004) 909–914

World Scientific
www.worldscientific.com

INFLUENCE OF THE N-DIFFUSION LAYER ON THE CHANNEL CURRENT AND THE BREAKDOWN VOLTAGE IN 4H-SiC SIT

YOUNG CHUL CHOI, HO-YOUNG CHA, LESTER F. EASTMAN, AND MICHAEL G. SPENCER

School of Electrical and Computer Engineering, Cornell University
Ithaca, NY 14853, USA

For a recessed gate (RG) silicon carbide (SiC) static induction transistor (SIT), an n-diffusion layer technique to maximize a maximum channel current (I_{MAX}) and realize an efficient breakdown voltage (BV) was presented. An n-diffusion layer having a doping profile of a Gaussian distribution was formed from the source to the channel region and introduced to elevate the doping concentration of the channel region. Through the simulation of a RG SiC SIT with an n-diffusion layer, it was verified that an n-diffusion layer technique was an excellent way to realize a very small width of a source region (W_s), resulting in an larger I_{MAX} and a higher BV simultaneously, and it was desirable to employ an n-diffusion layer having a slightly larger junction depth (X_j) than the depth of a gate trench (L_T) ($L_T < X_j < 1.3L_T$) into a RG SiC SIT with a small width of the source region ($W_s \approx$ 0.5 μm).

1 Introduction

The silicon carbide (SiC) static induction transistors (SITs) are based on the original concepts invented by J. Nishizawa [1]. They have been demonstrated as a highly efficient power microwave device because SiC has a high critical electric field and a high electron saturated velocity [2-3]. These properties coupled with an extremely high thermal conductivity and a good stability help SiC to be ideal for high power, high temperature, and harsh environment electronic device applications. Additionally, the SiC SITs offer many advantages over other wide bandgap high frequency devices for high power operation, and no requirements for semi-insulating wafers or susceptibility to back-gating effects as is the case for MESFETs. However, conventional SiC SITs have a fundamental trade-off relation between maximum channel current (I_{MAX}) and breakdown voltage (BV). Thus, it is difficult for conventional SiC SITs to exhibit high BV and large I_{MAX} simultaneously.

In this paper, an n-diffusion layer technique to maximize I_{MAX} and realize an efficient BV is presented and simulated by using the SILVACO atlas device simulator. The influence of an n-diffusion layer on I_{MAX} and BV is also investigated as a function of the width of a source region (W_s) and the junction depth of an n-diffusion layer (X_j).

2 Brief Structure Description

SITs usually have three basic structural types according to the pattern of structures: surface gate, buried gate and recessed gate (RG) [4-5]. Among them the RG structure, in which the gate is formed by the bottom Schottky contact of recessed regions, is favored because of high operating frequency and high power. The cross-sectional view of a simulated RG 4H-SiC SIT with an n-diffusion layer is shown in Figure 1, with the primary difference found in the doping profile of the channel region. The n-diffusion layer formed from the source to the channel region by deep ion implantation and diffusion at a high temperature has a doping profile of a Gaussian distribution, as shown in Figure 2.

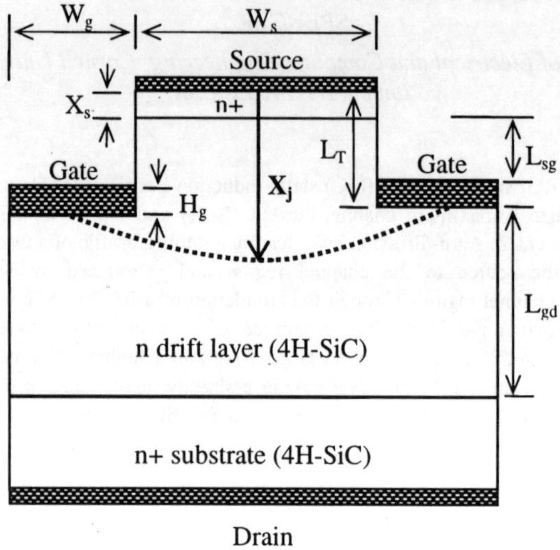

Figure 1 Cross-sectional view of a simulated RG 4H-SiC SIT including an n-diffusion layer (X_j is the junction depth of an n-diffusion layer)

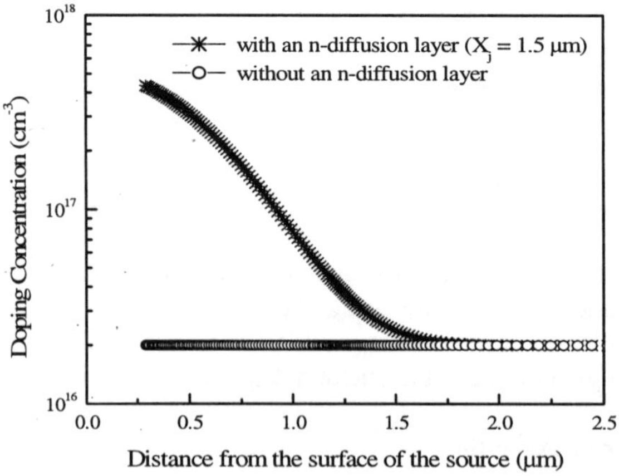

Figure 2 Doping profiles along a vertical line passing through the center of a device as shown in Figure 1

The doping concentration of the n drift region is 2 x 10^{16} cm^{-3} chosen as a value that is achievable with current technology. The n+ source is 0.3 μm thick (X_s) and has a doping concentration of 1 x 10^{19} cm^{-3}. The distance between the source and the gate (L_{sg}) is set to be 0.9 μm in order to avoid the source-to-gate punch-through when the depletion region from the gate extends to the source. The distance between the gate and the drain (L_{gd}) is set to be 3.5 μm in order to have a large BV. However, as L_{gd} increases, a series resistance in the n drift region also increases, resulting in the degradation of the frequency and channel current response of a device. The gate half-width (W_g) and the gate height (H_g) are set to be 0.5 μm and 0.3 μm, respectively. Increasing H_g improves gate control, DC voltage gain, and transconductance, while reducing the output resistance. The W_s and the X_j are variables and very critical factors in determining the BV and the I_{MAX} of a device.

3 Simulation Results and Discussions

A conventional RG 4H-SiC SIT has a gate trench structure where a high electric field near a gate corner is the main reason for instability and premature failure, resulting in a lower BV [6]. The SILVACO simulated breakdown in a conventional RG SiC SIT is shown in Figure 3(a), where a maximum electric field (E_{max}) occurs directly under the gate corner in the drift region. One way to reduce E_{max} is by use of a PN junction gate, in which a buried p+ junction is placed directly under the gate contact [7]. A PN junction gate offers less gate leakage, even at high temperatures, and potentially higher power density, but it is used with a higher gate resistance. It is especially tricky to realize the buried p+ junction directly under the gate contact because of the high degree of technological difficulty.

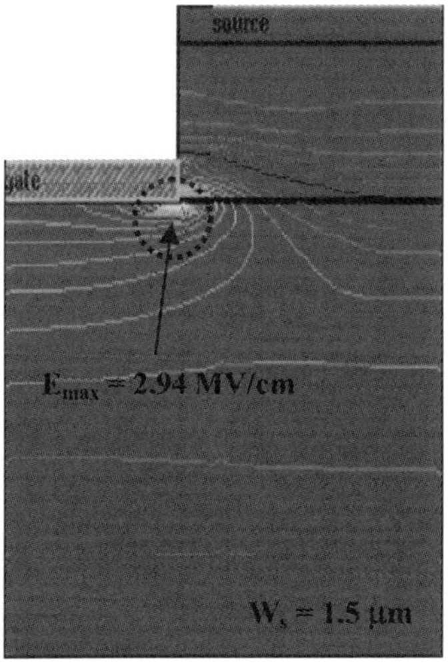

(a)

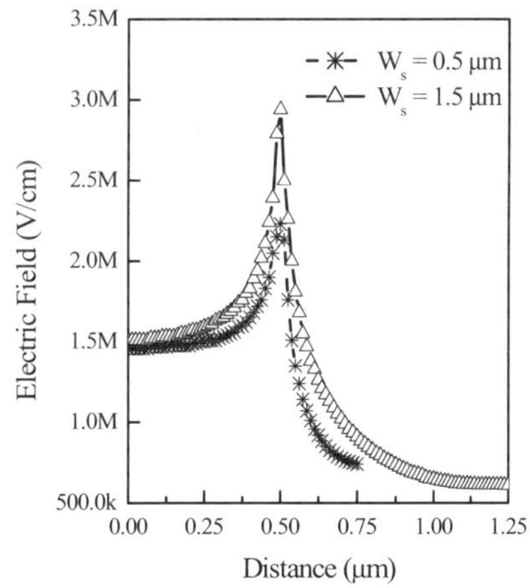

(b)

Figure 3 Electric field distribution from SILVACO simulation for conventional RG SiC SITs (V_{ds} = 500V, V_{gs} = -20V) (a) Electric field contour lines (b) Electric field along a horizontal line directly under the gate trench

Another simple way is by use of a smaller W_s. Because the crowding of the electric field directly under the gate corner is significantly reduced as shown in Figure 3(b), the BV of a device having a smaller W_s is higher. Furthermore, a smaller W_s are helpful to achieve an improved frequency response. However, the I_{MAX} becomes lower due to the increased ohmic resistance in the channel region. Figure 3 illustrates the examples of the simulated electric field distribution in a conventional RG 4H-SiC SIT, where one can easily observe the point of E_{max}. The simulation for figure 3 are performed at V_{ds} = 500V and V_{gs} = -20V. The E_{max} found directly under the gate corner is 2.94 MV/cm for W_s = 1.5 μm and 2.23 MV/cm for W_s=0.5 μm. This comparison implies that E_{max} directly under the gate corner can be relieved by decreasing W_s.

Figure 4 and Figure 5 show the simulation results of BV and I_{MAX} as a function of W_s. The BV is the static breakdown voltage between drain and source simulated at V_{gs} = -20V, and the I_{MAX} is the maximum current in the channel region at V_{gs} = 0V and V_{ds}=V_{knee}. As W_s decreases, BV increases abruptly and I_{MAX} decreases linearly, as mentioned earlier. In particular, for W_s = 0.5 μm, where a device provides a very high BV, considering the elevated doping concentration in the channel region of a device with an n-diffusion layer, the BV of a device with an n-diffusion layer is so high that it is hard to see any difference compared to that of a device without an n-diffusion layer. However, the I_{MAX} of a device with an n-diffusion layer for W_s=0.5 μm is definitely low even though the I_{MAX} is considerably higher than that of a device without an n-diffusion layer. This means that the X_j of 1 μm is not large enough to efficiently suppress the JFET effect which occurs between the adjacent gate electrodes [8]. Another important design consideration for a high power capability is the pinch-off voltage (V_P) of a device. V_P increases with increasing W_s. It is highly desirable that V_P be lower than 10V in order to achieve a blocking state with a low negative gate voltage. Based on this limitation, it is clear that a large W_s is not good choice to get a stable blocking state. Therefore, this makes it difficult to achieve high power, as shown in figures 4 and 5.

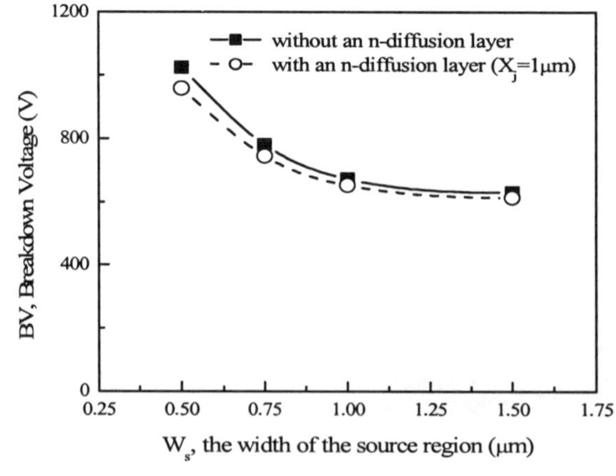

Figure 4 Comparison of the simulated BV as a function of W_s

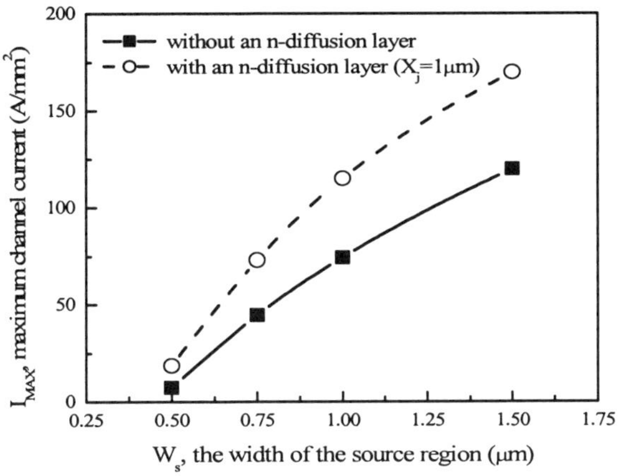

Figure 5 Comparison of the simulated I_{MAX} as a function of W_s

Figure 6 shows the simulated DC performances of a RG 4H-SiC SIT with an n-diffusion layer as a function of X_j for W_s = 0.5 μm. As X_j increases and becomes larger than L_T (= X_s + L_{sg} + H_g), I_{MAX} increases dramatically due to the effective suppression of the JFET effect and BV decreases gradually due to the elevated doping concentration near the gate corners. From Figure 6, the rapid fall-off of BV is obvious for larger X_j than 2 μm, but BV indicates a very small change over the simulated range.

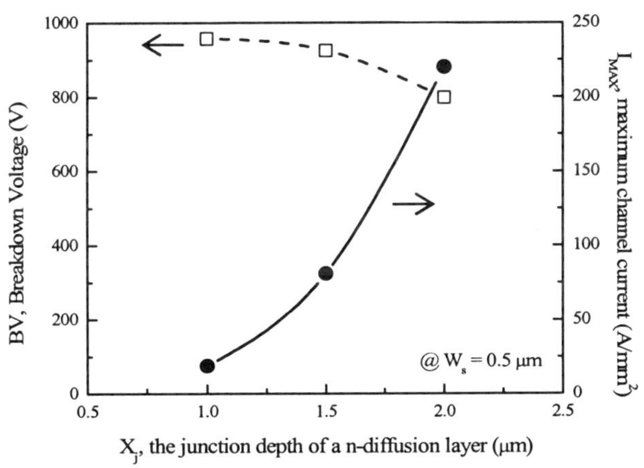

Figure 6 Simulated DC performances of a RG 4H-SiC SIT with an n-diffusion layer

Therefore, the n-diffusion layer technique is a good way to realize a very small W_s in a RG 4H-SiC SIT, resulting in a maximized I_{MAX} and an efficient BV. Considering the V_P limitation as mentioned before, it is highly desirable to introduce an n-diffusion layer having a slightly larger X_j than L_T ($L_T < X_j < 1.3L_T$) into a conventional device with a small W_s.

4 Conclusions

It has been demonstrated that the n-diffusion layer technique for a RG 4H-SiC SIT is effective for realizing a very small W_s, resulting in a maximized I_{MAX} and an efficient BV simultaneously. The influence of an n-diffusion layer on I_{MAX} and BV has been also observed as a function of W_s and X_j. As long as an n-diffusion layer having a slightly larger X_j than the L_T ($L_T < X_j < 1.3L_T$) is employed into a RG 4H-SiC SIT with a small W_s, both high BV and large I_{MAX} can be easily acquired.

Acknowledgments

This work is supported by the Office of Naval Research.

References

1. J. Nishizawa, T. Terasaki and J. Shibata, *IEEE Trans. Electron Devices* **22**, 185 (1975).
2. M. Bhatnagar and B. J. Baliga, *IEEE Trans. Electron Devices* **40**, 645 (1993).
3. I. A. Khan, J. A. Cooper, Jr., *IEEE Trans. Electron Devices* **47**, 269 (2000).
4. J. Nishizawa, K. Motoyan and A. Itioh, *IEEE Trans. Electron Devices* **47**, 482 (2000).
5. B. J. Baliga, *IEEE Trans. Electron Devices* **29**, 1560 (1982).
6. A. K. Agarwal, R. R. Siergiej, S. Seshadri, M. H. White, P. G. McMullin, A. A. Burk, L. B. Rowland, C. D. Brandt, and R. H. Hopkins, *Proc. 8th International Symp. Power Semiconductor Devices and ICs*, 119 (1996).
7. A. K. Agarwal, L. S. Chen, G. W. Eldridge, R. R. Siergiej, R. C. Clarke, *IEEE Device Research Conference Digest,* 94 (1998).
8. S. C. Sun and J. D. Plummer, *IEEE Trans. Electron Devices* **27**, 356 (1980).

International Journal of High Speed Electronics and Systems
Vol. 14, No. 3 (2004) 915–916

World Scientific
www.worldscientific.com

Participant Index

Anwar, M.	U Connecticut
Appenzeller, J.	IBM
Bhat, I.	RPI
Bilenko, Y.	Sensor Electronic Tech.
Brar, B.B.	Rockwell
Bu, G.g	RPI
Buttari, D.	UC Santa Barbara
Cha, H.-Y.	Cornell
Chandrashekhar, MVS	Cornell University
Chao, P.C.	BAE Systems
Chen, C.	U Delaware
Chen, M.	HRL Laboratories, LLC
Chen, X.	Cornell University
Chhajed, Sameer	RPI
Choi, Y. C.	Cornell University
Chow, T. P.	RPI
Daniels, D.	Booz Allen Hamilton
Das, M.K.	Cree, Inc.
De Knoop, L.	Cornell University
DeSalvo, G.	Northrop Grumman
Doolittle, W.A.	Georgia Tech
Eastman, L.F.	Cornell University
Evans, K.	Crystal Innovative Semi.
Faraclas, E.W.	U Connecticut
Fischer, A.	Sandia Labs
Foppes, John	Lockheed Martin
Fouad, B.K.	IAS, Germany
Ghori, A.	Syracuse University
Ghosh, S.	U Delaware
Giziewicz, W.P.	MIT
Gupta, M.	U Delhi, India
Gupta, R.S.	U Delhi, India
Huang, W.	RPI
Islam, S.S.	RIT
Jenkins, T.	AFRL/SNDM-WPAFB
Keogh, D.M.	UC San Diego
Kim, H.	Ohio State University
Kim, Jong Kyu	RPI
Kohn, Erhard	U Ulm-Germany
Kolodzey, J.	U Delaware
Krishnaswami, S.	Cree, Inc
Kumar, A.	RPI
Lai, J.-W.	Univ. of Illinois
Lee, J.	Ohio State University
Lee, K.-K.	Georgia Tech
Leoni, R.E.	Raytheon
Li, J.C.	UC San Diego
Liu, Q.	University of Notre Dame
Losee, P.	RPI
Lu, H.	Cornell University
Lu, W.	Ohio State University
Luck, C.R.	Virginia Tech
Luo, H.	RPI
Luukanen, A.	NIST/HRL Laboratories
McDonald, J.	RPI
Mishra, U.	UC Santa Barbara
Mukherjee, S.S.	RIT
Namkoong, G.	Georgia Tech
Neuberger, M.	U Ulm-Germany
Pala, N.	RPI
Pappaterra, R.	WPAFB
Paz, S.	Rafael Company, Israel
Pino, R.	RPI
Rai, S.S.	RIT
Rajan, S.	UC Santa Barbara
Roumiantsev, S.	RPI
Ryu, S.-H.	Cree, Inc
Sadler, R.	Northrop Grumman
Sawyer, S.	RPI
Schaff, W. J.	Cornell University
Schubert, E. F.	RPI
Schujman, S.B.	Crystal Innovative Semi.
Schulman, J.	HRL Laboratories, LLC
Seabaugh, A.	Univ. of Notre Dame
Shah, J.M.	RPI
Shahidi, G.G.	IBM
Shur, Michael	RPI
Siegel, Peter	CIT/JPL

Snow, E.	Naval Research Lab
Sommerhalter, C.	Aixtron
Spencer, M.	Cornell University
Sturm, M.	Princeton University
Stutzmann, M.	Walter Schottky Institute
Sun, Y.-J.	Cornell University
Susteric, N.	U Delaware
Thomas, C.	Cornell University
Troeger, R.T.	U Delaware
Turner, S.	U Maine
Vitusevich, S.	Institute fuer Schichten unde Grenzflaechen
Waldron, J.	Silicon Power Corporation
Washington, M.	RPI
Webster, R.T.	AFRL/SNHA
Wetzel, C.	RPI
Worley, R.	WPAFB
Wraback, M.	US Army Research Lab
Wu, H.	Cornell University
Wu, Y.	Cree, Inc
Xi, J.	RPI
Xi, Y.	RPI
Xie, S.	UC Santa Barbara
Xing, H.G.	University of Notre Dame
Xu, H.	UC Santa Barbara
Xuan, G.	U Delaware
Yang, J.	NRL
Ye, P.P.	Agere Systems
Zhu, L.	RPI
Zolper, J.	DARPA

International Journal of High Speed Electronics and Systems
Vol. 14, No. 3 (2004) 917–919

World Scientific
www.worldscientific.com

Author Index